Methods for Solving Incorrectly Posed Problems

V.A. Morozov

Methods for Solving Incorrectly Posed Problems

Translation editor: Z. Nashed

Translated by A.B. Aries

Springer-Verlag
New York Berlin Heidelberg Tokyo

V.A. Morozov
Computer Center
Moscow State University
Moscow 119899
U.S.S.R.

Z. Nashed (Translation editor)
Department of Mathematical Science
University of Delaware
Newark, Delaware 19711
U.S.A.

AMS Classification: 35R25

Library of Congress Cataloging in Publication Data
Morozov, Vladimir Alekseevich
Methods for solving incorrectly posed problems.
Translation of: Reguliarnye metody resheniia nekorrektno postavlennykh zadach.
Includes bibliography.
1. Differential equations, Partial–Improperly posed problems. I. Nashed, Z. II. Title.
QA377.M6813 1984 515.3'53 84-13961

Title of the original Russian edition: Reguliarnye metody resheniia nekorrektno postavlennykh zadach. Moskva: Izdatel'stvo MGU, 1974.

With 5 illustrations

Printed and bound by R.R. Donnelley & Sons, Harrisonburg, Virginia.
Printed in the United States of America.

9 8 7 6 5 4 3 2 1

ISBN 0-387-96059-7 Springer-Verlag New York Berlin Heidelberg Tokyo
ISBN 3-540-96059-7 Springer-Verlag Berlin Heidelberg New York Tokyo

Introduction

Some problems of mathematical physics and analysis can be formulated as the problem of solving the equation

$$Au = f, \quad f \in F, \tag{1}$$

where $A: D_A \subset U \to F$ is an operator with a non-empty domain of definition D_A, in a metric space U, with range in a metric space F. The metrics on U and F will be denoted by ρ_U and ρ_F, respectively. Relative to the twin spaces U and F, J. Hadamard [106] gave the following definition of correctness: the problem (1) is said to be *well-posed (correct, properly posed)* if the following conditions are satisfied:

(1) The range of the value Q_A of the operator A coincides with F ("solvability" condition);

(2) The equality $Au_1 = Au_2$ for any $u_1, u_2 \in D_A$ implies the equality $u_1 = u_2$ ("uniqueness" condition);

(3) The inverse operator A^{-1} is continuous on F ("stability" condition).

Any reasonable mathematical formulation of a physical problem requires that conditions (1)-(3) be satisfied. That is why Hadamard postulated that any "*ill-posed*" (improperly posed) problem, that is to say, one which does not satisfy conditions (1)-(3), is non-physical. Hadamard also gave the now classical example of an ill-posed problem, namely, the Cauchy problem for the Laplace equation. It became clear later that many branches of mathematics and/or natural sciences in general involve ill-posed problems; for example, the continuation problem for analytic and harmonic functions, geophysical problems, the supersonic body problem, biophysical problems, etc. In the 1950's the Cauchy problem for the Laplace equation became a canonical problem in numerous investigations [35], [110].

The general theory and methods for solving ill-posed (unstable) problems were developed soon after Hadamard. In this connection, we must mention the names of some outstanding Soviet mathematicians: A. N. Tikhonov, M. M. Lavrentiev, V. K. Ivanov, and also their disciples, who contributed much to the development of the theory and techniques which became a new and most fruitful area in numerical analysis. The intensive development of methods for solving unstable problems was spurred by the increasing utilization of digital computers in mathematics and engineering. The natural feedback between theory and practice produced a wide variety of problems requiring urgent solutions. It also necessitated the development of new approximate methods for solving a much wider class of problems free from the rigid constraints of their well-posedness. This facilitated the successful tackling of problems (mainly, unstable problems) in geophysics, spectroscopy, electronic microscopy, automatic control, thermophysics, gravimetrics, electrodynamic, optics, nuclear physics, plasma theory, and many other fields of science and technology.

The foundations of approximate methods for solving ill-posed problems were laid by A. N. Tikhonov [89], where he introduced an important class of the so-called "inverse" problems involving construction of qualitative characteristics of a medium from physical field measurements. Many theoretical and practical problems of data processing, and the estimation of unknown parameters in equations based on functionals of their solutions, can be formulated as inverse (as a rule, ill-posed) problems [46], [76].

A. N. Tikhonov gave the following generalization of the classical (in the Hadamard sense) concept of correctness (well-posedness) of the problem. The concept of *narrowing* the region of definition of the primary operator is at the heart of Tikhonov's generalization [89]. Thus, the problem (1) is said to be *well-posed in the Tikhonov sense* (conditionally well-posed) if:

(1') it is known *a priori* a solution u of the problem (1) exists for all f belonging to a given subset of F and, furthermore, it belongs to the specified set $M \subseteq D_A$;

(2') the solution u is unique in the class M;

(3) infinitely small variations of the solution u correspond to infinitely small variations of the right-hand side of (1), retaining the solution in the class M.

The set M, in the Tikhonov sense, is said to be a set of correctness. Tikhonov was the first to invoke a topological theorem providing sufficient conditions for the problem (1) to be well-posed in his sense. This theorem, given below, plays an essential role in the theory of methods for solving ill-posed problems.

Theorem (On the Stability of the Inverse Operator). Let A be a continuous operator on a non-empty compact set $M \subseteq D_A$, and assume that A is one-to-one on M. Then the inverse operator A^{-1} considered on $A[M] = N$ is continuous.

As Lavrentiev [37] noted, when the conditions of the theorem are satisfied, there exists a continuous non-decreasing function $\omega(\tau) = \omega(\tau;M)$, $\tau > 0$, $\omega(0) = 0$, such that for any $u,v \in M$ satisfying $\rho_F(Au,Av) \leq \tau$ the estimate $\rho_U(u,v) \leq \omega(\tau)$ holds. The function $\omega(\tau)$ is frequently called the *function of correctness* (or stability) of the problem (1) on the set M; the set M itself is called the *set of correctness*. It is seen that the function $\omega(\tau)$ is the modulus of continuity of the operator A^{-1} on N.

Generalizations of this theorem to metric and topological spaces for closed operators were obtained by Ivanov [22], [27], and in the case of a non-invertible operator by Liskovetz [44]. A local version of this theorem was proved by the author in [53].

Tikhonov's theorem on the stability of the inverse operator resolved, in principle, the question of the possibility of obtaining a stable solution of (1) but did not indicate a *method* for finding a solution. The latter is due to the fact that, in practice, the condition that the approximately specified right-hand side F of equation (1) belongs to the set N, cannot, as a rule, be satisfied, that is to say, equation (1) is *unsolvable* for the specified $\tilde{f}$. Lavrentiev [37] showed that under certain conditions on the operator A it is possible to replace problem (1) with a closely related problem which is solvable; for any $f \in F$. In this event, the *necessary* knowledge of both the accuracy of specifying $\tilde{f}$, in other words, the estimate of deviation $\rho_F(f,\tilde{f}) \leq \delta$, and the function of correctness $\omega(\tau)$, is an essential ingredient for the approximate solution of problem (1). This enabled Lavrentiev to determine, for a wide class of problems, an algorithm of construction of approximations $\tilde{u} \in U$

for which $\rho_U(\tilde{u},u) \to 0$, when $\rho_F(\tilde{f},f) \to 0$, where u is the solution of problem (1), belonging to the compact set M. John [108] investigated the problem under similar conditions.

V. K. Ivanov exploited some concepts of mathematical programming and in [20], [21] he was able to get rid of the specification of the correctness function $\omega(\tau)$ while solving problem (1) approximately. At the same time, the knowledge of δ characterizing the accuracy of specifying the right-hand side of (1) was not required. However, the method suggested by Ivanov *required* that the compact set M be specified, that is, the set of correctness of problem (1) be given. Ivanov defined approximate solutions (called *quasi-solutions*) as elements of the set

$$\left\{u_M \in M: \inf_{u \in M} \rho_F(Au,\tilde{f}) = \rho_F(Au_M,\tilde{f})\right\}.$$

The existence of quasi-solutions for each $f \in F$ for continuous A and compact M follows from the fact that the continuous function $\phi(u) = \rho_F(Au,\tilde{f})$, $u \in M$, attains its lower bound on the compact set M. Convergence follows immediately from the theorem on continuity of an inverse operator. The method of quasi-solutions has an obvious geometric interpretation, which was used as a starting point in some further investigations. In the case where U and F are Hilbert spaces and $M = \{u \in U: ||u|| \leq R\}$ (which is weakly compact), the method of quasi-solutions is reduced to solving the operator equation of the second kind

$$(\lambda E + A^*A)u_\lambda = A^*\tilde{f}, \quad \lambda > 0$$

(where E is the identity operator on U) and in addition, the determination of the Lagrange parameter (regularization parameter) from the condition: $||u_\lambda|| = R$. The weak convergence of u_λ to u is assured if $u \in M$.

Y. Douglas [104] applied the method of quasi-solutions in solving numerically Volterra-type integral equations of the first kind

$$Au \equiv \int_0^x k(x,\zeta)u(\zeta)d\zeta = f(x).$$

It is not hard to see that a quasi-solution generalizes the classical solution of the equation $Au = f$. It is also closely related to generalized inverses in the case of a linear operator.

Ivanov as well as others generalized the method of quasi-solutions in various ways. In particular, they required neither uniqueness of the solution of problem (1) nor continuity of the operator A. In [61], for example, the compactness of the set M was not required in constructing quasi-solutions. Also, see [28], [29], and [45].

Tikhonov suggested a new method which he called a "*regularization" method* for solving ill-posed problems [90], [91]. This method has been crucial in the development of methods for solving ill-posed problems and Tikhonov's works [90] and [91] were the basis on which a general theory was formulated. The Tikhonov regularization method is based on the radical idea that the minimum deviation of the values of Au from the specified right-hand side $\tilde{f}$ is stabilized by means of some auxiliary non-negative functional $\Omega(u)$, defined on some subset $U_0 \subseteq D_A$, which itself is a metric space. It is required that the sets $M_c = \{u \in U_0: \Omega(u) \leq c\}$ be compact in U for any C. With regard to the solution of (1), it is assumed that it is contained in M_C for some $C = \overline{C}$. Then the minimum of the parametric functional is to be sought:

$$M_\alpha[u,\tilde{f}] = \rho_F^2(Au,\tilde{f}) + \alpha\Omega(u),$$

for $\alpha > 0$.

Solutions of this problem $\tilde{u}_\alpha$ for some value $\alpha = \alpha(\delta)$ where $\delta: \rho_F(\tilde{f},f) \leq \delta$ are assumed to be the approximations to u. It is proved that if α is chosen, such that $\delta\alpha^{-1/2}$ is uniformly bounded in δ, then the elements $\tilde{u}_\alpha \to \overline{u}$ in the space U_0 as $\delta \to 0$ (later, Bakushinskii [3] and Morozov [50] proved the convergence of $\tilde{u}_\alpha \to \overline{u}$ in the *principal* space U, if $\delta/\sqrt{\alpha} \to 0$, $\delta \to 0$, and U is Hilbert).

The Tikhonov regularization method turned out to be important in applications because it does not require the knowledge of the compact set M containing the *desired* solution $\overline{u}$ of equation (1). For A linear, and U and F Hilbert spaces, the regularization method could be reduced, as in the case of Lavrentiev's method, to solving an equation of the second kind:

$$(\alpha E + A^*A)\tilde{u}_\alpha = A^*f.$$

In the case of non-linear A, the parametric functional defined earlier has to be minimized.

The main difficulty in applying the regularization method lies in deducing *algorithmic* principles for choosing the regularization parameter α. The author deals with this problem in a number of his works. For example, in [50], under the assumptions in the Tikhonov regularization, a technique was suggested for choosing the parameter α from the condition

$$M^{\alpha}[\tilde{u}_{\alpha};\ \tilde{f}] \simeq \delta^2.$$

This method was extended to the non-linear case in the author's "Candidat Dissertation."

In [51] and [52] the choice of the regularization parameter for linear operator equations was suggested and justified, based on the residual principle (the advantageous application of this principle was noted in [16] as well), carrying over the widely used criterion of accuracy of approximate solutions to ill-posed problems: namely, to choose the regularization parameter from the condition

$$\rho(\alpha) = \delta, \quad \rho(\alpha) \equiv ||A\tilde{u}_{\alpha} - \tilde{f}||_F.$$

This principle was also justified in the case where the operator A is only approximately satisfied, and a refinement of this principle has been obtained in [10].

Effective numerical algorithms for choosing the regularization parameters, in practical use, have been obtained by this author together with V. I. Gordonova and by this author in [72], [79]. This allowed to implement and use in computations a number of programs written in Fortran (see "Numerical Analysis in Fortran." Edited by V. V. Vojevodin. Vyp. 6-7. Moskva: Izdatel'stvo MGU, 1974.) Efficient solution was made possible by the use of Vojevodin's method for solving the algebraic problem [8], [9].

The regularization method was further developed in the following works: [3], [6], [7], [12], [18], [33], [44], [48], [79], [86-97]. The residual method represented in terms of an inequality is a special form of the method of the choice of the parameter in accordance with the residual principle. Some recipes for the application of this method are given in Kantorovich [32] and Phillips [111], without, however, indicating any theoretical justification. I. N. Domrovskaya [14] suggested a similar idea, which is inapplicable.

The residual method was investigated and justified in a sufficiently general form in V. K. Ivanov [23], I. N. Domrovskaya [15], and the author [52], and later in T. F. Dolgopolova and V. K. Ivanov [13], the author [55], V. V. Vasin and V. P. Tanana [7], and others.

The residual method in outline is as follows: Let $U_\delta := \{u: \rho_F(Au,\tilde{f}) \leq \delta\}$. Obviously, under the condition that equation (1) has a solution, the set U_δ is non-empty and contains all *formal* solutions to equation (1). Given some non-negative functional $\Omega(u)$, as in the regularization method, we can formulate principles for the choice of meaningful solutions $\tilde{u}_\delta$ satisfying the condition

$$\Omega(\tilde{u}_\delta) = \min_{u \in U_\delta} \Omega(u).$$

It turns out that this method yields convergent approximations to a solution of equation (1) under sufficiently general conditions on the operator A and the functional $\Omega(u)$. Using the method of Lagrange multipliers, it is often possible to reduce the problem to finding the (unconditional) minimum of the functional

$$\phi^\lambda[u;\tilde{f}] = \lambda\rho_F^2(Au,\tilde{f}) + \Omega(u), \quad u \in U_\delta,$$

and to defining an appropriate value of the Lagrange multiplier from the condition

$$\rho_0(\lambda) = \delta,$$

where $\rho_0(\lambda) \equiv \rho_F(Au^\lambda,\tilde{f})$, u^λ is the extreme of ϕ^λ, that is, it can be defined in accordance with the residual principle.

The relationship between various variational approaches was investigated extensively in V. V. Vasin [6], Some aspects of this problem were discussed in [23], [58].

The regularization of the solutions in a specific form occurs in the presence of random noise for a given $\tilde{f}$. The close relationship of the regularization method with the Wiener optimal filtering was emphasized by M. M. Lavrentiev and V. G. Vasiliev [39], and used by V. Ya. Arsenin [1], [2] for optimization of coefficients in the Fourier method, and the choice of the regularization parameter in solving integral equations of the convolution type. This problem was considered by the author in [60] in which optimal estimates for the approximation errors are given for different *classes* of solutions and random perturbations. In the same work the problem of realizing optimal algorithms was formulated, and, furthermore, it was shown that the regularization method yields an unimprovable (with respect to the order) accuracy in the class of admissible solutions. Statistical regularization of systems of linear algebraic equations on the basis of sequential Bayes solutions was investigated in Ye. L. Zhukovsky and the author [17]. Also, see [81], [105].

The problem of estimation of approximate solutions is very crucial in the general theory of the regularization method. Some results related to this problem were obtained in Ivanov [18], [19] and the author [52], [58], in which the influence of errors of the specification of $\tilde{f}$ as well as A was studied. General techniques for estimating the accuracy of the methods for solving ill-posed problems were suggested by Denchev [11] and by Ivanov [28], [31] while considering equation (1) on a compact set. These techniques were justified for a somewhat more general problem in [61], where numerous examples illustrating the application of Ivanov's method of estimation to various cases of the operator equation (1) are given.

With regard to the regularization problem, Ivanov stated the problem finding *maximal* sets of correctness of problem (1). He noted the principal difference between the behavior of approximate regularization algorithms "at a point" and the "uniform" behavior. As it turned out, the "point" regularization occurs in a wide class of Banach spaces (so-called E-spaces), namely, those possessing the uniform convexity property and the Efimov-Stechkin property [30]. The spaces L_p, W_p^ℓ, $p > 1$ are examples of such spaces. An important problem of regularization theory formulated by Ivanov is the clarification of the dependence $\alpha = \alpha(\delta)$ which is necessary and sufficient for regularized solutions to converge in different spaces [25], [26]. The author solved this problem for a specific ill-posed problem, namely, the problem of determining values of an unbounded operator [67]. This problem is the following.

Let a mapping L from U onto a metric space G be given on a set $D_L \subset U$. The problem of determining values of the operator L consists in defining the element

$$g = Lu \tag{2}$$

for a given element $u \in U$.

Problem (2) is called *well-posed* if: (a) the mapping L is an operator; (b) $D_L = U$; (c) the operator L is continuous on U.

In the contrary case problem (2) is called *ill-posed* (in [30], Ivanov considered the problem of determination involving a *multi-valued* operator L). It is not hard to see that the problem of determing values of an *unbounded* operator L is ill-posed. Various cases of this problem were considered by the author in [54], [55], [61]. In [62] and [76], the algorithms for stable determination of values of an unbounded operator, constructed on the basis of a *smoothness* tool developed by the author, were used to justify numerical methods for defining parameters entering the operator equation on the basis of its approximately specified solution (the so-called inverse coefficient problem). With regard to problem (2), Strakhov [85] formulated the problem of determining the optimal accuracy while reconstructing g as well as the problem of finding the appropriate *optimal* algorithm. Strakhov solved this problem for the choice of a regularization parameter for the case where the maximum *a priori* information was required about both the accuracy of specification of δ and its belonging to a class of correctness M. Under weaker *a priori* assumptions, V. A. Morozov indicated the *optimal* (in the order) computational algorithm [67]. In the same work he gave unimprovable errors of such an algorithm and formulated the general problem of constructing optimal (in the order) algorithms under *minimum a priori* assumptions. The author's approach was developed further by V. N. Strakhov, in particular, in [86]. We note that the problem of algorithm optimization was considered by N. S. Bakhvalov [5] and by S. B. Stechkin [84].

The so-called *spectral* problems (for instance, the Neumann problem for the Laplace equation) constitute an important class of ill-posed problems. In order to solve problems of this class, the author developed a special theory of *pseudosolutions* [56], [57].† The general scheme of regularization was essentially modified and the formulations were sharpened. Methods for solving unstable (ill-conditioned, degenerate) systems of linear algebraic equations constitute a sophisticated theory. In the general theory of pseudosolutions *it is not required* that equation (1) have a classical solution. Similar problems for systems of linear algebraic equations were studied by V. V. Vojevodin [9].

In the present book a generalized scheme of solving both equation (1) and the problem of computation (2) is considered, taking the approach due to the author in collaboration with N. N. Kirsanova [76]. We call this problem the problem of computing the values of operator (2), using the solution of the operator equation (1). This problem is, in general, the following.

Let $U_A = \{u \in D_A : Au = f\} \cap D_L \neq \emptyset$. We assume that the set U_A consists of more than one element. We need to compute the value $g = Lu$ on some element $u \in U_A$. Usually, some fixed element $g^* \in G$ is given and the element $g = Lu$ is computed for which

$$\inf_{u \in U_A} \rho_G(Lu,g^*) = \rho_G(Lu,g^*). \qquad (3)$$

Problem (3) is referred to as *well-posed* (correct) if both problems (1) and (2) are well-posed.

†Editors' Note: The theory of pseudosolutions and generalized inverse operators is widely developed in the Western literature.

We can give many reasons for the advantages of such a formulation of a mathematical problem (some of them are given in this book) even in the case when both problems (1) and (2) are well-posed. Some optimal control problems [41], as can easily be seen, reduce to solving problem (3). We have just noted that for $L = E$, that is, an identity operator, problem (3) coincides with problem (1); for $A = E$, it coincides with problem (2). The consideration of problem (3) eliminates the current duplication in scientific literature on ill-posed problems because of separate consideration of problems (1) and (2).

In the main text of this book it is required that problem (1) have a solution only in the sense of the "least squares method". This requirement makes formulations of numerical methods more specific. Sufficiently general conditions are pointed out (in our opinion, these conditions are close to necessary conditions) under which it is possible to construct *regular* approximate solutions of the *basic problem* (3). A wide range of regular methods is investigated, many have never appeared before in the literature. The influence of errors on the specification of both operator A and operator L is studied. Special attention is given to estimates of accuracy of the regularization methods considered for solving the basic problem. We do not assume in the general case that problems (1) and (2) are well-posed, which helps to define the results obtained more precisely. The consideration of inconsistent equations (1) enables us to pose and solve some problems involving *a priori* estimation of the *adequacy* of the mathematical model (1) on the basis of the observations performed [67], [74]. The solution of the latter problem is essential in practice, especially for estimating the adequacy of new mathematical models.

An important point in solving both problems (1) and (2), which is essential from the point of view of applications, is the case of specification of input information as values of functionals or, in the more general case, as values of some system of operators (for instance, "traces" of some function on manifolds of various dimensions). The method of splines [98] is very suitable in this case. Taking a unified approach, the author presents a function space interpretation of the method of splines, which differs essentially from the well-known works [99], [101], in that the problem of existence and uniqueness of splines as well as the *convergence* is considered. An effective tool of smoothing is constructed on the basis of various methods for constructing splines. Also, it is shown that the algorithms suggested are optimal in a wide sense. The application of the method of splines to solving well-posed equations as well as the basic problem is explained (without assuming that the latter is well-posed). The role of the method of splines as an effective algorithm for solving problem (2) is delineated. Algorithms of numerical differentiation of discrete information are considered, in particular, the application of algorithms of fast Fourier transformation [99], and also the method of splines [112], and some other methods.

In this book algorithms for solving the non-linear equations (1) are considered. In particular, the algorithms of choosing a regularization parameter on the basis of consistency between the values $M^{\alpha}[\tilde{u}_{\alpha};\tilde{f}]$ and δ^2 is justified.

The author also developed a new theory for estimating accuracy of solutions of the basic problem. This theory is based upon the introduction of an estimation function, the computation of which is, in our opinion, much simpler than that which is currently available. This approach enables us to formulate both sufficient and necessary conditions for the convergence of regular methods. We note here that the concept of a *regular* approximate method is distinguished by its constructive aspects. This property enables us to prove that the concept of regularity is equivalent to that of convergence of the method. Furthermore, this makes it possible to construct other regular methods, in particular, the deterministic Bayes method considered by the author, which in its formulation is close to the statistical *Bayes method* [17].

In general, the discussion in this book follows the investigations carried out by the author. We do not give here the results related to regular methods based on the method of iterations [4], [50], since we believe that in the present work this problem is solved effectively enough, using the Tithonov regularization method, including special cases of (1) where the operator A is symmetric and non-negative definite (the space $U = H$ is Hilbert). However, we do not touch upon the question of regular methods of minimization of functionals. The author believes that the ideas presented in this book are sufficient in order to construct the appropriate algorithms. The reader who is interested in the problem indicated may refer to works by A. N. Tikhonov, and also [74], [75]. In this book we do not discuss the regularization of some special classes of problems, such as integral equations with kernel of the δ-function type (for numerical solution of which a sufficiently effective method of self-regularization was suggested in [12]), solution of evolutionary problems using the method of quasi-inversion [40], [41], etc.

To conclude our introduction, we note that almost all results given in this book are formulated for Hilbert spaces. However, they can easily be carried over to more general spaces, for instance, uniformly convex spaces and spaces satisfying the Efimov-Stechkin conditions. A more complete survey of the extant methods for solving ill-posed problems can be found in the author's work [73]. See also the survey and expository research papers by Nashed and Payne.

Table of Contents

Chapter 1
The Regularization Method

Section 1. The Basic Problem for Linear Operators

1. Let H, F, and G be Hilbert spaces, and let $A: H \to F$, $L: H \to G$ be linear operators, with domains D_A and D_L, respectively. We assume that the set $D_A \cap D_L =: D_{AL} \neq \emptyset$ and that a non-empty set $D \subseteq D_{AL}$ is given a priori.

The operators A and L are said to be *jointly closed* on D if for any sequence $\{u_n\} \subset D$ such that

$$\lim_{n\to\infty} u_n = u_0 \text{ (in } H), \quad \lim_{n\to\infty} Au_n = f_0 \text{ (in } F), \quad \lim_{n\to\infty} Lu_n = g_0 \text{ (in } G) \tag{1}$$

we have that $u_0 \in D$ and $Au_0 = f_0$, $Lu_0 = g_0$.

Lemma 1. If the operators A and L are closed on D, then they are jointly closed on D.

Proof: The proof follows from the definition of a closed operator.

Remark. The conditions of the lemma are satisfied if one of the operators is closed and the other is bounded (continuous) on H.

Lemma 2. For any $f \in F$, $g \in G$ let the equations

$$Au = f, \quad Lu = g \tag{2}$$

have a common solution $\bar{u} \in D_{AL}$, and let there exist a constant $\gamma > 0$ not depending on $u \in D_{AL}$ such that for all $u \in D_{AL}$

$$|u|^2 := ||Au||_F^2 + ||Lu||_G^2 \geq \gamma^2 ||u||_H^2. \tag{3}$$

Then the operators A and L are jointly closed on D_{AL}.

Proof: It follows from (3) that (2) has a unique solution. Let u_n, $n = 1,2,\ldots$ be a sequence satisfying (1).

Let $\bar{u}_0$ be a solution of (2) corresponding to the right sides f_0 and g_0. By (3) we then have

$$\gamma^2 ||u_n-\bar{u}_0||_H^2 \le ||Au_n-A\bar{u}_0||_F^2 + ||Lu_n-L\bar{u}_0||_G^2 \to 0 \quad \text{as} \quad n \to \infty.$$

Therefore, $u_0 = \bar{u}_0$ and $Au_0 = f_0$, $Lu_0 = g_0$. We have proved the lemma. □

Example 1. Let Ω be a given regular two-dimensional region. Also let $W_2^{(2)}$ be a set of functions which have square-summable derivatives in Ω up to the second order. We define the operators A and L as

$$Au = u|_\Gamma, \qquad Lu = \Delta u,$$

where $u \in W_2^{(2)}$, Γ is the boundary of the region Ω, and Δ is the Laplacian operator. We let $F = L_2(\Gamma)$, $G = L_2(\Omega)$. Then the equations

$$\Delta u = g, \quad u|_\Gamma = f$$

correspond to the Dirichlet problem for the Poisson equation and are solvable for all $g \in L_2(\Omega)$ and $f \in L_2(\Gamma)$. Furthermore, we have

$$\gamma^2 ||u||^2_{L_2(\Omega)} \le ||\Delta u||_{L_2(\Omega)} + ||u|_\Gamma||^2_{L_2(\Gamma)}.$$

Therefore, the operators A and L considered on the set D_{AL} of functions in $W_2^{(2)}$ are jointly closed in $L_2(\Omega)$.

Later, we give other examples of jointly closed operators.

2. Let

$$\mu_A = \inf\{||A_u - f||_F : u \in D\} \tag{4}$$

where $f \in F$ is an element in F such that the set

$$U_A = \{u \in D: ||Au - f||_F = \mu_A\} \text{ is not empty.}$$

We consider now the variational problem of finding $\hat{u} \in D$ for which

$$\nu_L = \inf_{u \in U_A} ||Lu - g||_G = ||L\hat{u} - g||_G, \tag{5}$$

where $g \in G$ is an arbitrary element in G.

We call the problem of finding the elements $\hat{u} \in D$ satisfying (5) the *basic* problem; we call the corresponding elements $\hat{u}$ *solutions of the basic problem*.

Theorem 1. A solution of the basic problem (5) exists and is unique if the following *basic conditions* are satisfied:

(a) the set D is convex;

(b) the operators A and L are jointly closed on D;

(c) the operators A and L satisfy (3) (the *completion condition*).

Proof: We note that for any $u_1, u_2 \in D$ we have the identities

$$||A(\frac{u_1-u_2}{2})||^2 = \frac{1}{2}||Au_1-f||^2 + \frac{1}{2}||Au_2-f||^2 - ||A(\frac{u_1+u_2}{2}) - f||^2 \tag{6}$$

$$||L(\frac{u_1-u_2}{2})||^2 = \frac{1}{2}||Lu_1-g||^2 + \frac{1}{2}||Lu_2-g||^2 - ||L(\frac{u_1+u_2}{2}) - g||^2$$

$$\forall f \in F, \quad g \in G \tag{7}$$

which can be verified immediately.

Assuming that the basic problem is solvable, we show that its solution is unique. In fact, let u_1 and u_2 be two solutions of the basic problem. Then it follows from (6) that

$$||A(\frac{u_1-u_2}{2})||^2 = \mu_A^2 - ||A(\frac{u_1+u_2}{2}) - f||^2 \leq 0$$

because D is convex.

It is easy to see that the set U_A is convex. Similarly, we have

$$||L(\frac{u_1-u_2}{2})||^2 = \nu_L^2 - ||L(\frac{u_1+u_2}{2}) - g||^2 \leq 0.$$

Letting $u = u_1-u_2$ in (3) and also using the relations above, we arrive at $u_1 = u_2$. Thus we have proved that there is at most one solution of the basic problem. Next we show that a solution exists.

Let $u_n \in U_A$ be some minimizing sequence, for example such that

$$\nu_L^2 \leq ||Lu_n - g||^2 \leq \nu_L^2 + \frac{1}{n}, \quad n = 1,2,... \tag{8}$$

Then, letting $u_1 = u_n$ and $u_2 = u_{n+p}$, where p is any natural number, in the relations (6) and (7), we obtain

$$||A(\frac{u_n-u_{n+p}}{2})||^2 = \mu_A^2 - ||A(\frac{u_n+u_{n+p}}{2}) - f||^2 \leq \mu_A^2 - \mu_A^2 = 0$$

$$||L(\frac{u_n-u_{n+p}}{2})||^2 \leq \frac{1}{2}(\nu_L^2 + \frac{1}{n}) + \frac{1}{2}(\nu_L^2 + \frac{1}{n+p}) - ||L(\frac{u_n+u_{n+p}}{2}) - g||^2$$

$$\leq \nu_L^2 + \frac{1}{n} - \nu_L^2 = \frac{1}{n} \to 0$$

as $n \to \infty$ independently of p. Making use of the completion condition (3), we have that

$$||u_n - u_{n+p}|| \to 0 \quad \text{as} \quad n \to \infty$$

independently of p.

Therefore, $\{u_n\}$, $\{Au_n\}$, $\{Lu_n\}$ are Cauchy sequences in the Hilbert spaces H, F and G, respectively. Let u_0, f_0, and g_0 denote their respective limits. Since the operators A and L are jointly closed, $u_0 \in D$, $Au_0 = f_0$, and $Lu_0 = g_0$. In this case, due to the choice of the sequence $\{u_n\}$ we have

$$\mu_A = \lim_{n\to\infty} ||Au_n - f|| + ||Au_0 - f||,$$

that is, $u_0 \in U_A$.

Taking the limit in (8), we obtain $\nu_L = ||Lu_0 - g||$ which implies that the element $u_0 \in U_A$ is a solution of the basic problem (5). □

3. The completion condition (3) does not require that *each* of the quadratic forms: $||u||_A^2 = ||Au||^2$, $||u||_L^2 = ||Lu||^2$ be positive definite. If

$$||Au||^2 \geq \gamma_A^2 ||u||^2, \qquad ||Lu||^2 \geq \gamma_L^2 ||u||^2, \qquad u \in D,$$

where the constants γ_A and γ_L do not depend on $u \in D$, the condition (3) is satisfied; for example, for $\gamma_A^2 + \gamma_L^2 = \gamma^2 > 0$. We note, however, that it can happen that both $||Au||^2$ and $||Lu||^2$ are only non-negative while the *sum* is positive definite.

<u>Example 2</u>. Let $H = F = G = R^2$ and $u = (u_1, u_2) \in R^2$. We define the operators A and L as

$$Au = (u_1, 0), \qquad Lu = (0, u_2).$$

Then, $||Au||^2 = u_1^2$, $||Lu||^2 = u_2^2$ and both quadratic forms are only non-negative. However, the sum

$$||Au||^2 + ||Lu||^2 \equiv u_1^2 + u_2^2$$

is, obviously, positive definite.

It is not hard to see that in Example 1 the operators A and L define quadratic forms which are only non-negative.

<u>Example 3</u>. Let the range $R(A)$ of the operator A be all of F, and let the range $R(L)$ of the operator L be all of G. Finally, let the

completion condition (3) be satisfied.

Then, obviously, $\mu_A = 0$ for any $f \in F$. The system (2) assumes the form

$$Lu = g, \quad u \in U_A. \tag{9}$$

The condition $u \in U_A$ can be interpreted as the imposition of boundary (or initial) conditions on the function u. If $L(U_A) = G$, Eq. (9) is solvable for any $g \in G$ and the solution is unique due to Theorem 1.

The condition (3) insures in this case that an *a priori estimate* is satisfied for the solution (9), namely

$$||\hat{u}||^2 \leq \frac{1}{\gamma^2} (||f||^2 + ||g||^2)$$

which implies the stability of the solution by the hypothesis of the problem; that is, the problem (9) is well-posed in the Hadamard sense. As follows from Example 1, we may write the Dirichlet problem for the Poisson equation in the form (9).

4. If $\mu_\Lambda = 0$, we call the basic problem a '*consistent*' problem. In the general case we do not assume that the basic problem is consistent. However, we shall assume in the sequel that the basic conditions (1), (b), and (c) in Theorem 1 are satisfied.

Let $H = G$ and let the operator L be the identity operation E on H. Then the completion condition (3) is satisfied a priori for each A. The condition for a joint closure is satisfied if the operator A is closed. In this event the basic problem is reduced to the problem of solving the operator equation

$$Au = f, \quad f \in F. \tag{10}$$

If $H = F$ and $A = E$, the completion condition (3) can be satisfied for any operator L. The condition for joint closure can, obviously, be satisfied, if the operator L is closed. In this event the basic problem (for $g = 0$) is reduced to the problem of determining the element $\hat{g} = L\hat{u}$; we then call the basic problem a problem of evaluation.

It is not hard to see that the basic problem is a generalization of the two problems. In the general case, we do not assume that the basic problem is well-posed.

5. We give now examples leading to the basic problem.

Example 4. Let $H = W_2^n[a,b]$ be the space of functions $u = u(x)$ with square-summable derivatives up to the nth order. Let

$$||u||^2_{W^n_2} = \int_a^b [u^{(n)}]^2(x)dx + \int_a^b u^2(x)dx$$

Then the problem of solving the equation

$$Au := \int_a^b k(\cdot,\xi)u(\xi)d\xi = f, \quad f \in L_2[a,b] \tag{11}$$

where the kernel $k = k(x,\xi)$ is a continuous function of its arguments, is ill-posed.

If Eq. (11) is solvable, we may assume that the operator L is the identity operator E in $W^2_2[a,b]$. It is seen that the completion condition (3) is satisfied.

Example 5. Let the spaces $H = F = G = L_2[a,b]$, and define the operator L by $Lu := d^nu/dx^n$ on the set $D = W^n_2$. The basic problem consists of determining the element $\hat{g} = L\hat{u}$, where $\hat{u} \in W^n_2$. Since the approximations to u from the space $L_2[a,b]$ are admissible, the given problem of *differentiation* is, obviously, unstable.

Example 6. Let $H = W^n_2[a,b]$ and define the operator A by

$$Au := (\ell_1(u),\ell_2(u),\ldots,\ell_n(u)), \quad u \in W^n_2$$

where the ℓ_i are linear (and bounded in W^n_2) functionals which operate from W^n_2 into $F = R^n$ with norm $||r|| = (\Sigma^n_{i=1} r^2_i)^{1/2}$ for $r \in R^n$. The problem of finding a solution of the equation

$$Au = f, \quad f = (f_1,f_2,\ldots,f_n) \in R^n \tag{12}$$

is known as the problem of *interpolation*. It is easily seen that this problem is ill-posed (the uniqueness condition of the interpolating function is violated). If we let $Lu \equiv d^nu/dx^n$ the basic problem is reduced to the problem of *smooth interpolation*: to find the function $\hat{u} \in U^n$ for which

$$||L\hat{u}||_{L_2} = \inf\{||Lu||_{L_2} : Au = f\}. \tag{13}$$

The function $\hat{u}$ satisfying Eq. (13) is said to be an interpolating spline. The general theory of splines will be considered in Chapter 4.

Other examples will be considered as we proceed.

Section 2. The Approximation of the Solution of the Basic Problem

1. In order to find the solution $\hat{u}$ of the basic problem, it is necessary to know the set U_A, which may be ill-specified. It is natural that we should get rid of this shortcoming of the problem.

First, we define the parametric functional

$$\Phi_\alpha[u] := ||Au-f||^2 + \alpha||Lu-g||^2, \quad u \in D$$

where $\alpha > 0$ is called the regularization parameter, and $f \in F$ and $g \in G$ are given.

We consider the problem of finding *regularized solutions*, that is, elements $u_\alpha \in D$ such that

$$\mu_\alpha^2 := \inf_{u \in D} \Phi_\alpha[u] = \Phi_\alpha[u_\alpha].$$

We have the following theorem.

Theorem 2. For any $\alpha > 0$, the minimization problem (1) has a unique solution.

Proof: It is not hard to see that the identity

$$||A(\frac{u_1-u_2}{2})||^2 + \alpha||L(\frac{u_1-u_2}{2})||^2 = \frac{1}{2}\Phi_\alpha[u_1] + \frac{1}{2}\Phi_\alpha[u_2] - \Phi_\alpha[\frac{u_1+u_2}{2}]$$

can be satisfied for any $u_1, u_2 \in D$. If these are the solutions of (1), then

$$||A(\frac{u_1-u_2}{2})||^2 + \alpha||L(\frac{u_1-u_2}{2})||^2 = \mu_\alpha^2 - \Phi_\alpha[\frac{u_1+u_2}{2}] \leq 0,$$

since $u_1+u_2/\alpha \in D$. However, then $A(u_1-u_2) = 0$, $\ell(u_1-u_2) = 0$. Making use of the completion condition, we have $u_1-u_2 = 0$, that is, $u_1 = u_2$. Therefore, the problem (1) has a unique solution.

We can prove the existence of the regularized solution u_α in a similar way as in Theorem 1.

2. Let $\hat{u}_\alpha$ be the solution of (1).

Theorem 3. If $\alpha \to 0$, then $||\hat{u}_\alpha - \hat{u}|| \to 0$.

Remark. It is seen that $||\hat{u}_\alpha-\hat{u}||_H$ tends to zero as $\alpha \to 0$.

Theorem 3 shows that the regularized solutions $\hat{u}_\alpha$ for small α approximate the solution of the basic problem. Since in minimizing the functional in (1) we need not know the set U_A, the regularized solutions can be defined effectively.

In order to prove Theorem 3, we need to introduce the following definition. The operators A and L are said to be *jointly weakly closed* on D if it follows from the relations

$$u_n \xrightarrow{\text{weak}} u_0 \text{ (in } H), \quad Au_n \xrightarrow{\text{weak}} f_0 \text{ (in } F), \quad Lu_n \xrightarrow{\text{weak}} g_0 \text{ (in } G), \quad \text{as } n \to \infty, \tag{2}$$

where $\xrightarrow{\text{weak}}$ denotes the weak convergence in the corresponding space, that $u_0 \in D$, $Au_0 = f_0$, $Lu_0 = g_0$. In this case the set D is not necessarily convex; furthermore, we do not assume that the operators A and L are linear.

We prove Theorem 3 using the following lemma.

Lemma 3. If the operators A and L are linear and the set D is convex, a necessary and sufficient condition for the operators A and L on D to be jointly weakly closed is that they be jointly closed.

Necessity: If the operators A and L are jointly weakly closed on D, they are, obviously, jointly closed, since strong convergence implies weak convergence.

Sufficiency: Let the operators A and L be jointly closed. We shall show that they will also be jointly weakly closed on D.

Let the relations (2) be satisfied. By the Banach-Saks theorem there exist convex linear combinations of points of the sequence $\{u_k, k \geq n\}$ strongly converging to u_0; that is, there exist elements

$$v_{nm} = \sum_{i=n}^{n+m} \lambda_{inm} u_i, \quad \lambda_{inm} \geq 0, \quad \sum_{i=n}^{n+m} \lambda_{inm} = 1$$

such that for each $n = 1,2,\ldots$ the relation

$$\lim_{m\to\infty} ||v_{nm} - u_0|| = 0$$

can be satisfied. This implies that there exists a sequence $\{\bar{v}_n\}$, $\bar{v}_n = v_{nn}$, strongly converging to u_0 (in H). Due to convexity of D, $\bar{v}_n \in D$ for any n.

We show now that as $n \to \infty$ the relations

$$A\bar{v}_n \xrightarrow{\text{weak}} f_0, \quad L\bar{v}_n \xrightarrow{\text{weak}} g_0 \tag{3}$$

hold.

Let z be an arbitrary element in F. Then, using the linearity of A, we have:

$$|(A\overline{v}_n - f_0, z)_F| = |\sum_{i=n}^{n+m_n} \lambda_{inm_n}(Au_i - f_0, v)|$$

$$\leq \max_{n\leq i\leq n+m_n} |(Au_i - f_0, z)_F| \leq \sup_{i\geq n} |(Au_i - f_0, z)_F| = m_n$$

Since, obviously, $\lim_{n\to\infty} m_n = 0$, then

$$\lim_{n\to\infty} |(A\overline{v}_n - f_0, z)| = 0$$

for each $z \in F$; i.e., $A\overline{v}_n \xrightarrow{\text{weak}} f_0$.

We prove the second relation in (3) in a similar way.

Let $f_n = A\overline{v}_n$. We construct the sequence $\overline{f}_n$ in a similar way as the sequence $\overline{v}_n$. Let

$$\overline{f}_n := \sum_{i=n}^{n+k_n} \mu_{ink_n} A\overline{v}_i = A\left(\sum_{i=n}^{n+k_n} \mu_{ink_n} \overline{v}_i\right)$$

where $\mu_{ink_n} \geq 0$, $\Sigma_{i=1}^{n+k_n} \mu_{ink_n} = 1$. Furthermore, let

$$\hat{v}_n := \sum_{i=n}^{n+k_n} \mu_{ink_n} \overline{v}_i.$$

Then

$$A\hat{v}_n \to f_0$$

by construction.

Next, we show that $\hat{v}_n \to u_0$. In fact,

$$||\hat{v}_n - u_0|| = ||\sum_{i=n}^{n+k_n} \mu_{ink_n}(\overline{v}_i - u_0)|| \leq \sup_{i\geq n} ||\overline{v}_i - u_0|| \to 0$$

as $n \to \infty$ by construction of the sequence $\{\overline{v}_i\}$. We note that the elements $\hat{v}_n \in D$ are convex combinations of the elements of the sequence $\{\overline{v}_i\}$.

Thus, we have constructed the sequence $\{\hat{v}_n\} \subset D$ in a way that

$$\hat{v}_n \to u_0 \text{ (in } H), \quad A\hat{v}_n \to f_0 \text{ (in } F). \tag{4}$$

Carrying out similar constructions and keeping the same notation, we assume also that

$$L\hat{v}_n \to g_0 \quad \text{(in } G). \tag{5}$$

Taking advantage of the joint closure of the operators A and L on D, it follows from (4) and (5) that

$u_0 \in D$;

i.e., the operators A and L are jointly weakly closed. □

Proof of Theorem 3: Making use of the extremal property of the regularized solutions $\hat{u}_\alpha$, we have

$$||A\hat{u}_\alpha - f||^2 + \alpha||L\hat{u}_\alpha - g||^2 \le ||A\hat{u} - f||^2 + \alpha||L\hat{u} - g||^2, \tag{6}$$

where $\hat{u}$ is the solution of the basic problem. Since $\hat{u}_\alpha \in D$, the extremal property of the solution of the basic problem implies that

$$||A\hat{u} - f|| \le ||A\hat{u}_\alpha - f||. \tag{7}$$

From (6) and (7) it follows that

$$||A\hat{u}_\alpha - f||^2 \le \mu_A^2 + \alpha\nu_L^2, \quad ||L\hat{u}_\alpha - g|| \le ||L\hat{u} - g|| = \nu_L \tag{8}$$

for any $\alpha > 0$.

Therefore, the families $A\hat{u}_\alpha$, $L\hat{u}_\alpha$ for α: $0 < \alpha \le \bar{\alpha}$ are bounded and, consequently, weakly compact, since F and G are Hilbert spaces. Taking advantage of the completion condition, we see that the family $\hat{u}_\alpha$ is weakly compact as well.

Let $\alpha_n > 0$, $n = 1,2,\ldots$ be a sequence tending to zero. We may assume without loss of generalty that the sequences $u_n = \hat{u}_{\alpha_n}$, $f_n = A\hat{u}_{\alpha_n}$, $g_n = L\hat{u}_{\alpha_n}$ are weakly convergent;

$$u_n \xrightarrow{\text{weak}} u_0, \quad Au_n \xrightarrow{\text{weak}} f_0, \quad Lu_n \xrightarrow{\text{weak}} g_0, \tag{9}$$

in the corresponding spaces.

Making use of the joint weak closure of the operators A and L on D (Lemma 3), we have:

$$u_0 \in D, \quad Au_0 = f_0, \quad Lu_0 = g_0. \tag{10}$$

Since the norm in a Hilbert space is weakly lower semi-continuous, we have from Eqs. (8), (9) and (10):

$$||Au_0 - f|| \le \underline{\lim}_{n\to\infty} ||Au_n - f|| \le \overline{\lim}_{n\to\infty} ||Au_n - f|| \le \mu_A, \tag{11}$$

$$||Lu_0 - g|| \le \underline{\lim}_{n\to\infty} ||Lu_n - g|| \le \overline{\lim}_{n\to\infty} ||Lu_n - g|| \le \nu_L. \tag{12}$$

It follows from (11) that $u_0 \in U_A$, and also from (12) that u_0 is the solution of the basic problem and, therefore, coincides with $\hat{u}$. Thus, it follows that

$$||A\hat{u}-f|| = \lim_{n\to\infty} ||Au_n-f|| = \mu_A$$
$$||L\hat{u}-g|| = \lim_{n\to\infty} ||Lu_n-g|| = \nu_L. \tag{13}$$

Noting that in a Hilbert space weak convergence together with convergence of the norms implies strong convergence, we deduce from (9) and (13) that

$$\lim_{n\to\infty} Au_n = A\hat{u}, \quad \lim_{n\to\infty} Lu_n = L\hat{u}.$$

From the completion condition it also follows that

$$\lim_{n\to\infty} u_n = \hat{u}.$$

Since the sequence $\alpha_n \to 0$ was arbitrary, similar relations hold for the entire family of regularized solutions. □

Corollary 1. If $\mu_A = 0$, that is, the basic problem is consistent, it follows from (8) that

$$||A\hat{u}_\alpha-\bar{f}|| \le \sqrt{\alpha}\,\nu_L \to 0 \quad \text{as} \quad \alpha \to 0.$$

If the operator A has, in addition, a bounded inverse, the estimate

$$||\hat{u}_\alpha-\hat{u}|| \le \sqrt{\alpha}\,\nu_L||A^{-1}||$$

holds, characterizing the rate of convergence of the regularized solutions $\hat{u}_\alpha$ to the solution of the basic problem.

Corollary 2. If $\mu_A = \nu_L = 0$ simultaneously, the regularized solutions $\hat{u}_\alpha \equiv \hat{u}$.

Section 3. The Euler Variation Inequality. Estimation of Accuracy

1. The rate of convergence of the deviation $||\hat{u}_\alpha-\hat{u}||$ to zero may be arbitrarily slow if we make no additional assumptions. Hence it is desirable to distinguish those cases where a prescribed rate of convergence can be guaranteed. Before solving this problem, we derive one essential property of regularized solutions.

Theorem 4. In order that $u_\alpha \in D$ be a solution of the problem

$$\Phi_\alpha[u_\alpha] = \inf_{u\in D} \Phi_\alpha[u]$$

where

$$\Phi_\alpha[u] \equiv ||Au-f||^2 + \alpha||Lu-g||^2$$

it is necessary and sufficient that the *Euler variational inequality* be satisfied:

$$(Au_\alpha - f, A(v-u_\alpha))_F + \alpha(Lu_\alpha - g, L(v-u_\alpha))_G \geq 0, \quad \forall v \in D \tag{1}$$

Proof: It is easy to show that the functional Φ_α is convex for each $\alpha > 0$. Hence a necessary and sufficient condition for $u_\alpha \in D$ to minimize Φ_α over D is that the (one-sided) directional derivative be non-negative, i.e.,

$$\lim_{t\to 0^+} \frac{\Phi_\alpha[u+tv] - \Phi_\alpha[u]}{t} \geq 0 \tag{2}$$

Let $\psi(t) := \Phi_\alpha(u+tv)$. Then $\psi'(0^+) \geq 0$ and this implies (1) by simple computations. □

2. Let D be a linear space. Assuming that $v = u_\alpha \pm w$ in (1), we have that the element u_α is characterized by the relation

$$(Au_\alpha - f, Aw)_F + \alpha(Lu_\alpha - g, Lw)_G = 0 \quad \forall w \in D, \tag{3}$$

which is the *Euler identity*.

We consider now the case where $D = D_{AL}$. We assume that $\overline{D}_{AL} = H$, where the bar denotes closure. This condition ensures the existence and uniqueness of the adjoint operators A^* and L^* as single-valued maps.

Let either A or L be bounded on D; to be specific, A. Then the functional $(Lu_\alpha - g, Lw)_G$ is bounded on D_{AL}. Therefore, the element $Lu_\alpha - g \in D_{L^*}$; that is, it is representable in the form $(L^*(Lu_\alpha - g), w)_H$, all $w \in D_{AL}$. Then the Euler identity (3) assumes the form

$$(A^*Au_\alpha - A^*f, w)_H + \alpha(L^*(Lu_\alpha - g), w) = 0.$$

Since, by hypothesis, $\overline{D}_{AL} = H$, the above relation is equivalent to the following *Euler equation:*

$$A^*Au_\alpha + \alpha L^*(Lu_\alpha - g) = A^*f. \tag{4}$$

In particular, if $A = E$ (that is, we consider the problem of determining the range of the operator L), the Euler equation becomes:

$$u_\alpha + \alpha L^*(Lu_\alpha - g) = f. \tag{5}$$

If the element g is sufficiently smooth, that is, $g \in D_{L^*}$, (5) becomes

$$(E + \alpha L^*L)u_\alpha = f + \alpha L^*g \tag{6}$$

Similarly, in the case where the operator L is bounded, for determining the regularized solutions u_α we have the Euler equation of the form

$$A^*(Au_\alpha - f) + \alpha L^*Lu_\alpha = \alpha L^*g. \tag{7}$$

If $L = E$ (that is, we consider the problem of solving the operator equation $Au = f$), we may write Eq. (7) as

$$A^*(Au_\alpha - f) + \alpha u_\alpha = \alpha g. \tag{8}$$

If $f \in D_{A^*}$, the Euler equation becomes:

$$(\alpha E + A^*A)u_\alpha = A^*f + \alpha g \tag{9}$$

We note that Eqs. (4)-(9) have unique solutions for any $\alpha > 0$, which follows from the fact that they are equivalent to the corresponding Euler identities.

The following assertion characterizes the solution $\hat{u}$ of the basic problem.

Theorem 5. Let $\hat{u}$ be the solution of the basic problem. Then the Euler inequality

$$(A\hat{u}-f,\ A(v-\hat{u}))_F \geq 0 \quad \forall v \in D. \tag{10}$$

is satisfied. If D is a linear space, the Euler identity

$$(A\hat{u}-f,\ Aw)_F = 0 \quad \forall w \in D.$$

holds.

Proof: It suffices to take the limit in Eq. (1) as $\alpha \to 0$ and use the theorem on the convergence of regularized solutions.

3. In Section 2 we proved the rate of convergence of the images of the regularized solutions $\hat{u}_\alpha$ under the operator A to $A\hat{u}$, provided the basic problem is consistent. We consider now the general case.

Let $\hat{u}_\beta \in D$ be regularized solutions for the parameters $\beta > 0$. By Theorem 4 we have

$$(A\hat{u}_\beta - f,\ A(w-u_\beta))_F + \beta(Lu_\beta - g,\ L(w-\hat{u}_\beta))_G \geq 0 \quad \forall w \in D. \tag{11}$$

Letting $v = \hat{u}_\beta$ in (1), $w = \hat{u}_\alpha$ in (11), and adding the resulting inequalities, we have

$$-(A(\hat{u}_\beta-\hat{u}_\alpha),\ A(\hat{u}_\beta-\hat{u}_\alpha))_F - \alpha(L(\hat{u}_\beta-\hat{u}_\alpha),\ L(\hat{u}_\beta-\hat{u}_\alpha))_G$$
$$+ (\alpha-\beta)(L\hat{u}_\beta-g,\ L\hat{u}_\beta-L\hat{u}_\alpha)_G \geq 0,$$

or

$$||A(\hat{u}_\beta-\hat{u}_\alpha)||^2 + \alpha||L\hat{u}_\beta-L\hat{u}_\alpha||^2 \leq (\alpha-\beta)(L\hat{u}_\beta-g,\ L\hat{u}_\beta-L\hat{u}_\alpha)_G.$$

Taking the limit as $\beta \to 0^+$ and applying the Cauchy-Schwarz inequality, we have

$$||A\hat{u}-A\hat{u}_\alpha||^2 + ||L\hat{u}-L\hat{u}_\alpha||^2\alpha \leq \alpha||L\hat{u}-g||\,||L\hat{u}-L\hat{u}_\alpha||, \tag{12}$$

which yields the following theorem.

Theorem 6. Let $\hat{u}$ be a solution of the basic problem and let $\hat{u}_\alpha$ be regularized solutions. Then we have

$$||L\hat{u}-L\hat{u}_\alpha|| \leq \nu_L, \qquad ||A\hat{u}_\alpha-A\hat{u}|| \leq \sqrt{\alpha}\,\nu_L. \tag{13}$$

Corollary. If $\nu_L = 0$, $A\hat{u}_\alpha \equiv A\hat{u}$, $L\hat{u}_\alpha \equiv L\hat{u}$ for $\alpha > 0$, as we see from (13). It also follows from the completion condition that $\hat{u}_\alpha \equiv u$.

The proof is immediate. In fact, the extremal properties of regularized solutions and the solution of the basic problem for $\nu_L = 0$ imply the inequalities

$$\begin{aligned} ||A\hat{u}_\alpha-f||^2 + \alpha||L\hat{u}_\alpha-g||^2 &\leq ||A\hat{u}-f||^2 + \alpha||L\hat{u}-g||^2 \\ &= ||A\hat{u}-f||^2 \\ &\leq ||Au_\alpha-f||^2 \qquad \forall\alpha > 0 \end{aligned}$$

which, in turn, imply

$$||A\hat{u}_\alpha-f|| = ||A\hat{u}-f|| = \mu_A, \qquad ||L\hat{u}_\alpha-g|| = 0 \qquad \forall\alpha > 0.$$

Therefore, by Theorem 1, on the uniqueness of solution of the basic problem, $\hat{u}_\alpha \equiv \hat{u}$, which was to be proved.

4. We shall assume now that the set D is a linear space and that it is dense in H. We then have the Euler identity (4). Assuming in the latter that $\hat{u}_\alpha = \hat{u}+z_\alpha$, where $\hat{u}$ is the solution of the basic problem, we have, using Theorem 5:

$$(Az^\alpha, Aw)_F + \alpha(Lz^\alpha+L\hat{u}-g,\ Lw)_G = 0 \qquad \forall w \in D. \tag{14}$$

The following theorems deal with the rate of convergence of the regularized solutions $\hat{u}_\alpha$ to the solution $\hat{u}$ of the basic problem if the latter solution possesses appropriate smoothness properties.

Theorem 7. Let A be a bounded linear operator, let $L\hat{u}-g \in D_{L*}$, and let there exist at least one element $\bar{y} \in F$ such that

$$L^*(L\hat{u}-g) = A^*\bar{y}. \tag{15}$$

Then we have the following:

$$||L\hat{u}_\alpha - L\hat{u}||_G \le \sqrt{\alpha}||\bar{y}||_F, \quad ||A\hat{u}_\alpha - A\hat{u}||_F \le 2\alpha||\bar{y}||_F,$$
$$||\hat{u}_\alpha - \hat{u}||_H \le 1/\gamma(\alpha+4\alpha^2)^{1/2}||\bar{y}||_F. \tag{16}$$

Proof: Under the condition (15) we obtain from the identity (14):

$$(Az^\alpha+\alpha\bar{y}, Aw)_F + \alpha(Lz^\alpha, Lw)_G = 0, \quad \forall w \in D.$$

By Theorem 4, we see that the element z^α attains the infimum of the functional

$$||Az+\alpha\bar{y}||^2 + \alpha||Lz||^2, \quad z \in D.$$

Choosing $z = 0$ for comparison, we obtain

$$||Az^\alpha+\alpha\bar{y}||^2 + \alpha||Lz^\alpha||^2 \le \alpha^2||\bar{y}||.$$

Therefore,

$$||L\hat{u}_\alpha - L\hat{u}|| \le \sqrt{\alpha}||\bar{y}||, \quad ||A\hat{u}_\alpha - A\hat{u}|| \le 2\alpha||\bar{y}||.$$

It remains only to use the completion condition in order to estimate the deviation $||u_\alpha - \hat{u}||$. □

Remark. The estimates (16) hold for unbounded A as well, provided only that the operator L is bounded.

Theorem 8. Let the operator A be bounded, let $L\hat{u}-g \in D_{L^*}$, and let there exist at least one element $\bar{h} \in D_L$ such that

$$L^*(L\hat{u}-g) = A^*A\bar{h}. \tag{17}$$

Then we have the following:

$$||L\hat{u}_\alpha - L\hat{u}||_G \le \alpha||L\bar{h}||_G,$$
$$||A\hat{u}_\alpha - A\hat{u}||_F \le \alpha||A\bar{h}||_F + \alpha^{3/2}||L\bar{h}||_G, \tag{18}$$
$$||\hat{u}_\alpha - \hat{u}|| \le \frac{1}{\gamma}\alpha[||L\bar{h}||^2 + (||A\bar{h}||_F + \sqrt{\alpha}||L\bar{h}||^2)]^{1/2}.$$

Proof: From Eq. (14) and the condition (17) we obtain:

$$(Az_\alpha+\alpha A\bar{h}, Aw)_F + \alpha(Lz_\alpha, Lw)_G = 0 \quad \forall w \in D.$$

By Theorem 4 it then follows that the element z_α attains the infimum of the functional

$$||Az+\alpha A\bar{h}||^2 + \alpha||Lz||^2, \quad z \in D.$$

Choosing $z = -\alpha\bar{h}$ for comparison, we obtain

$$||Az_\alpha+\alpha A\bar{h}||_F^2 + \alpha||Lz_\alpha||_G^2 \leq \alpha^3||L\bar{h}||^2.$$

Using the completion condition, we obtain (18), thus proving the theorem.

Remark. The estimates (18) hold for unbounded A as well, provided only that the operator L is bounded.

Let $L = E$. Then the conditions (15) and (17) assume the form

$$\hat{u} - g = A^*\bar{y} \tag{19}$$

or

$$\hat{u} - g = A^*A\bar{h}. \tag{20}$$

Let $A = E$. Then the conditions (15) and (17) become smoothness conditions of the solution of the basic problem:

$$L\hat{u} - g \in D_{L^*}, \tag{21}$$

$$L^*(L\hat{u}-g) \in D_L. \tag{22}$$

We note that the conditions (15) and (17) are satisfied a priori if the element g is chosen by

$$g = L\hat{u}.$$

Then the conditions (15), (17) are satisfied for $\bar{y} = \bar{h} = 0$. It is easily seen that

$$L\hat{u}_\alpha \equiv L\hat{u}, \quad A\hat{u}_\alpha \equiv A\hat{u}, \quad u_\alpha \equiv \hat{u} \quad \text{for} \quad \alpha > 0.$$

Sufficient conditions under which the relations (19) and (20) are satisfied are given in Chapter 5.

Section 4. Stability of Regularized Solutions

1. We consider the regularized solution as a function of the elements f and g, the operators A and L, and the set D.

It is very easy to estimate the stability of regularized solutions under perturbation of the elements f and g. We assume that instead of the elements f and g we are given the approximations f_δ and g_τ such that

$$||f_\delta-f||_F < \delta, \quad ||g_\tau-g||_G < \tau,$$

where δ and τ characterize the accuracy of approximation of these elements. We note that the element f_δ *need not* satisfy the condition for the problem (4) in Section 1 to be solved. Thus, f_δ may be an arbitrary element in the δ-neighborhood of the element f.

For brevity, we denote by $\tilde{u}_\alpha$ the solution of the *regularized* problem (1) in Section 2 for $f = f_\delta$, $g = g_\tau$ and for fixed values of the remaining parameters. By Theorem 2 the elements $\tilde{u}_\alpha \in D$ are uniquely defined for all $\alpha > 0$. By Theorem 4 the variational inequality

$$(A\tilde{u}_\alpha - f_\delta,\ A(w-\tilde{u}_\alpha))_F + \alpha(L\tilde{u}_\alpha - g_\tau,\ L(w-\tilde{u}_\alpha)) \geq 0, \quad \forall w \in D \tag{1}$$

holds. We write a similar inequality for the regularized solutions $\hat{u}_\alpha$:

$$(A\hat{u}_\alpha - f,\ A(v-\hat{u}_\alpha))_F + \alpha(L\hat{u}_\alpha - g,\ L(v-\hat{u}_\alpha))_G \geq 0, \quad \forall v \in D. \tag{2}$$

Assuming $w = \hat{u}_\alpha$ in (1), $v = \tilde{u}_\alpha$ in (2) and, furthermore, adding both inequalities obtained, we have

$$(A(\hat{u}_\alpha - \tilde{u}_\alpha) - \delta f,\ A(\tilde{u}_\alpha - \hat{u}_\alpha))_F + \alpha(L(\hat{u}_\alpha - \tilde{u}_\alpha) - \delta g,\ L(\tilde{u}_\alpha - \hat{u}_\alpha))_G \geq 0,$$

$$\delta f = f - f_\delta, \quad \delta g = g - g_\tau.$$

Therefore,

$$||A(\tilde{u}_\alpha - \hat{u}_\alpha)||^2 + \alpha||L(\tilde{u}_\alpha - \hat{u}_\alpha)||^2 \leq (\delta f,\ A(\hat{u}_\alpha - \tilde{u}_\alpha))_F + \alpha(\delta y,\ L(\hat{u}_\alpha - \tilde{u}_\alpha))_G. \tag{3}$$

We thus arrive at the following theorem.

Theorem 9. We have the following estimates for the deviations of $\tilde{u}_\alpha$ from $\hat{u}_\alpha$:

$$||L\tilde{u}_\alpha - L\hat{u}||_G \leq \frac{\delta}{\sqrt{\alpha}} + \tau, \quad ||A\tilde{u}_\alpha - A\hat{u}||_F \leq \delta + \sqrt{\alpha}\,\tau$$
$$||\tilde{u}_\alpha - \hat{u}||_H \leq \frac{1}{\gamma}\left(\frac{\delta}{\sqrt{\alpha}} + \tau + \delta + \sqrt{\alpha}\,\tau\right). \tag{4}$$

Proof: Applying the Cauchy-Schwarz inequality, we obtain from (3)

$$||A(\tilde{u}_\alpha - \hat{u}_\alpha)||^2 + \alpha||L(\tilde{u}_\alpha - \hat{u}_\alpha)||^2$$
$$\leq \delta||A(\tilde{u}_\alpha - \hat{u}_\alpha)||_F + \alpha\tau||L(\tilde{u}_\alpha - L\hat{u}_\alpha)||_G$$
$$\leq \frac{\delta^2}{2} + \frac{1}{2}||A(\tilde{u}_\alpha - \hat{u}_\alpha)||^2 + \alpha\left(\frac{\tau^2}{2} + \frac{1}{2}||L(\tilde{u}_\alpha - \hat{u}_\alpha)||^2\right)$$

(here we have used the elementary inequality $2ab \leq a^2 + b^2$).

Therefore,

$$||A(\tilde{u}_\alpha-\hat{u}_\alpha)||^2 + \alpha||L(\tilde{u}_\alpha-\hat{u}_\alpha)||^2 \leq \delta^2 + \alpha\tau^2$$

which together with the completion condition for the operators A and L proves (4). □

Remark. For small values of the regularization parameter α the perturbed regularized solution depends less on the perturbations of the element g than on those of the element f.

The convergence theorem for regularized solutions, (3), and Theorem 9 imply the following theorem.

Theorem 10. Let the regularization parameter $\alpha > 0$ be chosen such that the consistency condition

$$\lim_{\delta,\alpha\to 0} \frac{\delta}{\sqrt{\alpha}} = 0 \tag{5}$$

is satisfied. Then

$$||A\tilde{u}_\alpha-A\hat{u}|| + ||L\tilde{u}_\alpha-L\hat{u}|| + ||\tilde{u}_\alpha-\hat{u}|| \to 0$$

for $\delta,\tau,\alpha \to 0$.

Remark 1. The condition (5) is sufficient for the perturbed regularized solutions to converge to the solution $\hat{u}$ of the basic problem. Further, in investigating the problem of determining the range of an unbounded operator (Chapter 4) we shall show that this condition is also necessary for the convergence mentioned.

Remark 2. Let the conditions of Theorem 7 be satisfied. Then the deviations of the images of the regularized solutions from those of the solution of the basic problem with respect to the operators A and L are seen to be

$$||A\tilde{u}_\alpha-A\hat{u}|| \leq 2\alpha||\bar{y}|| + \delta + \sqrt{\alpha}\,\tau$$
$$||L\tilde{u}_\alpha-L\hat{u}|| \leq \sqrt{\alpha}\,||\bar{y}|| + \frac{\delta}{\sqrt{\alpha}} + \tau.$$

We find $\alpha_{k,opt}$ by minimizing the function

$$\sqrt{\alpha} + \frac{\delta}{\sqrt{\alpha}}. \tag{6}$$

We then have $\alpha_{k,opt} = \delta$. The corresponding estimates are seen to be

$$\begin{aligned} ||A\tilde{u}_{\alpha_{k,opt}}-A\hat{u}|| &\leq 2||\bar{y}||\delta + \tau\sqrt{\delta} + \delta \\ ||L\tilde{u}_{\alpha_{k,opt}}-L\hat{u}|| &\leq (||\bar{y}|| + 1)\sqrt{\delta} + \tau. \end{aligned} \tag{7}$$

Remark 3. Let the conditions of Theorem 8 be satisfied. Then, as in the preceding case, we have

$$||L\tilde{u}_\alpha - L\hat{u}|| \leq \alpha C_1 + \frac{\delta}{\sqrt{\alpha}} + \tau$$

$$||A\tilde{u}_\alpha - A\hat{u}|| \leq \alpha C_2 + \alpha^{3/2} C_1 + \delta + \sqrt{\alpha}\,\tau$$

where $C_1 = ||L\bar{h}||$, $C_2 = ||A\bar{h}||$. We find $\alpha_{k,opt}$ by minimizing the function

$$\frac{\alpha}{2} + \frac{\delta}{\sqrt{\alpha}}\,. \tag{8}$$

It is seen that $\alpha_{k,opt} = \delta^{2/3}$. For this value of the regularization parameter we have the following error estimate:

$$\begin{aligned} ||A\tilde{u}_{\alpha_{k,opt}} - A\hat{u}|| &\leq \delta C_1 + \delta^{1/3} C_2 + \delta + \delta^{1/3}\tau \\ ||L\tilde{u}_{\alpha_{k,opt}} - L\hat{u}|| &\leq C_1 \delta^{2/3} + \delta^{2/3} + \tau. \end{aligned} \tag{9}$$

We call the choice of the regularization parameter $\alpha_{k,opt}$ on the basis of the conditions similar to (6) and (8) the *minimum principle of majorant estimates*. It is easy to see that this principle yields the simplest sufficient conditions for the perturbed regularized solutions to converge to the solution of the basic problem. It is also seen that this principle ensures the accuracy estimates of these approximations.

2. We consider now the case where the elements f and g as well as the operators A and L are given only approximately.

Thus, we assume that on D_{AL} there are given linear operators A_h and L_t (where $0 < h \leq h_0$, $0 < t \leq t_0$) into F and G, respectively, and jointly closed on the set D. Let the following approximation conditions on the operators A and L be satisfied: for any $u \in D$

$$||Au - A_h u|| \leq h|u|, \qquad ||Lu - L_t u|| \leq t|u|, \tag{10}$$

where

$$|u|^2 := ||Au||^2 + ||Lu||^2.$$

Lemma 4. Let $|u|^2_{h,t} = ||A_h u||^2 + ||L_t u||^2$, $\forall u \in D$. For sufficiently small h_0 and t_0 we have

$$(1-\varepsilon)|u|_{h,t} \leq |u| \leq (1+\varepsilon)|u|_{h,t}, \quad \forall u \in D \tag{11}$$

where $0 \leq \varepsilon = \varepsilon(t,h) \to 0$ as $h,t \to 0$.

Proof: The proof is elementary.

<u>Corollary</u>. For sufficiently small h_0 and t_0 the operators A_h and L_t satisfy the completion conditions; that is, there exists a constant $\bar{\gamma} > 0$, not depending on $u \in D$, such that

$$|u|_{h,t}^2 \geq \bar{\gamma}^{-2}||u||_H^2, \quad \forall u \in D.$$

It follows from the above that if we put $f = f_\delta$, $g = g_\tau$, $A = A_h$, $L = L_t$ in the functional Φ_α, the conditions of Theorem 2 on existence as well as uniqueness of regularized solutions are satisfied. We denote these solutions by $\tilde{\tilde{u}}_\alpha$. Further, we estimate the deviations of the values of the operators A and L on the elements $\tilde{\tilde{u}}_\alpha$ from their values on the elements $\tilde{u}_\alpha$.

We write the Euler variational inequality for $\tilde{\tilde{u}}_\alpha$:

$$(A_h\tilde{\tilde{u}}_\alpha - f_\delta, A_h(v-\tilde{\tilde{u}}_\alpha))_F + \alpha(L_t\tilde{\tilde{u}}_\alpha - g_\tau, L_t(v-\tilde{\tilde{u}}_\alpha))_G \geq 0, \quad \forall v \in D. \tag{12}$$

Assuming $w = \tilde{\tilde{u}}_\alpha$ in (1) and $v = \tilde{u}_\alpha$ in (12), we add the inequalities obtained. Letting $z = \tilde{\tilde{u}}_\alpha - \tilde{u}_\alpha$, we obtain after some elementary manipulations

$$||A_hz||^2 + \alpha||L_tz||^2 \leq (A\tilde{u}_\alpha - f_\delta, (A-A_h)z)_F + ((A-A_h)\tilde{u}_\alpha, A_hz)_F$$
$$+ \alpha[(L\tilde{u}_\alpha - g_\tau, (L-L_t)z)_G + ((L-L_t)\tilde{u}_\alpha, L_tz)_G]$$

Applying the Cauchy-Schwarz inequality and the approximation conditions (10), we have

$$||A_hz||^2 + ||L_tz||^2 \leq h||A\tilde{u}_\alpha - f_\delta||\ |z| + h|\tilde{u}_\alpha|\ |z|_{h,t}$$
$$+ \alpha t\{||L\tilde{u}_\alpha - g_\tau||\ |z| + |\tilde{u}_\alpha|\ |z|_{h,t}\}.$$

Using (11), we then have

$$||A_hz||^2 + \alpha||L_tz||^2 \leq h[||A\tilde{u}_\alpha - f_\delta||(1+\varepsilon) + |\tilde{u}_\alpha|$$
$$+ \alpha t||L\tilde{u}_\alpha - g_\tau||(1+\varepsilon) + \alpha t|\tilde{u}_\alpha|]|z|_{h,t}$$

and therefore

$$||A_hz||^2 \leq h[||A\tilde{u}_\alpha - f_\delta||(1+\varepsilon) + |\tilde{u}_\alpha| + \alpha t(||\tilde{u}_\alpha - g_\tau||(1+\varepsilon) + |\tilde{u}_\alpha|]|z|_{h,t}$$

$$||L_tz||^2 \leq \frac{h}{\alpha}[||A\tilde{u}_\alpha - f_\delta||(1+\varepsilon) + ||\tilde{u}_\alpha|] + t[||L\tilde{u}_\alpha - g_\tau||(1+\varepsilon) + |\tilde{u}_\alpha|]|z|_{h,t}$$

Adding the above inequalities and dividing out the common factor, we have, using (11) again:

$$|\tilde{u}_\alpha - \tilde{u}_\alpha| = |z| \le (1+\varepsilon)|z|_{h,t}$$
$$\le (1+\varepsilon)h + \frac{h}{\alpha})[||A\tilde{u}_\alpha - f_\delta||(1+\varepsilon) + |\tilde{u}_\alpha|] \tag{13}$$
$$+ t(1+\varepsilon)(1+\alpha)[||L\tilde{u}_\alpha - g_\tau||(1+\varepsilon) + |\tilde{u}_\alpha|]$$

We next estimate the quantities in the square brackets. We have

$$||A\tilde{u}_\alpha - f_\delta|| \le \mu_A + ||A\tilde{u}_\alpha - A\hat{u}|| + \delta \tag{14}$$

and, similarly,

$$||L\tilde{u}_\alpha - g_\tau|| \le \nu_L + ||L\tilde{u}_\alpha - L\hat{u}|| + \tau, \tag{15}$$

where μ_A and ν_L have been defined earlier. For the value $|\tilde{u}_\alpha|$ we have

$$|\tilde{u}_\alpha| \le ||A\tilde{u}_\alpha|| + ||L\tilde{u}_\alpha|| \le ||A\tilde{u}_\alpha - A\hat{u}|| + ||A\hat{u}|| + ||L\tilde{u}_\alpha - L\hat{u}|| + ||L\hat{u}||. \tag{16}$$

The following assertion is a corollary of Theorem 10 and the estimates (13)-(16).

Theorem 11. Let the approximation conditions (10) be satisfied. Also, let the operators A_h and L_t be jointly closed on D. If the consistency condition

$$\left(\frac{\delta}{\sqrt{\alpha}} + \frac{h}{\alpha}\right) = 0 \tag{17}$$

is satisfied, then

$$\lim_{\delta,\tau,h,t\to 0} |\tilde{\tilde{u}}_\alpha - \hat{u}| = 0.$$

3. The consistency condition (17) imposes certain restrictions on the choice of the regularization parameter α as a function of the perturbations of the element f and the operator A. For the element g and the operator L the consistency condition is not necessary. This fact can be employed in stating the initial problem. In particular, the formulation of the problem as that of determining the values of the operator may be more acceptable than that as the problem of solving the operator equation. (We should note here that some problems, for instance, the problem of differentiating empirically derived information, may be formulated as either problem. Moreover, some special operator equations, for instance, the Abel equation, permit inversion; i.e., there exists an inverse operator A^{-1} and, therefore, the problem of solving the equation

$Au = f$ can be reformulated as the problem of determining the corresponding unbounded operator $L = A^{-1}$.)

4. Let $\nu_L = 0$ in the primary problem. As was noted in the corollary to Theorem 6, the regularized solution is

$$\hat{u}_\alpha \equiv \hat{u}$$

for *any* value of the regularization parameter. In this event some assertions proved above become more precise.

In fact, the estimates (4) assume the form:

$$\begin{aligned} ||L\tilde{u}_\alpha - L\hat{u}||_G &\le \frac{\delta}{\sqrt{\alpha}} + \tau \\ ||A\tilde{u}_\alpha - A\hat{u}|| &\le \delta + \sqrt{\alpha}\,\tau \end{aligned} \tag{19}$$

and therefore

$$|\tilde{u}_\alpha - \hat{u}| \le \delta + \tau + \sqrt{\alpha}\,\tau + \frac{\delta}{\sqrt{\alpha}}\,. \tag{20}$$

For $\alpha = 1$ we have

$$||L\tilde{u}_\alpha - L\hat{u}|| \le \delta + \tau, \quad ||A\tilde{u}_\alpha - A\hat{u}|| \le \delta + \tau$$

$$|\tilde{u}_\alpha - \hat{u}| \le 2(\delta + \tau).$$

The estimates (13)-(16) become simpler as well. Assuming $\alpha = 1$, we have

$$\begin{aligned} |\tilde{\tilde{u}}_\alpha - \tilde{u}_\alpha| &\le 2h(1+\varepsilon)[||A\tilde{u}_\alpha - f_\delta||(1+\varepsilon) + |\tilde{u}_\alpha|] \\ &\quad + 2(1+\varepsilon)t[||L\tilde{u}_\alpha - g_\tau||(1+\varepsilon) + |\tilde{u}_\alpha|] \end{aligned} \tag{21}$$

$$||A\tilde{u}_\alpha - f_\delta|| \le \mu_A + 2\delta + \tau$$

$$||L\tilde{u}_\alpha - g_\tau|| \le 2\tau + \delta, \quad |\tilde{u}_\alpha| \le \sqrt{2}\,|\hat{u}| + 2(\delta + \tau).$$

From the estimates (19)-(21) the following assertion follows.

Theorem 12. Let $\nu_L = 0$. For $\alpha = 1$ we then have the accuracy estimate

$$|\tilde{\tilde{u}}_\alpha - \hat{u}| = O(\delta + h + \tau + t). \tag{22}$$

The conditions of the theorem are satisfied for well-posed problems, as in (9) in Section 1.

The estimate (22) is similar to accuracy estimates in the case of approximate solution of well-posed problems.

Section 5. Approximation of the Admissible Set. Choice of the Basis.

1. Let $\delta = h = \tau = t = 0$, that is, let all the data be exact. We consider the question of the influence of regularized solutions on the accuracy of approximation of the solution of the basic problem in the case of "approximation" of the admissible set D. We assume that the space H is separable.

We are given in H a family of finite-dimensional linear subspaces S_n, $n = 1,2,\ldots$. We call the family S_n *approximating* if:

1. $S_n \subset D_{AL}$ for all n;
2. for each $u \in D_{AL}$,

$$\inf_{v \in S_n} |u-v| \to 0 \quad \text{as} \quad n \to \infty.$$

Let D_n denote a sequence of sets approximating the set D, given

$$D_n = \{P_D v \in D: \inf_{u \in D} |u-v| = |P_D v - v|, \quad \forall v \in S_n\}.$$

The appropriateness of this definition of the 'approximating' sets D_n follows from the following lemma.

Lemma 5. For each $v \in S_n$ there exists a unique element $P_D v \in D$ for which

$$\inf_{u \in D} |u-v| = |P_D v - v|. \tag{2}$$

Proof: We can prove this lemma in the same way as Lemma 2 on existence and uniqueness of regularized solutions. Hence we omit the proof.

Lemma 5 defines the *projection* operator P_D on the space S_n, with range in D. Therefore, we can write $D_n = P_n S_n$; that is, D_n is the image of the subspace S_n under its projection onto D.

Lemma 6. Let $u \in D$ be arbitrary and let $\varepsilon > 0$ be an arbitrary number. There exists $n_0 = n_0(\varepsilon;u)$ such that for all $n \geq n_0$ there is an element $u_n \in D_n$ for which

$$|u - u_n| \leq \varepsilon.$$

(This lemma shows that the set D_n possesses approximating properties with regard to the set D.)

Proof: By virtue of (1) there exists $n_0 = n_0(\varepsilon;u)$ and an element $v_n' \in S_n$, $n \geq n_0$, such that

$$|u - v_n'| \le \varepsilon, \quad \forall n \ge n_0.$$

Since $|u-P_Dv_n'| \le |u-v_n| \le \varepsilon$, we may, obviously, assume that $u_n = P_Dv_n'$. □

Lemma 7. The operators A and L are jointly closed on the set D_n, $n = 1,2,\ldots$.

Proof: Let the sequence $\{u_k\} \subset D_n$ be such that

$$u_k \to u_0 \text{ (in } H), \quad Au_k \to f_0 \text{ (in } F), \quad Lu_k \to g_0 \text{ (in } G) \tag{3}$$

as $k \to \infty$. We shall show that $u_0 \in D_n$, $Au_0 = f_0$, $Lu_0 = g_0$.

Due to the joint closure of the operators A and L on D, it follows from the foregoing relations that $u_0 \in D$, $Au_0 = f_0$, $Lu_0 = g_0$. Therefore, we need only to prove that $u_0 \in D_n$. Since $u_k \in D_n$, there exist elements $v_k \in S_n$ such that

$$u_k = P_Dv_k, \quad k = 1,2,\ldots .$$

This condition, in general, defines the elements v_k non-uniquely. Hence we may define the projection operator Q_n onto the subspace S_n as follows:

$$|Q_nv-v| = \inf_{y\in S_n} |y-v|, \quad v \in D_{AL}.$$

We take as elements v_k the projections of the elements u_k onto the subspace S_n, that is, $v_k = Q_nu_k$. It is well-known (see below, Lemma 9) that the operation of projection onto a subspace is continuous. Therefore, taking the limit in k, we have that there exists an element $v_0 \in S_n$ such that

$$v_0 = Q_nu_0,$$

which, as is easily seen, implies the equality

$$u_0 = P_Dv_0,$$

thus completing the proof of the lemma. □

Using these results, we can prove the next theorem.

Theorem 13. The variational problem of finding the element $u_\alpha^n \in D_n$ such that

$$\inf_{u\in D_n} \Phi_\alpha[u] = \Phi_\alpha[u_\alpha^n] \tag{4}$$

has a unique solution in D_n.

Proof: The proof follows that of Theorem 2 on existence and uniqueness of regularized solutions. We need to take into consideration that the subspace S_n plays the role of D_{AL} and the set D_n plays the role of the set D.

Similarly, we can prove that in order that the element $u_\alpha^n \in D_n$ be the solution of Eq. (4), it is *necessary and sufficient* that the *Euler variational inequality* be satisfied for each $w \in D_n$:

$$(Au_\alpha^n-f,\ A(w-u_\alpha^n))_F + \alpha(Lu_\alpha^n-g,\ L(w-u_\alpha^n))_G \geq 0 \qquad \forall w \in D_n. \tag{5}$$

In order to estimate the deviation of the elements u_α^n from the regularized solutions $\hat{u}_\alpha$, we assume $v = u_\alpha^n$ in the variational inequality (1) in Section 3, $w = w_\alpha^n = P_D Q_n \hat{u}_\alpha$ in the inequality (5), and add these inequalities. After simple manipulations we have the estimate

$$||A(\hat{u}_\alpha-u_\alpha^n)||^2 + \alpha||L(\hat{u}_\alpha-u_\alpha^n)||^2$$
$$\leq (Au_\alpha^n-f,\ A(\hat{u}_\alpha-w_\alpha^n))_F + \alpha(Lu_\alpha^n-g,\ L(\hat{u}_\alpha-w_\alpha^n))_G$$

in which we apply the Cauchy-Schwarz inequality to obtain

$$||A(\hat{u}_\alpha-u_\alpha^n)||^2 + \alpha||L(\hat{u}_\alpha-u_\alpha^n)||^2$$
$$\leq ||Au_\alpha^n-f||\ ||A(\hat{u}_\alpha-w_\alpha^n)|| + \alpha||Lu_\alpha^n-g||\ ||L(\hat{u}_\alpha-w_\alpha^n)||$$
$$\leq (||A(u_\alpha^n-\hat{u}_\alpha)|| + ||A\hat{u}_\alpha-f||)||A(\hat{u}_\alpha-w_\alpha^n)||$$
$$+ \alpha(||L(u_\alpha^n-\hat{u}_\alpha)|| + ||L\hat{u}_\alpha-g||)||L(\hat{u}_\alpha-w_\alpha^n)||.$$

Using the elementary inequality $2ab \leq a^2 + b^2$, we derive from the foregoing estimate:

$$||A(\hat{u}_\alpha-u_\alpha^n)||^2 + \alpha||L(\hat{u}_\alpha-u_\alpha^n)||^2 \leq ||A(\hat{u}_\alpha-w_\alpha^n)||^2$$
$$+ \alpha||L(\hat{u}_\alpha-w_\alpha^n)||^2 + 2||A\hat{u}_\alpha-f||\ ||A(\hat{u}_\alpha-w_\alpha^n)|| \tag{6}$$
$$+ 2\alpha||Lu_\alpha-g||\ ||L(\hat{u}_\alpha-w_\alpha^n)||.$$

We note that in Lemma 6 we may take as the element u_n the element $u_n = P_D Q_n u$. By Lemma 6, we have

$$\lim_{n\to\infty} |u-u_n| = 0, \qquad \forall u \in D.$$

Similarly,

$$\lim_{n\to\infty} |\hat{u}_\alpha - w_\alpha^n| = 0.$$

Therefore, the following theorem holds.

Theorem 14. Let the regularization parameter $\alpha = \alpha(n)$, as $n \to \infty$, be chosen so that the *consistency condition* is satisfied:

$$\lim_{n\to\infty} \frac{|\hat{u}_\alpha - Q_n\hat{u}_\alpha|}{\alpha(n)} = 0. \tag{7}$$

Assuming $\hat{u}_n \equiv u_\alpha^n$ for the choice of $\alpha = \alpha(n)$ mentioned, we obtain

$$\lim_{n\to\infty} |\hat{u}_n - \hat{u}| = 0, \tag{8}$$

where $\hat{u}$ is the solution of the basic problem.

This theorem shows that it is possible, in principle, to construct approximate solutions for the solution of the basic problem by minimizing the functional $\Phi_\alpha[u]$ on the set D_n. It is not hard to show that the problem of minimizing $\Phi_\alpha[u]$ on D_n is reduced to a *finite-dimensional* convex programming problem and, therefore, it can be readily solved by techniques well known in convex programming.

We can prove a similar assertion under another perturbation condition on the elements f, g and the operators A and L. In particular, let $\tilde{\tilde{u}}_\alpha^n$ be a solution of the problem (4) for the perturbed parameters in the sense indicated above. Then we have the following theorem.

Theorem 15. Let the parameter $\alpha = \alpha(\delta,h)$ be consistent with δ and h as required in Theorem 11. The following limiting relations then hold:

$$\lim_{\delta,h,t,\tau\to 0} \lim_{n\to 0} |\tilde{\tilde{u}}_\alpha^n - \hat{u}| = 0. \tag{9}$$

2. In practice, it is hard to verify the consistency condition (7) because it depends on the regularized solution $\hat{u}_\alpha$. We consider a particular but important case, $\nu_L = 0$. In this case, the regularized solutions $\hat{u}_\alpha \equiv \hat{u}$, the solution of the basic problem for all $\alpha > 0$. Further, the relations (6) transform into

$$||A(\hat{u}-u_\alpha^n)||^2 + \alpha||L(\hat{u}-u_\alpha^n)||^2 \le ||A(\hat{u}-\hat{u}_n)||^2$$
$$+ \alpha||L(\hat{u}-\hat{u}_n)||^2 + 2\mu_A||A(\hat{u}-\hat{u}_n)||,$$

where $\hat{u}_n = P_D Q_n \hat{u}$, yielding for $\alpha = 1$

$$|\hat{u}-u_\alpha^n|^2 \le |\hat{u}-\hat{u}_n|^2 + 2\mu_A|\hat{u}-\hat{u}_n|.$$

Since $(a^2 + b^2)^{1/2} \le |a| + |b|$,

$$|\hat{u}-u_\alpha^n| \le |\hat{u}-\hat{u}_n| + \sqrt{2\mu_A}\,|\hat{u}-\hat{u}_n|^{1/2}. \tag{10}$$

Thus, the rate of convergence of the approximations u_α^n (for $\alpha = 1$) to the solution $\hat{u}$ of the basic problem can be completely defined by the rate of convergence of the elements $\hat{u}_n$ to $\hat{u}$, that is to say, by the properties of the solution of the basic problem as well as the approximating spaces S_n.

Next, we assume that $r_n \to 0$, $n \to \infty$, are known, such that

$$|\hat{u} - Q_n\hat{u}| \le r_n, \qquad n = 1,2,\dots . \tag{11}$$

Then the following lemma holds.

Lemma 8. If the estimate (11) is known, we have

$$|\hat{u} - \hat{u}_n| \le r_n, \tag{12}$$

where $\hat{u}_n = P_D Q_n \hat{u}$.

The proof of Lemma 8 is based on the following lemma.

Lemma 9. For any u and v in D_{AL} we have the estimate

$$|P_D u - P_D v| \le |u-v|; \tag{13}$$

that is, the operator P_D is a contraction.

In fact, it is not hard to show that in order that the element $P_D u \in D$ be a solution of the problem (2), it is necessary and sufficient that the Euler variational inequality be satisfied:

$$(AP_D u-Au,\ A(z-P_D u))_F + (LP_D u-Lu,\ L(z-P_D u))_G \ge 0, \qquad \forall z \in D. \tag{14}$$

We now write a similar inequality for the element $P_D v$, $v \in D_{AL}$:

$$(AP_D v-Av,\ A(w-P_D v))_F + (LP_D v-Lv,\ L(w-P_D v))_G \ge 0, \qquad \forall w \in D. \tag{15}$$

Assuming that $z = P_D v$ in (14) and $w = P_D w$ in (15), and then adding the resulting inequalities, we have

$$||A(P_D u-P_D v)||^2 + ||L(P_D u-P_D v)||^2$$

$$\le (A(P_D u-P_D v),\ A(v-u))_F + (L(P_D v-P_D u),\ L(v-u))_G.$$

Applying the Cauchy-Schwarz inequality, we obtain:

$$|P_Du-P_Dv|^2 \leq ||A(P_Dv-P_Du)|| \; ||A(v-u)|| + ||L(P_Dv-P_Du)|| \; ||L(v-u)||$$
$$\leq \frac{1}{2}|P_Du-P_Dv|^2 + \frac{1}{2}|u-v|^2,$$

yielding (13). Lemma 9 is proved.

Noting that $\hat{u} = P_D\hat{u}$ and also using the condition (11), we have

$$|\hat{u}-\hat{u}_n| = |P_D\hat{u}-P_DQ_n\hat{u}| \leq |\hat{u}-Q_n\hat{u}| \leq \tau_n$$

as required. This proves Lemma 8. □

We note now that we have obtained the error estimate of approximation of the element $\hat{u}$ by the element $\hat{u}_n$ in the set D_n on the basis of the accuracy of approximation only of the subspaces S_n. This fact is even more noteworthy because this estimate depends neither on the nature of the elements of the set D nor on the set D itself.

The following theorem is a corollary of the above arguments as well as the estimate (10).

Theorem 16. Let $\nu_L = 0$ and let the approximating spaces S_n be such that the error estimate (14) is known for the solution $\hat{u}$ of the basic problem. Then, for the deviation of the regularized solutions u_α^n (for $\alpha = 1$) from the solution of the basic problem we have the error estimate

$$|\hat{u}-u_\alpha^n| \leq \tau_n + \sqrt{2\mu_A}\,\tau_n^{1/2}. \tag{16}$$

Remark. If $\mu_A = 0$ as well as $\nu_L = 0$, that is, the basic problem is consistent, the estimate (16) implies the better estimate

$$|\hat{u}-u_\alpha^n| \leq \tau_n;$$

i.e., the rate of convergence of the approximate solutions u_α^n to the element $\hat{u}$ is given by the accuracy of approximation of the element $\hat{u}$ by the elements of the subspace S_n.

3. We wish to draw the reader's attention to the following. In the case $\nu_L = 0$, the basic problem is reduced to that of finding the operator equation

$$Lu = g \tag{17}$$

provided

$$u \in U_A. \tag{18}$$

The latter may be interpreted as a generalized "boundary" condition.

We showed before under the condition for existence of the (unique) solution $\hat{u}$ satisfying the "boundary" conditions (18) and Eq. (17) that the problem of finding this solution is reduced to that of minimizing the functional

$$\Phi[u] = ||Au-f||^2 + ||Lu-g||^2$$

on the set D.

The remarkable feature of this approach lies in the fact that we minimize the functional Φ without, in general, concerning ourselves whether the elements of the set D satisfy the "boundary" conditions or not. The same fact is reflected in the situation that the approximating spaces S_n satisfy only rather general completion conditions (1), thus enabling us to construct the subspaces S_n as the linear spans of the linearly independent elements $w_1^n, w_2^n, w_3^n, \ldots$ *not satisfying* the "boundary" conditions, which essentially simplifies the construction of approximating subspaces.

A similar situation develops in the general case when we are finding the regularized solutions $\hat{u}_\alpha$ by means of minimization of the functional Φ_α on the set D.

We shall illustrate our statement with an example.

Example. Let $H = W_2^{(n)}[a,b]$ with norm

$$||u||_{W_2} = (||u^{(n)}||_{L_2}^2 + ||u||_{L_2}^2)^{1/2}.$$

Also, let $A_0u(x) \equiv \int_a^b k(x,\xi)u(\xi)d\xi$, $a \le x \le b$, where $\int_a^b\int_a^b k^2(x,\xi)dxd\xi < \infty$, be an operator from H into $L_2[a,b]$. We now consider a solution of the Fredholm integral equation of the first kind

$$A_0u = f_0, \quad f_0 \in L_2, \tag{19}$$

the latter being a given function. In practice, the boundary conditions are frequently known which the required solution itself, or values of its derivatives, satisfy at some points of the interval $[a,b]$. In the general case, we may assume that we know values of some linear functionals of the required solution:

$$a_1(u) = f_1, \quad a_2(u) = f_2, \ldots, a_s(u) = f_s \tag{20}$$

where a_i are the functionals given on H and the f_i are the values of

a_i. We define now the space F as the Cartesian product

$$L_2 \times \underbrace{R_1 \times R_1 \times \dots R_1}_{s}$$

with norm

$$||f||_F = (||f_0||^2_{L_2} + f_1^2 + f_2^2 + \dots + f_s^2)^{1/2},$$

$$f = (f_0, f_1, \dots, f_s).$$

Furthermore, let the operator A be defined by

$$Au = (A_0u, \ a_1(u), \dots, a_s(u)).$$

Then, solving Eq. (19) under the conditions (20) is reduced to solving the operator equation

$$Au = f, \quad f = (f_0, f_1, \dots, f_s) \in F. \tag{21}$$

Let Eq. (21) have a unique solution (for this, it is sufficient that Eq. (19) has a unique solution) $\hat{u} \in W_2^{(n)}$. Let $Lu \equiv d^nu/dx^n$. Then the functional

$$\Phi_\alpha[u] \equiv |||Au-f||^2_F + \alpha||Lu-g||^2_{L_2}, \quad g \in W_2^{(n)} \tag{22}$$

is minimized on the entire space $W_2^{(n)}$, in spite of the additional conditions (20).

Our approach does not depend in this case on the specific form of operator A_0. Hence it can be applied more generally.

We also note that the minimization of Eq. (22) always leads us to the Euler equation with a self-adjoint operator.

4. We consider now the problem of choosing the approximating subspaces S_n. The consistency condition between the operator A and the operator L implies that the combined quadratic form $||Au||^2 + ||Lu||^2$ is positive definite on D_{AL}. Let a positive definite self-adjoint unbounded operator T be given on D_{AL}. We assume that the quadratic form (Tu,u), $u \in D_{AL}$, given by the operator T is *equivalent* to the quadratic form $||Au||^2 + ||Lu||^2$; that is to say, there exist constants γ_0 and γ_1, $0 < \gamma_0 \leq \gamma_1$, such that the relations

$$\gamma_0^2(Tu,u)_H \leq ||Au||^2_F + ||Lu||^2_G \leq \gamma_1^2(Tu,u)_H \qquad \forall u \in D_{AL} \tag{23}$$

are satisfied. We also assume that the operator T has a complete orthonormal (on H) system of eigenvectors w_i:

$$Tw_i = \lambda_i w_i, \qquad i = 1,2,\ldots,$$

where $\lambda_1 \le \lambda_2 \le \ldots$ are the corresponding eigenvalues (for this, it is sufficient that every set of elements in D_{AL} which is bounded in the sense of the quadratic form (Tu,u) be compact on H) tending to $+\infty$ as $i \to \infty$.

Then, we may assume that S_n is equal to the linear subspace $L(w_i, w_2, \ldots, w_n)$ spanned by the first n eigenvectors. Let $||u||_T^2 = (Tu,u)_H$ and let $u = \Sigma_{i=1}^n u_i w_i$, where $u_i = (u,w_i)_H$ are the Fourier coefficients of the element $u \in D_{AL}$ in the coordinate system w_i, $i = 1,2,\ldots$. Then

$$Q_n u = u_n = \sum_{i=1}^{n} u_i w_i$$

and

$$||u-u_n||_T^2 = \sum_{i=n+1}^{\infty} \lambda_i u_i^2 \to 0 \quad \text{as} \quad n \to \infty.$$

If the element $u \in D_{T^2}$, the series $\Sigma_{i=1}^{\infty} \lambda_i^2 u_i^2 = ||Tu||_H^2$ is convergent. In this event, it is seen that

$$||u-u_n||_T^2 \le \frac{1}{\lambda_{n+1}} \sum_{i=n+1}^{\infty} \lambda_i^2 u_i^2 \le \frac{||Tu||_H^2}{\lambda_{n+1}} \to 0 \quad \text{as} \quad n \to \infty.$$

By virtue of (23)

$$|u-u_n| \le \gamma_1 \frac{||Tu||_H}{\lambda_{n+1}^{1/2}} = \tau_n \to 0 \quad \text{as} \quad n \to \infty$$

i.e., the condition (11) is satisfied.

We note that there can be an infinite number of operators T satisfying the equivalence condition (23). It is quite natural to choose T to be the simplest possible operator and, in particular, with known asymptotic behavior of its eigenvalues. The latter problem has been extensively investigated and we shall not dwell upon it here.

If $u \in D_{T^2}$, it is possible to estimate the rate of convergence for the deviation of u_n from u in the H-norm in a similar way. In fact, we have then:

$$||u-u_n||_H \le ||Tu||_H/\lambda_{n+1}.$$

Chapter 2
Criteria for Selection of Regularization Parameter

Section 6. Some Properties of Regularized Solutions

1. Many computational techniques for solving operator equations of various types were developed based on minimizing the *residual:* the discrepancy between the value of a specified operator at the approximate solution and the given right-hand side. If the corresponding problem is well-posed in the Hadamard sense, the operator is given exactly and it is known a priori that the problem of solving the equation in which the exact right-hand side is *consistent* (that is to say, the consistency measure $\mu_A = 0$), this approach is natural. The picture changes if the operator is given only approximately, or if the problem is ill-posed. In the first case, as is well-known in the approximation of differential equations by difference schemes, an "exact" solution of the approximate problem, that is, a solution with zero residual, does not always reduce to an exact solution of the basic problem if there is no requirement for the difference scheme to be *stable*. Therefore, even for well-posed problems, minimizing the residual may turn out to be an erroneous criterion while solving these problems approximately. In the case of approximate solution of ill-posed problems or simply inconsistent equations, this approach without appropriate modifications has little to recommend it.

Nevertheless, we may make the following assumption which we call *the residual principle*. As will be seen below, this assumption resotres the role of the residual as the accuracy index for approximate solution of a wide class of problems (ill-posed problems included): the magnitude of the residual must be commensurate with the inconsistency measure and the accuracy of specification of input information of the problem.

Depending on the nature of the basic problem and the "quality" of the input information, this assumption may be necessary. Moreover, it can take different forms.

The residual principle seems to be unfeasible because it is paradoxical. In effect, the following scheme is commonly invoked for proving that a stable solution of ill-posed operator equations is impossible in a classical way; that is to say, seeking a solution of the operator equation

$$Au = \tilde{f}, \quad \tilde{f} \in F: ||f-\tilde{f}||_F < \delta,$$

with the approximate right-hand side $\tilde{f}$. Let the operator A be completely continuous. Then, there exist finite variations of the solution $\hat{u}$ of the equation $Au = f$ which lead to infinitely small variations of the right-hand side. Now the idea of using the residual as the basis for obtaining algorithms for constructing stable solutions looks impossible. It can be done nevertheless. In particular, our immediate problem consists of proving that it is possible to construct stable approximations based on choosing a regularization parameter, using the a priori residual.

2. Before solving the problem stated above, we shall investigate the behavior of some auxiliary functions which we will need in order to formulate the principles of selecting a regularization parameter and to justify it.

For $\alpha > 0$ let

$$\rho(\alpha) = ||A\hat{u}_\alpha - f||_F, \quad \phi(\alpha) = \Phi_\alpha[\hat{u}_\alpha], \qquad \gamma(\alpha) = ||L\hat{u}_\alpha - g||_G, \tag{1}$$

where $\hat{u}_\alpha$ is the solution of the regularized problem ((1), Section 2). We shall now establish some additional properties of regularized solutions.

We assume that the following additional conditions are satisfied. Let

$$M_L = \inf_{u \in D} ||Lu-g||_G. \tag{2}$$

Also, let the set $U_L = \{u \in D: ||Lu-g|| = \mu_L\}$ be non-empty. We note that this condition is a priori satisfied if $g = 0$ and the zero element belongs to D.

Then, as follows from Theorem 1, there exists a unique solution, u_∞, of the problem

$$\nu_A = \inf_{u \in U_L} ||Au-f||_F = ||Au_\infty - f||_F. \tag{3}$$

We call Problem (3) an *auxiliary problem*. The main condition for this problem to have a solution is that the set U_L be non-empty. The latter condition is always satisfied if, for instance, the element $g = Lu^*$, where $u^* \in D$ is fixed. It is easily seen that always

$$\mu_A \leq \nu_A, \quad \mu_L \leq \nu_L. \tag{4}$$

3. We establish next the limiting properties of the functions (1) as $\alpha \to 0+$.

Lemma 10. We have the following limit relations:

$$\lim_{\alpha\to 0^+} \rho(\alpha) = \lim_{\alpha\to 0^+} \phi^{1/2}(\alpha) = \mu_A, \quad \lim_{\alpha\to 0^+} \gamma(\alpha) = \nu_L. \tag{5}$$

Proof: Using the extremal properties of the elements $\hat{u}$ and $\hat{u}_\alpha$, we obtain a chain of inequalities:

$$\mu_A^2 \leq ||A\hat{u}_\alpha - f||^2 = \rho^2(\alpha) \leq \phi(\alpha) = \Phi_\alpha[\hat{u}_\alpha]$$
$$\leq \Phi_\alpha[\hat{u}] = \mu_A^2 + \alpha\nu_L^2 \leq \rho^2(\alpha) + \alpha\nu_L^2, \tag{6}$$

yielding

$$\mu_A \leq \rho(\alpha) \leq \mu_A + \sqrt{\alpha}\,\nu_L,$$
$$\mu_A \leq \phi^{1/2}(\alpha) \leq \mu_A + \sqrt{\alpha}\,\nu_L, \quad \gamma(\alpha) \leq \nu_L. \tag{7}$$

The first two limit relations (5) follow from the first two estimates (7). The limit relation

$$\lim_{\alpha\to 0^+} \gamma(\alpha) = \nu_L$$

is a corollary of Theorem 3 on the convergence of regularized solutions. □

Remark. It readily follows from Eq. (7) that

$$0 \leq \rho(\alpha) - \mu_A \leq \sqrt{\alpha}\,\nu_L,$$
$$0 \leq \phi^{1/2}(\alpha) - \mu_A \leq \sqrt{\alpha}\,\nu_L,$$

i.e., as $\alpha \to 0^+$, the functions $\rho(\alpha)$ and $\phi^{1/2}(\alpha)$ behave identically.

Next, we investigate the behavior of the functions ρ, ϕ and γ as $\alpha \to +\infty$.

Lemma 11. We have the limit relations

$$\lim_{\alpha\to+\infty} \rho(\alpha) = \nu_A, \quad \lim_{\alpha\to\infty} \gamma(\alpha) = \mu_L, \quad \lim_{\alpha\to\infty}[\phi(\alpha)-\nu_A^2-\alpha\mu_L^2] = 0. \tag{8}$$

Proof: Using the extremal properties of the elements $\hat{u}_\alpha$ and u_∞, we have

$$\phi(\alpha) = \rho^2(\alpha) + \alpha\gamma^2(\alpha) = \Phi_\alpha[\hat{u}_\alpha] \le \Phi_\alpha[u_\infty]$$
$$= \nu_A^2 + \alpha\mu_L^2 \le \nu_A^2 + \alpha\gamma^2(\alpha), \tag{9}$$

yielding

$$\gamma(\alpha) \le \frac{\nu_A}{\sqrt{\alpha}} + \mu_L.$$

Since $\mu_L \le \gamma(\alpha)$ a priori, the inequalities

$$\mu_L \le \gamma(\alpha) \le \mu_L + \frac{\nu_A}{\sqrt{\alpha}}$$

are satisfied and therefore

$$\lim_{\alpha\to\infty} \gamma(\alpha) = \mu_L. \tag{10}$$

At the same time, we obtain from Eq. (9),

$$\rho^2(\alpha) \le \nu_A^2 + \alpha(\mu_L^2 - \gamma^2(\alpha)) \le \nu_A^2. \tag{11}$$

Reasoning similarly as in proving Theorem 3 on the convergence of regularized solutions as $\alpha \to 0$ to a solution of the basic problem, we derive from Eqs. (10), (11) the following assertions.

Let a regularization parameter $\alpha \to \infty$. Then

$$\lim_{\alpha\to\infty} |\hat{u}_\alpha-u_\infty| = \lim_{\alpha\to\infty} ||\hat{u}_\alpha-u_\infty||_H = 0. \tag{12}$$

Therefore, the desired limit relations for the functions ρ and ϕ are simple corollaries of Eq. (12). □

Corollary. Let $\mu_L = 0$. Then, obviously,

$$\lim_{\alpha\to\infty} \rho(\alpha) = \lim_{\alpha\to\infty} \phi^{1/2}(\alpha) = \nu_A, \quad \lim_{\alpha\to\infty} \gamma(\alpha) = 0.$$

The condition $\mu_L = 0$ is satisfied a priori if the element $g = Lu^*$, $u^* \in D$. In this case we have

$$\nu_A \le ||Au^* - f||_F.$$

If, furthermore, the operator L is invertible, then

$$\nu_A = ||Au^* - f||_F.$$

Under these conditions we have $\nu_A = ||f||_F$ for $g = 0$.

4. We now investigate the continuity of the functions ρ, ϕ, and γ.

Let $\alpha > 0$ and $\beta > 0$ be two regularization parameters. Also, let $\hat{u}_\alpha$ and $\hat{u}_\beta$ be the corresponding regularized solutions. We write the corresponding variational inequalities:

$$(A\hat{u}_\alpha - f, A(v-\hat{u}_\alpha))_F + \alpha(L\hat{u}_\alpha - g, L(v-\hat{u}_\alpha))_G \geq 0, \quad \forall v \in D$$

$$(A\hat{u}_\beta - f, A(w-\hat{u}_\beta))_F + \beta(L\hat{u}_\beta - g, L(w-\hat{u}_\beta))_G \geq 0, \quad \forall w \in D.$$

Assuming $v = \hat{u}_\beta$ in the first inequality and $w = \hat{u}_\alpha$ in the second inequality, we have, after a little algebra, that

$$||A(\hat{u}_\beta - \hat{u}_\alpha)||^2 + \alpha||L(\hat{u}_\beta - \hat{u}_\alpha)||^2 \leq (\alpha-\beta)(L\hat{u}_\beta - g, L(\hat{u}_\alpha - \hat{u}_\beta))_G.$$

Applying the Cauchy-Schwarz inequality and the estimates

$$\mu_L \leq \gamma(\beta) \leq \nu_L$$

established when proving Lemmas 10 and 11, we have

$$||A(\hat{u}_\beta - \hat{u}_\alpha)||^2 + \alpha||L(\hat{u}_\beta - \hat{u}_\alpha)||^2 \leq |\alpha-\beta|\nu_L||L(\hat{u}_\alpha - \hat{u}_\beta)|| \leq 2|\alpha-\beta|\nu_L^2$$

and therefore,

$$||A(\hat{u}_\beta - \hat{u}_\alpha)|| \leq \sqrt{2}\ \nu_L|\alpha-\beta|^{1/2}, \quad ||L(\hat{u}_\beta - \hat{u}_\alpha)|| \leq \nu_L \frac{|\alpha-\beta|}{\alpha}. \tag{13}$$

Lemma 12. The functions $\rho(\alpha)$, $\gamma(\alpha)$ and $\phi(\alpha)$ are continuous for all α, $0 < \alpha < \infty$. If $\mu_A < \nu_A$, the values of the functions $\rho(\alpha)$ and $\phi^{1/2}(\alpha)$ exhaust the interval (μ_A, ν_A). Similarly, for $\mu_L < \nu_L$ the values of the function $\gamma(\alpha)$ exhaust the interval (μ_L, ν_L).

Proof: The continuity (even Lipschitz property) of the functions ρ, ϕ, and γ follows immediately from Eq. (13). The second part of the assertion of the lemma follows from the limiting properties of the functions ρ, ϕ, and γ proved in Lemmas 10 and 11. □

Remark. Let the parameter $\beta = \tau\alpha$, where $\tau \leq 1$. Then it follows from the estimates (13) that

$$||A(\hat{u}_{\tau\alpha} - \hat{u}_\alpha)|| \leq \sqrt{2\alpha}\ |1-\tau|^{1/2}, \quad ||L\hat{u}_{\tau\alpha} - L\hat{u}_\alpha|| \leq \nu_L|1-\tau|.$$

For $0 < \alpha \leq \bar{\alpha}$, the last estimates show that the two families $\hat{u}_{\tau\alpha}$ and $\hat{u}_\alpha$ are uniformly close in the sense that

$$|\hat{u}_{\tau\alpha}-\hat{u}_\alpha| \leq \sqrt{2\bar{\alpha}}\,|1-\tau|^{1/2} + \nu_L|1-\tau|$$

independently of α. This property is essential in constructing numerical algorithms for the regularized problem which involve the idea of *descent* in the regularization parameter. Such algorithms include iteration methods of minimization of functionals: gradient, Newton, Newton-Gauss, etc. In [70], a similar idea was considered for regularized variational problems.

5. We shall now define the properties of these functions more precisely.

Lemma 13. Let $\mu_A < \nu_A$. Then, for any $\alpha > 0$ the functions ρ and ϕ are strictly increasing while the function γ is strictly decreasing.

Proof: Let α and β be positive. Also, let $\hat{u}_\alpha$ and $\hat{u}_\beta$ be the corresponding regularized solutions. Using the extremal property of the elements $\hat{u}_\alpha$, we have

$$\phi(\alpha) = \Phi_\alpha[\hat{u}_\alpha] \leq \Phi_\alpha[\hat{u}_\beta]$$

and therefore

$$\phi(\alpha) - \phi(\beta) \leq \Phi_\alpha[\hat{u}_\beta] - \Phi_\beta[\hat{u}_\beta] = (\alpha-\beta)||L\hat{u}_\beta-g||^2. \tag{14}$$

For $\beta > \alpha$, this yields

$$\phi(\alpha) \leq \phi(\beta),$$

i.e., the function ϕ is non-decreasing.

Interchanging α and β in Eq. (14), we obtain:

$$\phi(\beta) - \phi(\alpha) \leq (\beta-\alpha)||L\hat{u}_\alpha-g||^2. \tag{15}$$

Now let $\beta > \alpha$. It follows from Eqs. (14), (15) that

$$\gamma^2(\beta) = ||L\hat{u}_\beta-g||^2 \leq \frac{\phi(\beta)-\phi(\alpha)}{\beta-\alpha} \leq ||L\hat{u}_\alpha-g||^2 = \gamma^2(\alpha), \tag{16}$$

i.e., the function γ is non-increasing.

We shall show that the function ρ is non-decreasing. Let $\beta > \alpha$. Using Eq. (16), we obtain

$$\rho^2(\alpha) + \alpha\gamma^2(\alpha) = \Phi_\alpha[\hat{u}_\alpha] \leq \Phi[\hat{u}_\beta] = \rho^2(\beta) + \alpha\gamma^2(\beta) \leq \rho^2(\beta) + \gamma^2(\alpha)\alpha \tag{17}$$

and hence

$$\rho(\alpha) \le \rho(\beta), \quad \beta > \alpha,$$

i.e., the function ρ is non-decreasing.

In order to prove the strict monotonicity of the functions, we show that

$$||L\hat{u}_\beta - g|| \ne 0, \quad \forall\beta > 0.$$

We suppose to the contrary that there is a $\beta > 0$ for which

$$||L\hat{u}_\beta - g|| = 0. \tag{18}$$

Writing the variational inequality for the element $\hat{u}_\beta$ and also using (16), we have:

$$(A\hat{u}_\beta - f, A(v-\hat{u}_\beta))_F \ge 0, \quad \forall v \in D.$$

By Theorem 5, the last condition implies the inclusion $u_\beta \in U_A = \{u \in D: \mu_A = ||Au-f||_F\}$ and, therefore,

$$||Au_\beta - f|| = \mu_A.$$

Since $\mu_L \le \gamma(\beta)$, Eq. (18) implies $\mu_L = 0$ and $u_\beta \in U_L$. By definition,

$$\nu_A = \inf_{u \in U_L} ||Au - f||_F$$

and therefore,

$$\nu_A \le ||\ u_\beta - f||_F = \mu_A,$$

contradicting $\mu_A < \nu_A$.

This shows that

$$||Lu_\beta - g|| \ne 0, \quad \beta > 0$$

and hence

$$\phi(\alpha) < \phi(\beta)$$

for $\beta > \alpha$, i.e., the function ϕ is strictly increasing.

We shall show that the function γ is strictly decreasing. In fact, suppose there exists $\beta > \alpha > 0$ such that

$$\gamma(\alpha) = \gamma(\beta) = s > 0.$$

Then, it follows from Eq. (16) that

$$s^2 = \gamma^2(\alpha) = \gamma^2(\beta) = \frac{\phi(\beta)-\phi(\alpha)}{\beta-\alpha} = \frac{\rho(\beta)-\rho(\alpha)}{\beta-\alpha} + s^2$$

i.e., $\rho(\alpha) = \rho(\beta)$, $\phi(\beta) = \phi(\alpha)$. The last relation immediately implies that $\beta = \alpha$, thus proving the assertion.

It then readily follows from Eq. (17) that the function γ is strictly increasing. □

In Figure 1 we illustrate the behavior of the functions ρ, ϕ, and γ which we have proved above $(\mu_A < \nu_A,\ \mu_L < \nu_L)$:

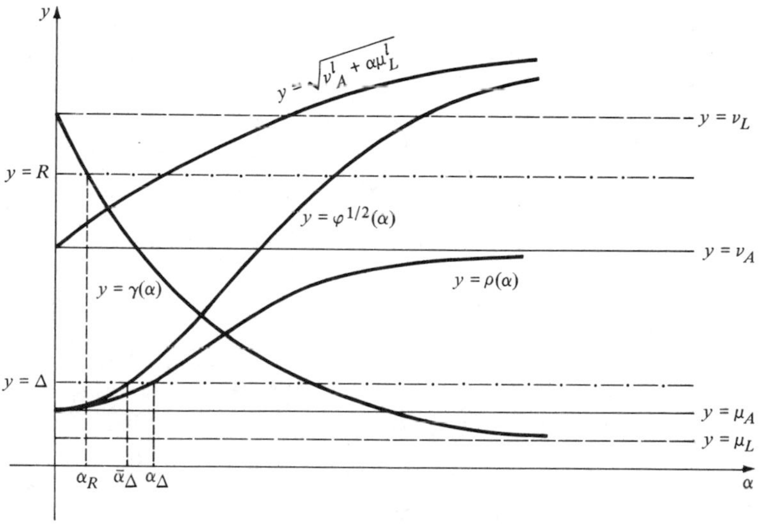

Figure 1

6. As was noted in the Corollary to Theorem 6, for $\nu_L = 0$ the regularized solutions $\hat{u}_\alpha \equiv \hat{u}$; that is, they solve the basic problem. We define more precisely the conditions under which this property is a priori satisfied. We need to do this because there is no problem of choosing "appropriate" values of the regularization parameter in this case, at least for the case of exact specification of the information. Moreover, in the case of approximate specification of the information, one can hope that the problem of choice is much simplified.

Lemma 14. Let either $\mu_A = \nu_A$ or $\mu_L = \nu_L$. Then, the regularized solutions $\hat{u}_\alpha \equiv \hat{u}$; that is, they solve the basic problem.

Proof: For example, let $\mu_A = \nu_A$. Using the extremal properties of the elements $\hat{u}_\alpha$, we have:

$$\begin{aligned}\Phi_\alpha[\hat{u}_\alpha] &\le \Phi_\alpha[u_\infty] = \nu_L^2 + \alpha\mu_L^2 \le \nu_A^2 + \alpha||L\hat{u}_\alpha - g||^2 \\ &= \mu_A^2 + \alpha||L\hat{u}_\alpha - g||^2 \le ||A\hat{u}_\alpha - f||^2 + \alpha||L\hat{u}_\alpha - g||^2 \\ &= \Phi_\alpha[\hat{u}_\alpha]\end{aligned}$$

that is, $\Phi_\alpha[\hat{u}_\alpha] \equiv \Phi_\alpha[u_\infty]$.

By Theorem 2, this implies $\hat{u}_\alpha \equiv u_\infty$. Since, on the other hand, $\lim_{\alpha\to 0^+} \hat{u}_\alpha = \hat{u}$ we have finally

$$\hat{u}_\alpha \equiv \hat{u} = u_\infty,$$

as required.

Now, let $\mu_L = \nu_L$. We have similarly:

$$\begin{aligned}\Phi_\alpha[\hat{u}_\alpha] &\le \Phi_\alpha[\hat{u}] = \mu_A^2 + \alpha\nu_L^2 \le ||A\hat{u}_\alpha - f||^2 + \alpha\nu_L^2 \\ &= ||A\hat{u}_\alpha - f||^2 + \alpha\mu_L^2 \le ||A\hat{u}_\alpha - f||^2 + \alpha||L\hat{u}_\alpha - g||^2 \\ &= \Phi_\alpha[\hat{u}_\alpha]\end{aligned}$$

and therefore, $\Phi_\alpha[\hat{u}_\alpha] \equiv \Phi_\alpha[\hat{u}]$. Hence $\hat{u}_\alpha = \hat{u} = u_\infty$. □

<u>Corollary</u>. The conditions $\mu_A < \nu_A$ and $\mu_L < \nu_L$ are satisfied simultaneously, as well as the conditions $\mu_A = \nu_A$ and $\mu_L = \nu_L$.

<u>Remark</u>. If $\mu_L = \nu_L$ (or $\mu_A = \nu_A$), then, as follows from Lemma 14, $u_\alpha \equiv \hat{u}$. We can thus see that the results obtained in Subsection 5.2 and the beginning of Subsection 5.3 still hold, widening the range of application of these results.

Section 7. Methods for Choosing the Parameter: Case of Exact Information

1. We consider now the most interesting case from the point of view of applications to ill-posed problems, namely, $\mu_A < \nu_A$.

Let a quantity Δ be given, $\mu_A < \Delta < \nu_A$. We consider the equation

$$\rho(\alpha) = \Delta, \quad 0 < \alpha < \infty. \tag{1}$$

From Lemmas 12 and 13 it follows that Eq. (1) always has a unique root. We shall denote this root by α_Δ. It is seen that $0 < \alpha_\Delta < \infty$.

We shall refer to the choice of the regularization parameter based on (1) as *criterion* ρ.

The following assertion holds.

<u>Theorem 17</u>. Let the regularization parameter α_Δ be chosen by criterion ρ; see (1). Then

$$\lim_{\Delta\to\mu_A+} |u_\Delta-\hat{u}| = \lim_{\Delta\to\mu_A+} ||u_\Delta-\hat{u}||_H = 0, \tag{2}$$

where $u_\Delta = \hat{u}_{\alpha_\Delta}$, the regularized solution corresponding to the chosen value of the parameter α_Δ.

Proof: Let $\alpha = \alpha_\Delta$. Due to the extremal property of the element u_Δ we have

$$||Au_\Delta-f||_F^2 + \alpha_\Delta||Lu_\Delta-g||^2 \le ||A\hat{u}-f||^2$$
$$+ \alpha_\Delta||L\hat{u}-g||^2 \le ||Au_\Delta-f||^2 + \alpha_\Delta||L\hat{u}-g||^2$$

which together with Eq. (1) implies

$$||Au_\Delta-f|| = \Delta, \quad ||Lu_\Delta-g|| \le \nu_L. \tag{3}$$

We note that the relations (3) are completely similar to the relations (9) in Section 2 which implied the convergence of the regularized solutions $\hat{u}_\alpha$ to $\hat{u}$; that is, the solutions of the basic problem (Theorem 3). Reasoning in a similar way as in Section 2 (which we shall not repeat here), we can establish (2). □

Remark. The importance of Theorem 17 lies in the transition from a formal parametric family of regularized solutions $\hat{u}_\alpha$ to a parametric family of approximate solutions u_Δ in which the parameter Δ has an obvious geometric meaning (see Figure 1).

2. Let a quantity Δ, $\mu_A < \Delta < \nu_A$ be given. We consider the equation

$$\phi(\alpha) = \Delta^2, \qquad 0 < \alpha < \infty. \tag{4}$$

From Lemmas 12 and 13 it follows that Eq. (4) always has a unique root. We shall denote this root by $\bar{\alpha}_\Delta$. It is seen that $0 < \bar{\alpha}_\Delta < \infty$ (see Figure 1), where α_Δ is defined according to Eq. (1).

We call the choice of the regularization parameter as the root of Eq. (4) the choice of the parameter by the functional $\Phi_\alpha[u]$ on the regularized solutions $u = \hat{u}_\alpha$, or, in short, *criterion* ϕ.

Theorem 18. Let the regularization parameter $\bar{\alpha}_\Delta$ be chosen by criterion ϕ; see (4). Then

$$\lim_{\Delta\to\mu_A+} |\bar{u}_\Delta-\hat{u}| = \lim_{\Delta\to\mu_A+} ||\bar{u}-\hat{u}||_H = 0, \tag{5}$$

where $\hat{u}_\Delta = \hat{u}_{\overline{\alpha}_\Delta}$, the regularized solution corresponding to the chosen value of the parameter $\overline{\alpha}_\Delta$.

Proof: From (4) we have:

$$\Delta^2 = ||A\overline{u}_\Delta - f||^2 + \overline{\alpha}_\Delta ||L\overline{u}_A - g||^2 \leq ||A\hat{u} - f||^2 + \overline{\alpha}_\Delta ||L\hat{u} - g||^2 \leq ||A\overline{u}_\Delta - f||^2 + \overline{\alpha}_\Delta ||L\hat{u} - g||^2.$$

We derive from this chain of inequalities that

$$||A\overline{u} - f|| \leq \Delta, \qquad ||L\overline{u}_\Delta - g|| \leq \nu_L. \tag{6}$$

Reasoning in a similar way as in Theorem 3, we derive (5) from (6). □

Remark. If the information is exact, criterion ρ is equivalent to criterion ϕ; namely, the regularized solutions approximate arbitrarily closely in the strong sense the solution $\hat{u}$ of the basic problem in H and the values of the operators A and L on the element $\hat{u}$ in the spaces F and G, respectively. As will be seen from the sequel, the behavior of these solutions for the case of approximate information is quite different.

3. Let R, $\mu_L < R < \nu_L$ be given. Consider the equation

$$\gamma(\alpha) = R, \qquad 0 < \alpha < \infty. \tag{7}$$

From Lemmas 12 and 13 it follows that Eq. (17) always has a unique root. We shall denote this root by α_R. It is seen that $0 < \alpha_R < \infty$ (see Figure 1).

The choice of the regularization parameter as the root of Eq. (7) will be referred to as *criterion* γ.

Theorem 19. Let the regularization parameter α_R be chosen by criterion γ; see (7). Then

$$\lim_{R \to \nu_L -} |u_R - \hat{u}| = \lim_{R \to \nu_L -} ||u_R - \hat{u}||_H = 0, \tag{8}$$

where $u_R = \hat{u}_{\alpha_R}$, that is, the regularized solution corresponding to the chosen value of the parameter α_R.

Proof: Let $\alpha = \alpha_R$. Using the extremal property of the elements u_R, we have:

$$||Au_R - f||^2 + \alpha_R ||Lu_R - g||^2 \leq ||Au - f||^2 + \alpha_R ||L\hat{u} - g||^2.$$

Therefore,

$$||Au_R-f||^2 \le \mu_A^2 + \alpha_R(\nu_L^2-R^2), \qquad ||Lu_R-g||^2 = R^2. \tag{9}$$

We show that the values of the parameter α_R have an upper bound.

Let $R_0 = (\mu_L+\nu_L)/2$ and let the parameter $\alpha_0 = \alpha_R$. Then, by Theorems 12 and 13 we have that $\alpha_R \le \alpha_0$ for $R_0 \le R < \nu_L$. We have further from the relations (9):

$$\overline{\lim_{R\to\nu_L-}} ||Au_R-f|| \le \mu_A, \quad \overline{\lim_{R\to\nu_L-}} ||Lu_R-g|| \le \nu_L. \tag{10}$$

Also, reasoning as in proving Theorem 3, we derive from the inequalities (10) the desired limits (8). □

4. Thus, the choice of the regularization parameter by criterion ρ, ϕ, or γ guarantees the convergence of the solutions obtained to a solution of the basic problem. We need however to give a priori either the desired level of the residual Δ or the quantity R. It is not hard to see (see Figure 1) that the specification of one of these values, for instance Δ, is equivalent to specification of the quantity $R = R(\Delta) = ||Lu -g||$, if the element u_Δ is defined by criterion ρ, or the quantity $\overline{R} = \overline{R}(\Delta) = ||L\overline{u}_\Delta-g||$, if the element $\overline{u}_\Delta$ is defined by criterion ϕ. In this case, the functions $R = R(\Delta)$ and $\overline{R} = \overline{R}(\Delta)$ are one-to-one. The methods for choosing the regularization parameter by criteria ρ and ϕ are related in a similar manner.

The assertion above are immediate corollaries of Lemmas 12 and 13 proved earlier.

<u>Remark</u>. Let α_Δ be defined by criterion ρ and let $\Delta \to \nu_A^-$. Then, as in Theorem 17, we prove the following.

<u>Theorem 17'</u>. If α_Δ satisfies (1), the following relations hold:

$$\lim_{\Delta\to\nu_A-} |u_\Delta-u_\infty| = \lim_{\Delta\to\nu_A-} ||u_\Delta-u_\infty||_H = 0, \tag{2'}$$

where $u_\Delta = \hat{u}_{\alpha_\Delta}$.

Also, the following assertions hold.

<u>Theorem 18'</u>. Let $\mu_L = 0$. If $\overline{\alpha}_\Delta$ is defined by criterion ϕ as stated in (4), then

$$\lim_{\Delta\to\nu_A-} |\overline{u}_\Delta-u_\infty| = \lim_{\Delta\to\nu_A-} ||\overline{u}_A-u_\infty||_H = 0, \tag{5'}$$

where $\bar{u}_\Delta = \hat{u}_{\bar{\alpha}_\Delta}$.

Theorem 19'. If the parameter α_R is defined by criterion γ as stated in (7), then

$$\lim_{R\to\mu_L+} |u_R - u_\infty| = \lim_{R\to\mu_L+} ||u_R - u_\infty||_H = 0, \tag{8'}$$

where $u_R = \hat{u}_{\alpha_R}$.

Section 8. The Residual Method and the Method of Quasi-solutions: Case of Exact Information

1. Let $\Delta \geq \mu_A$ be given. We define the set $u_\Delta = \{u \in D: ||Au-f|| \leq \Delta\}$. The condition $\Delta \geq \mu_A$ ensures that the set U_Δ is non-empty; the element $\hat{u}$, the solution of the basic problem, is obviously in U_Δ. Furthermore, the set $U_A \subset U_\Delta$.

We consider now a variational problem: find the elements $u^\Delta \in U_\Delta$ for which

$$\nu_L^\Delta = \inf_{u \in U_\Delta} ||Lu-g|| = ||Lu^\Delta - g||. \tag{1}$$

Let $\hat{U}_\Delta = \{u^\Delta \in U_\Delta: \nu_L^\Delta = ||Lu^\Delta - g||\}$. It is seen that $\hat{U}_\Delta$ is the set of all solutions to Problem (1).

The residual method yields $u^\Delta \in \hat{U}_\Delta$.

Theorem 20. For any $\Delta \geq \mu_A$ the set $\hat{U}_A \neq \emptyset$, i.e., Problem (1) has a solution. Furthermore,

(a) for $\Delta = \mu_A$ the element $\hat{u} \in \hat{U}_\Delta$, where $\hat{u}$ is a solution of the basic problem;

(b) for $\mu_A < \Delta < \nu_A$, the element $u_\Delta \in \hat{U}_\Delta$, where u_Δ is chosen by criterion ρ;

(c) for $\Delta \geq \nu_A$, the element $u_\infty \in \hat{U}_\Delta$, where u_∞ is a solution of the auxiliary problem (6.3).

Proof: (a) Let $\Delta = \mu_A$; then, obviously, the set $U_\Delta = U_A$ and Problem (1) coincides with the basic problem.

(b) Let $\mu_A < \Delta < \nu_A$. Also, let the element u_Δ be defined by criterion ρ.

Let $\varepsilon > 0$ be given and let $u_\varepsilon \in U_\Delta$ be such that

$$||Lu_\varepsilon - g|| \leq \nu_L^\Delta + \varepsilon.$$

Using the extremal property of the element u_Δ, we have:

$$||Au_\Delta-f||^2 + \alpha_\Delta||Lu_\Delta-g||^2 \leq ||Au_\varepsilon-f||^2 + \alpha_\Delta||Lu_\varepsilon-g||^2 \leq \Delta^2+\alpha_\Delta(\nu_L^\Delta+\varepsilon)^2,$$

where α_Δ is a solution of Eq. (1) in Section 7.

Since $||Au_\Delta-f|| = \Delta$ we obtain from the above inequality that

$$||Lu_\Delta-g|| \leq \nu_L^\Delta + \varepsilon.$$

Because ε is arbitrary and the element $u_\Delta \in U_\Delta$, we have that u_Δ is a solution of Problem (1), that is, $u_\Delta \in \hat{U}_\Delta$.

(c) Let $\Delta \geq \nu_A$. Then the element $u_\infty \in U_\Delta$. Since the set $U_\Delta \leq D$, $\nu_L^\Delta \geq \mu_L$ for any $\Delta \geq \mu_A$. However, by definition, $\mu_L = ||Lu_\infty-g||$, yielding

$$||Lu_\infty-g|| \leq \Delta_L^\Delta,$$

and therefore, $u_\infty \in \hat{U}_\Delta$. □

We shall next establish the conditions under which the residual method yields a unique solution. Let the operator L be invertible; that is to say, the homogeneous equation $Lu = 0$ has only the trivial solution $u = 0$. Then the residual method yields a unique solution for any $\Delta \geq \mu_A$.

In fact, the set U_Δ is, obviously, convex. Let $u_1 \in \hat{U}_\Delta$ and $u_2 \in \hat{U}_\Delta$. Then

$$||L(\frac{u_2-u_1}{2})||^2 = \frac{1}{2}||Lu_2-g||^2 + \frac{1}{2}||Lu_1-g||^2 - ||L(\frac{u_1+u_2}{2})-g||^2$$

$$= \frac{1}{2}(\nu_L^\Delta)^2 + \frac{1}{2}(\nu_L^\Delta)^2 - ||L(\frac{u_1+u_2}{2})-g||^2 \leq 0$$

since $(u_1+u_2)/2 \in U_\Delta$, this proving the assertion. It follows from the theorem that the residual method is a variational analog of the ρ-method. Moreover, if $U_\Delta = \{u^\Delta\}$ and $\mu_A < \Delta < \nu_A$, these methods coincide.

2. Suppose $\mu_L \leq R < \infty$. This condition ensures that the set $U_R = \{u \in D: ||Lu-g|| \leq R\}$ is non-empty, since, obviously, the element $u_\infty \in U_R$, where u_∞ is a souution of the auxiliary problem (3) in Section 6.

We consider another variational problem: find the elements $u^R \in U_R$ for which

$$\mu_A^R = \inf_{u \in U_R} ||Au-f|| = ||Au^R-f||. \tag{2}$$

Let $\hat{U}_R = \{u^R \in U_R: \mu_A^R = ||Au^R-f||\}$. It is seen that $\hat{U}_R$ is the set of all solutions to Problem (2).

The method of quasi-solutions leads to solutions u^R of Problem (2).

<u>Theorem 21</u>. For any $R \geq \mu_L$ the set $\hat{U}_R \neq \emptyset$, i.e., Problem (2) has a solution. Furthermore,

(a) for $R = \mu_L$ the element $u_\infty \in \hat{U}_R$, where u_∞ is a solution of the auxiliary problem (3) in Section 6;

(b) for $\mu_L < R < \nu_L$ the element $u_R \in \hat{U}_R$, where u_R is chosen by criterion γ;

(c) for $R \geq \nu_L$ the element $\hat{u} \in \hat{U}_R$, where u is a solution of the basic problem.

<u>Proof</u>: (a) Let $R = \mu_L$. Then $U_R = U_L$ and Problem (2) coincides with the auxiliary problem (3) in Section 6, u_∞ being a solution of Problem (3).

(b) Let R: $\mu_L < R < \nu_L$. Let the elements u_R be defined by the criterion γ. Let an arbitrary $\varepsilon > 0$ be given and, finally, let $u_\varepsilon \in U_R$ be such that

$$||Au_\varepsilon-f|| \leq \mu_A^R + \varepsilon.$$

Using the extremal property of the element u_R, we obtain:

$$\begin{aligned}||Au_R-f||^2 + \alpha_R||Lu_R-g||^2 &\leq ||Au_\varepsilon-f||^2 + \alpha_R||Lu_\varepsilon-g||^2 \\ &\leq (\mu_A^R + \varepsilon)^2 + \alpha_R R^2.\end{aligned}$$

Since $||Lu_R-g|| = R$ by the choice of the element u_R, this implies:

$$||Au_R-f|| \leq \mu_A^R + \varepsilon.$$

Because ε is arbitrary and the element $u_R \in U_R$, we have that u_R is a solution of Problem (2).

(c) Let R: $R \geq \nu_L$. By the definition of ν_L the element $\hat{u}$ is a solution of the basic problem which belongs to the set U_R. Since the set $U_R \subseteq D$, $\mu_A^R \geq \mu_A$ for any $R \geq \mu_L$. However, by definition,

$$\mu_A = ||A\hat{u}-f||,$$

and therefore,

$$||A\hat{u}-f|| \leq \mu_A^R, \qquad \hat{u} \in U_R,$$

that is, $\hat{u}$ is a solution to Problem (2).

We establish now the conditions under which the method of quasi-solutions yields a unique solution. We have the following theorem.

Theorem 22. If the operator A is invertible; that is to say, the homogeneous equation $Au = 0$ has only a trivial solution $u = 0$, quasi-solutions given by the method (2) are defined uniquely for any $R \geq \mu_L$.

Proof: This proof follows the proof of uniqueness of solutions of the residual method.

Remark. It follows from Theorems 21 and 22 that the method of quasi-solutions is a variational analog of the γ-method. Furthermore, under the conditions of Theorem 22 and for $\mu_L < R < \nu_L$ the γ-method coincides with the method of quasi-solutions.

We can see now that solutions of the residual method for the variation Δ within the range from μ_A to $+\infty$ and solutions of the method of quasi-solutions for the variation R within the range from μ_L to $+\infty$ exhaust completely the set of regularized solutions $\hat{u}_\alpha$ as the regularization parameter α varies from 0 to $+\infty$, including its limits $\hat{u}$ and u_∞.

Section 9. Properties of the Auxiliary Functions.

1. In determining an approximate solution of the basic problem, let the element f and the operator A be given approximately, that is, $f_\delta \in F$ and linear operators A_h, $0 < h \leq h_0$, be known, such that

$$||f_\delta - f|| < \delta, \qquad ||Au - A_h u|| \leq h|u|, \qquad \forall u \in D, \tag{1}$$

and the operators A_h and L are jointly closed. These conditions ensure existence and uniqueness of the regularized solutions $\tilde{u}_\alpha$.

Let

$$\tilde{\mu}_A = \inf_{u \in D} ||A_h u - f_\varepsilon||, \qquad \tilde{\nu}_A = \inf_{u \in U_L} ||A_h u - f_\varepsilon||, \tag{2}$$

where the set U_L was defined in Subsection 6.1. Since we assumed that the set $U_L \neq \emptyset$, there exists a unique element $\tilde{u}_\infty \in U_L$ for which

$$\tilde{\nu}_A = ||A_h \tilde{u}_\infty - f_\varepsilon||.$$

This assertion can be proved in a similar way as Theorem 2 on the existence and uniqueness of a solution of the basic problem.

Theorem 23. Let $\sigma = (h, \delta) \to 0$. Then

$$\lim_{\sigma\to 0} |\tilde{u}_\infty - u_\infty| = 0. \tag{3}$$

Proof: We have the element $\tilde{u}_\infty \in U_L$, that is,

$$||L\tilde{u}_\infty - g|| = \mu_L. \tag{4}$$

On the other hand, we have the following obvious chain of inequalities:

$$\nu_A = \inf_{u\ U_L} ||Au-f|| \le ||A\tilde{u}_\infty - f|| \le ||A_h\tilde{u}_\infty - f_\delta||$$

$$+ h|\tilde{u}_\infty| + \delta = \tilde{\nu}_A + h|\tilde{u}_\infty| + \delta \le ||A_h u_\infty - f_\delta||$$

$$+ h|\tilde{u}_\infty| + \delta \le ||Au_\infty - f|| + h(|u_\infty| + |\tilde{u}_\infty|) + 2\delta$$

$$= \nu_A + h(|u_\infty| + |\tilde{u}_\infty|) + 2\delta. \tag{5}$$

Since

$$||A_h\tilde{u}_\infty - f_\delta|| \le ||A_h u_\infty - f_\delta|| \le \nu_A + h|u_\infty| + \delta,$$

we have

$$||A\tilde{u}_\infty|| \le ||A\tilde{u}_\infty - f_\delta|| + ||f_\varepsilon|| \le ||A_h u_\infty^2 - f_\delta|| + h|\tilde{u}_\infty| + ||f_\varepsilon||$$

$$\le \nu_A + h(|u_\infty| + |\tilde{u}_\infty|) + 2\varepsilon + ||f||.$$

It follows from (4) that

$$||L\tilde{u}_\infty|| \le \mu_L + ||g||.$$

Adding the last two inequalities, we obtain

$$|\tilde{u}_\infty| \le ||A\tilde{u}_\infty|| + ||L\tilde{u}_\infty|| \le \nu_A + \mu_L + ||f|| + ||g|| + h(|\tilde{u}_\infty| + |u_\infty|).$$

Thus, for sufficiently small h the quantities $|\tilde{u}_\infty|$ are uniformly bounded in σ.

From (5) it follows that

$$\lim_{\sigma\to 0} ||A\tilde{u}_\infty - f|| = \nu_A. \tag{6}$$

Arguing in the same way as in proving Theorem 3 on the convergence of regularized solutions, we can convince ourselves that the relations (4) and (6) yield the desired limit (3). □

Corollary. It follows from (5) that

$$\lim_{\sigma\to 0} \tilde{\nu}_A = \nu_A. \tag{7}$$

Since

$$\tilde{\mu}_A \le ||A_h\hat{u}-f_\delta|| \le ||A\hat{u}-f|| + h|\hat{u}| + \delta,$$

we have

$$\overline{\lim_{\sigma\to 0}}\, \tilde{\mu}_A \le \mu_A. \tag{8}$$

We note that Eq. (8) holds uniformly for all $\hat{u}$ with $|\hat{u}| \le C$, where C is a given constant, since, obviously,

$$\tilde{\mu}_A \le \mu_A + hC + \delta. \tag{9}$$

2. Now we define the auxiliary functions

$$\tilde{\rho}(\alpha) = ||A_h\tilde{u}_\alpha - f_\delta||, \qquad \tilde{\gamma}(\alpha) = ||L\tilde{u}_\alpha - g||,$$

$$\tilde{\phi}(\alpha) = \tilde{\rho}^2(\alpha) + \alpha\tilde{\gamma}^2(\alpha) = \Phi_\alpha[\tilde{u}_\alpha], \quad \alpha > 0, \tag{10}$$

where $\tilde{u}_\alpha$ is the regularized solution corresponding to the given $f = f_\varepsilon$, $A = A_h$.

<u>Lemma 15</u>. For all $0 < \alpha < \infty$, the functions $\tilde{\rho}(\alpha)$, $\tilde{\gamma}(\alpha)$ and $\tilde{\phi}(\alpha)$ are continuous.

<u>Proof</u>: Let $\alpha \ge \underline{\alpha} > 0$. It follows from the inequality

$$\Phi_\alpha[\tilde{u}_\alpha] \le \Phi_\alpha[\hat{u}]$$

that

$$||L\tilde{u}_\alpha - g|| \le \nu_L + \frac{||A_h\hat{u}-f_\delta||}{\sqrt{\underline{\alpha}}} \le \nu_L + \frac{\mu_A+hC+\delta}{\sqrt{\underline{\alpha}}} = R$$

uniformly for all $\alpha \ge \underline{\alpha}$.

Continuity can be proved in the same way as for the functions ρ, γ, and ϕ (see Section 6). □

Using Theorem 23, we have, similarly to Lemma 11, the following limit relations:

$$\lim_{\alpha\to\infty} \tilde{\rho}(\alpha) = \nu_A, \quad \lim_{\alpha\to\infty} \tilde{\gamma}(\alpha) = \mu_L, \quad \lim_{\alpha\to\infty}[\tilde{\phi}(\alpha) - \nu^2 - \alpha\mu_L^2] = 0. \tag{10}$$

<u>Lemma 16</u>. Let $\mu_A < \nu_A$. Then (for sufficiently small h and δ) the values of the functions $\tilde{\rho}$ and $\tilde{\phi}^{1/2}$ exhaust the semi-interval

$$[\mu_A + hC + \delta, \tilde{\nu}_A),$$

where $||\hat{u}|| \le C < +\infty$, as the regularization parameter α varies from 0 to $+\infty$.

Proof: For sufficiently small h and δ we have:

$$\tilde{\mu}_A \le \mu_A + hC + \delta < \tilde{\nu}_A \tag{11}$$

which follows readily from the conditions $\mu_A < \nu_A$ and the relation (7) proved earlier.

Furthermore, since

$$\begin{aligned}||A_h\tilde{u}_\alpha - f_\varepsilon||^2 + \alpha||L\tilde{u}_\alpha - g||^2 &\le ||A_h\hat{u} - f_\varepsilon||^2 + \alpha||L\hat{u} - g||^2 \\ &< (\mu_A + hC + \delta)^2 + \alpha\nu_A^2,\end{aligned}$$

for sufficiently small values of the parameter α the function $\tilde{\rho}(\alpha)$ (as well as the function $\tilde{\phi}^{1/2}(\alpha)$) assumes smaller values than $\mu_A + hC + \delta$.

Making use of the relations (10) and the continuity of the functions $\tilde{\rho}$ and $\tilde{\phi}$, we derive the assertion of the lemma from the remark made above. □

Remark. It is not hard to show that in fact more exact relations hold, namely:

$$\lim_{\alpha\to 0^+} \tilde{\rho}(\alpha) = \lim_{\alpha\to 0^+} \tilde{\phi}^{1/2}(\alpha) = \tilde{\mu}_A. \tag{12}$$

Indeed, let $\varepsilon > 0$ be arbitrary. Let $u_\varepsilon \in D$ be such that

$$\tilde{\mu}_A \le ||A_h u_\varepsilon - f_\delta|| \le \tilde{\mu}_A + \varepsilon.$$

Using the extremal property of the element $\tilde{u}_\alpha$, we obtain

$$\begin{aligned}||A_h\tilde{u}_\alpha - f_\delta||^2 + \alpha||L\tilde{u}_\alpha - g||^2 &\le ||A_h u_\varepsilon - f||^2 + \alpha||Lu_\varepsilon - g||^2 \\ &\le (\tilde{\mu}_A + \varepsilon)^2 + \alpha||Lu_\varepsilon - g||^2.\end{aligned}$$

Therefore

$$\tilde{\rho}(\alpha) \le \tilde{\phi}^{1/2}(\alpha) \le \mu_A + \varepsilon + \sqrt{\alpha}\, ||Lu_\varepsilon - g||.$$

Let α be such that

$$\alpha \le \frac{\varepsilon}{||Lu_\varepsilon - g||}.$$

Then, obviously,

$$\tilde{\rho}(\alpha) = \tilde{\phi}^{1/2}(\alpha) \leq \tilde{\mu}_A + 2\varepsilon, \quad 0 < \alpha \leq \frac{\varepsilon}{||Lu_\varepsilon - g||}.$$

Since ε is arbitrary, and $\tilde{\mu}_A \leq \tilde{\rho}(\alpha) \leq \tilde{\phi}^{1/2}(\alpha)$ for all $\alpha > 0$, we obtain (12).

3. Next, we assume that the condition

$$\mu_A < \nu_A$$

is satisfied. For sufficiently small h and δ we have also the inequality

$$\tilde{\mu}_A < \tilde{\nu}_A.$$

We prove the following lemma (similar to Lemma 13).

Lemma 17. Let $\mu_A < \nu_A$ and let h and δ be sufficiently small. On $[0,\infty)$, the functions $\tilde{\rho}$ and $\tilde{\phi}$ are strictly increasing and the function $\tilde{\gamma}$ is strictly decreasing.

The fact that the function $\tilde{\gamma}$ is strictly decreasing for $\alpha > 0$ implies that the (finite or infinite) limit

$$\lim_{\alpha\to 0^+} \tilde{\gamma}(\alpha) = \tilde{\gamma}_0 \tag{13}$$

exists.

The following assertion is of some interest.

Let $\tilde{\gamma}_0 < +\infty$ in Eq. (13). Then Problem (2) has a solution.

Proof: It follows from Eqs. (12) and (13) that the values $|\tilde{u}_\alpha|$ are bounded uniformly with respect to α as $\alpha \to 0$. Due to the completeness condition, the values $||\tilde{u}_\alpha||_H$ are bounded uniformly with respect to α. Hence the sequence $\tilde{u}_\alpha$ is weakly compact in H. We take $\tilde{u}_\alpha$ as weakly convergent in H to some element $\tilde{u}_0$ (if need be, we distinguish a weakly convergent subsequence); that is,

$$\lim_{\alpha\to 0} \tilde{u}_\alpha = \tilde{u}_0 \quad (\text{in } H). \tag{14}$$

It also follows from Eqs. (12) and (13) that $\{A_h\tilde{u}_\alpha\}$ and $\{L\tilde{u}_\alpha\}$ are weakly compact in F and G, respectively. We assume that they are weakly convergent, using subsequences if necessary:

$$\begin{aligned} &\lim_{\alpha\to 0} A_h\tilde{u}_\alpha = \tilde{f}_0 \quad (\text{in } F), \\ &\lim_{\alpha\to 0} L\tilde{u}_\alpha = \tilde{g}_0 \quad (\text{in } G). \end{aligned} \tag{15}$$

From the fact that the operators A_h and L are mutually weakly closed on D it follows that $\tilde{u}_0 \in D$, $A_h\tilde{u}_0 = \tilde{f}_0$, $L\tilde{u}_0 = \tilde{g}_0$.

Using the weak lower semi-continuity of the norm in a Hilbert space, we derive from Eq. (12)

$$||A_h\tilde{u}_0-f_\delta|| \leq \lim_{\alpha\to 0} ||A_h\tilde{u}_\alpha-f_\delta|| = \tilde{\mu}_A;$$

that is, the element $\tilde{u}_0$ is a solution of Problem (2). □

<u>Corollary</u>. The fact that Problem (2) has a solution under the condition $\tilde{\gamma}_0 < +\infty$, together with Theorem 3 on the convergence of regularized solutions, implies that the element $\tilde{u}_0$ constructed in proving Lemma 18 is a solution of the following problem: find the element $\tilde{u}_0 \in \tilde{U}$ for which

$$\tilde{\nu}_L = \inf_{u\in U} ||Lu-g|| = ||L\tilde{u}_0-g|| \tag{16}$$

where the set $\tilde{U} = \{u \in D\colon \tilde{\mu}_L = ||A_hu-f_\delta||\} \neq \emptyset$. In this case, we obviously have the limit relation

$$\lim_{\alpha\to 0} |\tilde{u}_\alpha-\tilde{u}_0| = 0.$$

We have, in fact, proved the following lemma.

<u>Lemma 18</u>. Let $\lim_{\alpha\to 0} \tilde{\gamma}(\alpha) = \tilde{\gamma}_0 < +\infty$. Then Problem (16) has a unique solution.

<u>Remark</u>. We do not assume here that Problem (2) has a solution a priori, as we did in formulating the basic problem. We derived our problem on the basis of the behavior of regularized solutions when the regularization parameter tends to zero. From the foregoing and Theorem 3 on the convergence, the following assertion follows.

<u>Theorem 24</u>. In order that the set $U_A \neq \emptyset$; (that is, the basic problem have a solution), it is necessary and sufficient that

$$\lim_{\alpha\to 0} \gamma(\alpha) < +\infty.$$

We note, however, that the elements $\tilde{u}_0$ need not, generally speaking, converge to a solution $\hat{u}$ of the basic problem even if $\lim_{\alpha\to 0} \tilde{\gamma}(\alpha) < +\infty$ for all admissible A_h and f_δ.

<u>Example</u>. Let $H = F = G = R_2 = \{u\colon u = (u_1,u_2)\}$,

$$Au = \begin{pmatrix} 1 & 0 \\ 0 & 0 \end{pmatrix}\begin{pmatrix} u_1 \\ u_2 \end{pmatrix}.$$

$Lu = u$, $f = (1, 0)$. It is easily seen that the general solution of $Au = f$ can be written as

$$u = (1,0) + t(0,1), \qquad t \in (-\infty,\infty).$$

If $g = 0$, the solution of the basic problem is $\hat{u} = (1,0)$.

Let $A_h \equiv \begin{pmatrix} 1 & 0 \\ h & h \end{pmatrix}$. Then, the vector $u_h := (1,-1)$, $h > 0$, is a solution of $A_h u = f$. Since, obviously, $\tilde{u}_0 = u_h$,

$$\tilde{u}_0 - \hat{u} = (0,-1) \not\to 0,$$

as $h \to 0$.

Section 10. Criteria for the Choice of a Parameter: Case of Inexact Data

1. We assume again that $\mu_A < \nu_A$. We denote by S_C the set of all $u \in D$ such that $|u| \leq C$. Also, we assume that for some $C > 0$ the solution $\hat{u}$ of the basic problem belongs to the set S_C.

Analogous to the choice of the regularization parameter by criterion ρ (Section 7), we assume that we know the upper estimate of the inconsistency measure μ_A; that is to say, we know the quantity $\hat{\mu}_A$, $\mu_A \leq \hat{\mu}_A$,

$$\Delta = \hat{\mu}_A + hC + \delta < \tilde{\nu}_A.$$

We now consider the equation for the regularization parameter α_Δ:

$$\tilde{\rho}(\alpha_\Delta) = \Delta. \tag{1}$$

By Lemmas 16 and 17, Eq. (1) always has a unique root.

We call the choice of the regularization parameter α_Δ by the condition (1) *criterion* $\tilde{\rho}$.

The following ltheorem holds.

Theorem 25. Let the regularization parameter α_Δ be chosen by criterion $\tilde{\rho}$ (see (1)). Then the following limit relations hold:

$$\lim_{\Delta\to\mu_A} |\tilde{u}_\Delta - \hat{u}| = \lim_{\Delta\to\mu_A} ||\tilde{u}_\Delta - \hat{u}||_H = 0, \tag{2}$$

where $\tilde{u}_\Delta = \tilde{u}_{\alpha_\Delta}$, the regularized solution $\tilde{u}_{\alpha_\Delta}$ corresponding to the value of the parameter α_Δ chosen by criterion $\tilde{\rho}$, and u is a solution of the basic problem.

Proof: Let $\alpha = \tilde{\alpha}_\Delta$. By the extremal property of tbe element $\tilde{u}_\Delta$, we have

$$||A_h\tilde{u}_\Delta - f_\delta||^2 + \alpha_\Delta||L\tilde{u}_\Delta - g||^2 \le ||A_h\hat{u} - f_\delta||^2 + \alpha_\Delta||L\hat{u} - g||^2. \tag{3}$$

Since, obviously,

$$||A_h\hat{u} - f_\delta|| < ||A\hat{u} - f|| + h|\hat{u}| + \delta \le \hat{\mu}_A + hC + \delta = \Delta,$$

it follows from (3) that

$$\Delta^2 + \alpha_\Delta||L\tilde{u}_\Delta - g||^2 < \Delta^2 + \alpha_\Delta\nu_L^2,$$

that is,

$$||L\tilde{u}_\Delta - g|| \le \nu_L. \tag{4}$$

On the other hand, we have

$$||A\tilde{u}_\Delta - f|| \le ||A_h\tilde{u}_\Delta - f_\delta|| + h|\tilde{u}_\Delta| + \delta = \Delta + h|\tilde{u}_\Delta| + \delta. \tag{5}$$

It follows from (4), (5) that for Δ sufficiently close to μ_A, that is, for h, δ, and $\hat{\mu}_A - \mu_A$ sufficiently small, the values $|\tilde{u}_\Delta|$ are uniformly bounded with respect to these parameters. We therefore have

$$\overline{\lim_{\Delta\to\mu_A}} ||A\tilde{u}_\Delta - f|| \le \mu_A, \qquad \overline{\lim_{\Delta\to\mu_A}} ||L\tilde{u}_\Delta - g|| \le \nu_L.$$

Arguing as in proving Theorem 3 on the convergence of regularized solutions for the case of exact information to a solution of the basic problem, we obtain (2). □

Remark. If $\delta = h = 0$, $\hat{\mu}_A = \mu_A$, criterion $\tilde{\rho}$ coincides with criterion ρ considered in Section 7.

Remark 2. If it is known a priori that the consistency measure $\mu_A = 0$, it is appropriate to assume that $\hat{\mu}_A = 0$. In this case the residual level $\Delta = hC + \delta$. For $h \ll \delta$, that is, when one can ignore an approximation error of the operator A compared to the specification error of the element f, it is appropriate in practice to assume that $\Delta = \delta$.

Remark 3. In the general case of the application of criterion $\tilde{\rho}$, it is necessary to know a priori the sphere S_C which contains the desired solution of the basic problem. This makes the application of criterion $\tilde{\rho}$ more difficult in practice if the condition $h \ll \delta$ is not satisfied. We shall suggest a modified criterion $\tilde{\rho}$ free from this shortcoming (also, see [13]).

2. We construct now an analog of the choice of the regularization parameter by criterion ϕ for the case of approximate specification of the information. We shall assume that the inconsistency measure $\mu_A = 0$, and in addition, $\nu_L > 0$.

Let

$$\Delta = \sqrt{2}\,(hC + \delta), \qquad C\colon \hat{u} \in S_C.$$

Furthermore, we assume that h and δ are small so that the condition

$$\Delta < \tilde{\nu}_A$$

is satisfied.

Now we consider the equation for the regularization parameter $\bar{\alpha}_\Delta$:

$$\tilde{\phi}(\bar{\alpha}_\Delta) = \Delta^2, \qquad 0 < \bar{\alpha}_\Delta < \infty. \tag{6}$$

By Lemmas 16 and 17, Eq. (6) always has a unique root $\bar{\alpha}_\Delta$, $0 < \bar{\alpha}_\Delta < \infty$.

We call the choice of the regularization parameter by the condition (6) *criterion* $\tilde{\phi}$.

Let $\tilde{u}_{\bar{\alpha}_\Delta} = \bar{u}_\Delta$, where $\bar{\alpha}_\Delta$ is defined in accord with Eq. (6). Using the extremal property of regularized solutions and of the solution of the basic problem, we have

$$\Delta^2 = ||A_h\bar{u}_\Delta - f_\delta||^2 + \bar{\alpha}_\Delta||L\bar{u}_\Delta - g||^2 \le ||A_h\hat{u} - f_\delta||^2$$
$$+ \bar{\alpha}_\Delta\nu_L^2 < (hC + \delta)^2 + \bar{\alpha}_\Delta\nu_L^2 = \frac{\Delta^2}{2} + \bar{\alpha}_\Delta\nu_L^2 \tag{7}$$

which yield the lower estimate for the root of Eq. (6), namely:

$$\bar{\alpha}_\Delta > (hC + \delta)^2/\nu_L^2 = \Delta^2/2\nu_L^2.$$

Using the last inequality, we derive from Eq. (7)

$$||L\bar{u}_\Delta - g||^2 \le \frac{\Delta^2}{2\bar{\alpha}_\Delta} + \nu_L^2 \le \nu_L^2 + \nu_L^2 = 2\nu_L^2. \tag{8}$$

Furthermore, since obviously $||A_h\bar{\alpha}_\Delta - f_\delta|| \le \Delta$, we have

$$||A\bar{u}_\Delta - f|| \le h|\bar{u}_\Delta| + \delta + \Delta,$$

and therefore

$$\lim_{\Delta \to 0} ||A\bar{u}_\Delta - f|| = 0. \tag{9}$$

We show that in the sense to be defined below, the approximations $\bar{u}_\Delta$ obtained approximate the set

$$\hat{U}_A = \{u \in U_A : ||Lu-g|| \leq \sqrt{2}\,\nu_L\}.$$

It is seen that $\hat{U}_A$ is the set of solutions of $Au = f$ which belong to the set D.

In fact, we prove the following result.

<u>Theorem 26</u>. Let $\mu_A = 0$, $\nu_{\tilde{L}} > 0$ and let the regularization parameter $\bar{\alpha}_\Delta$ be chosen by criterion ϕ. Then, for any element $u \in H$ and $g \in G$ we have:

$$\lim_{\Delta\to 0} \inf_{u\in\hat{U}_A} \{|(L\bar{u}_\Delta - L\bar{u},\ g)| + |(\bar{u}_\Delta - \bar{u}, u)| + ||A\bar{u}_\Delta - f|| = 0 \tag{10}$$

<u>Proof</u>: From Eqs. (8), (9) and the completeness condition it follows that the family of approximate solutions $\bar{u}_\Delta$ is bounded in the space H; therefore, it is weakly compact. At the same time, the family $L\bar{u}_\Delta$ is, obviously, weakly compact in the space G. Therefore, it is always possible to distinguish a subfamily $\{\bar{u}_{\Delta^i}\} \subseteq \{\bar{u}_\Delta\}$ such that

$$\bar{u}_{\Delta^i} \xrightarrow{\text{weakly}} u_0 \ \ (\text{in } H),\quad A\bar{u}_{\Delta^i} \longrightarrow \bar{f}\ \ (\text{in } F),\quad L\bar{u}_{\Delta^i} \xrightarrow{\text{weakly}} g_0\ \ (\text{in } G) \tag{11}$$

as $\Delta^i \to 0$.

Due to the fact that the operators A and L are jointly weakly closed, it follows from Eq. (11) that

$$u_0 \in D,\quad Au_0 = f,\quad Lu_0 = g_0. \tag{12}$$

Therefore, the element $u_0 \in U_A$. We show that $u_0 \in \hat{U}_A$. In fact, using the weak lower semi-continuity of the norm in a Hilbert space, we obtain from Eqs. (8), (11) and (12):

$$||Lu_0 - g|| \leq \underline{\lim}_{\Delta'\to 0}\ ||L\bar{u}_\Delta - g|| \leq \overline{\lim}_{\Delta'\to 0}\ ||L\bar{u}_{\Delta'} - g|| \leq \sqrt{2}\,\nu_L;$$

that is, $u_0 \in \hat{U}_A$.

Thus, we have proved that given any subfamily $\{\bar{u}_{\Delta^i}\}$ satisfying the relations (11), one can obtain an element $u_0 \in \hat{U}_A$ such that

$$\lim_{\Delta'\to 0} \{|(L\bar{u}_{\Delta'} - Lu_0, g)| + |(\bar{u}_{\Delta'} - u_0, u)| + ||A\bar{u}_{\Delta'} - f|| = 0.$$

As we can see, the relation (10) will be satisfied as well. □

Remark. If the set $U_A = \{\hat{u}\}$ (that is, it consists of a single element), under the conditions of Theorem 26, the following relation holds:

$$\lim_{\Delta\to 0} \{||A\bar{u}_\Delta - A\hat{u}|| + |(L\bar{u}_\Delta - L\hat{u}, g)| + |(\bar{u}_\Delta - u, u)|\} = 0, \quad \forall g \in G, \quad u \in H.$$

Let a linear operator B be given on D_{AL}, acting from H into some Banach space V and satisfying the following condition (similar to the completeness condition):

$$||Bu||_V \leq \gamma_B |u|, \qquad \forall u \in D_{AL} \tag{13}$$

We call the condition (13) *the B-completeness condition.*

We demand that for each sequence $u_n \in H$ such that

$$Au_n \xrightarrow{\text{weakly}} Au_0, \quad Lu_n \xrightarrow{\text{weakly}} Lu_0, \quad u_0 \in D_{AL},$$

the following is satisfied (the operator B is *strongly subordinate to the operators* A *and* L*):* $\lim_{n\to\infty} ||Bu_n - Bu_0||_V = 0$.

Example. Let $H = F = G = L_2[a,b]$. Let $A = E$, the identity operator. Let $Lu = d^n u/dx^n$, the operator of n-fold differentiation, whose domain D_L consists of all functions which are (n-1)-times continuously differentiable (on $[a,b]$) with derivative order n square-integrable on $[a,b]$. Then we have:

$$|u|^2 = \int_a^b u^2 dx + \int_a^b [u^{(n)}]^2 dx = ||u||^2_{W_2^n}$$

By well-known theorems on imbedding [80], weak convergence of the elements in $W_2^{(n)}[a,b]$ implies their strong convergence in any of the spaces $C^{(k)}[a,b]$, $k = 0,1,\dots,n-1$. In this case, we have the inequality

$$||u||_{C^{(k)}} \leq \gamma_k ||u||_{W_2^{(n)}}, \qquad \forall u \in W_2^{(n)}. \tag{14}$$

As the operator B we can, obviously, take the operator of imbedding of the space $W_2^{(n)}$ into any of the spaces $C^{(k)}$.

In the genral case the following theorem holds.

Theorem 27. Let the conditions of Theorem 26 be satisfied. Also, let the operator B be strongly subordinate to the operators A and L. Then we have

$$\lim_{\Delta\to 0} \inf_{u\in\hat{U}_A} ||Bu - B\bar{u}_\Delta||_V = 0. \tag{15}$$

Proof: This proof follows the proof of Theorem 26.

Remark 1. If the set $U_A = \{\hat{u}\}$, it follows from the condition (27) of the theorem that

$$\lim_{\Delta\to 0} ||B\bar{u}_\Delta - B\hat{u}||_V = 0.$$

Remark 2. Criterion $\tilde{\phi}$ for the choice of the regularization parameter, as well as criterion $\tilde{\rho}$, depends on the a priori knowledge of the sphere S_C containing the solution $\hat{u}$ of the basic problem, and thus is not useful enough in this sense. Under the condition $h \ll \delta$ one may assume $\Delta = \sqrt{2\delta}$; then the choice of the regularization parameter by criterion $\tilde{\phi}$ does not depend on such a priori information. The numerical realizations of algorithms for the choices $\tilde{\rho}$ and $\tilde{\phi}$ will be considered in Section 26.

3. We now change the approximation conditions for the operator A. We assume that the operators A_h satisfy an approximation condition of the form

$$||Au - A_h u||_F \leq h||Lu - g||_G, \qquad \forall u \in D, \tag{16}$$

with the remaining assumptions being unchanged.

Under this modification, we can construct an algorithm for the choice of the regularization parameter which does not require a priori knowledge of the sphere S_C containing the solution $\hat{u}$ of the basic problem.

We note that (16) implies the earlier conditions. Hence the results related to the behavior of the functions $\tilde{\rho}$, $\tilde{\gamma}$, and $\tilde{\phi}$ continue to hold under the new approximation condition (16) as well.

We consider the following equation for the regularization parameter:

$$\tilde{\rho}(\alpha) = h\tilde{\gamma}(\alpha) + \hat{\mu}_A + \delta, \tag{17}$$

where, as before, $\hat{\mu}_A = \mu_A$ is an approximation from above to the inconsistency measure μ_A. Our objective is to find the conditions under which Eq. (17) has a unique solution.

Lemma 19. Let $\mu_A < \nu_A$ (then also $\mu_L < \nu_L$). If h and δ are such that

$$h(\nu_L - \mu_L) < \delta - \delta_0, \tag{18}$$

where δ_0: $||f_\delta - f|| \leq \delta_0 < \delta$, for sufficiently small σ, $\sigma = (h,\delta,\hat{\mu}_A - \mu_A)$ Eq. (17) has a unique root α_σ, $0 < \alpha_\sigma < \infty$.

Proof: By definition, $\tilde{\mu}_A \le ||A_h\hat{u}-f_\delta||$. Therefore, by the condition (18), we have

$$\begin{aligned}\tilde{\mu}_A &\le \mu_A + h\nu_L + \delta_0 < \mu_A + h\mu_L + \delta \\ &\le \hat{\mu}_A + h\mu_L + \delta.\end{aligned} \tag{19}$$

It follows from the properties of the function $\tilde{\rho}$ proved in Section 9 that the values of this function exhaust the interval $(\tilde{\mu}_A, \tilde{\nu}_A)$ for $\alpha \in (0,\infty)$. Hence it follows from (19) that the graph of the function $\tilde{\rho}$ intersects the line $f_0 = \hat{\mu}_A + h\mu_L + \delta$ for sufficiently small positive values of the regularization parameter α (see Figure 2).

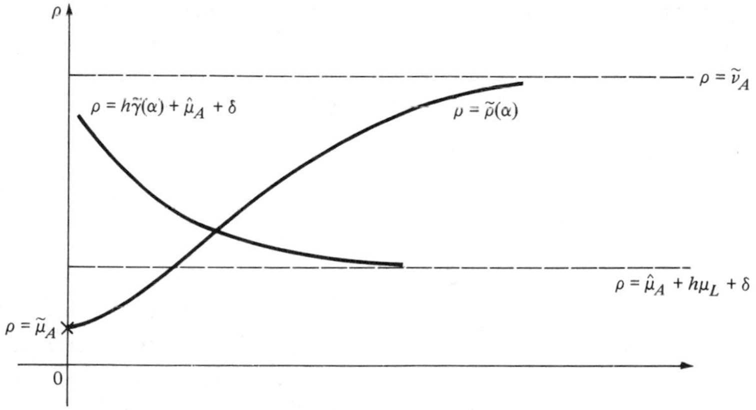

Figure 2

On the other hand, for any $\alpha > 0$ we have

$$h\tilde{\gamma}(\alpha) + \hat{\mu}_A + \delta \ge h\mu_L + \hat{\mu}_A + \delta = \rho_0;$$

that is, the graph of the right-hand side of Eq. (17) is above the line $\rho = \rho_0$. Since

$$\lim_{\alpha\to\infty} \tilde{\gamma}(\alpha) = \mu_L,$$

the line $\rho = \rho_0$ is a horizontal asymptote for the right-hand side of Eq. (17). Taking into consideration that by the conditions of the lemma, the right-hand side of Eq. (17) is a strictly decreasing function of the parameter α and the left-hand side of this equation is a strictly increasing function, and in addition that both functions are continuous, we can see that Eq. (17) has a unique solution. □

Next, we assume that the conditions of Lemma 19 are satisfied.

We call the choice of the regularization parameter α as the root α_σ of Eq. (17) *generalized criterion* $\tilde{\rho}$.

Theorem 28. Let the regularization parameter α_σ be chosen by generalized criterion $\tilde{\rho}$ (17). Then we have

$$\lim_{\sigma\to 0} |u_\sigma - \hat{u}| = \lim_{\sigma\to 0} ||u_\sigma - \hat{u}||_H = 0 \tag{20}$$

where $u_\sigma = \tilde{u}_{\alpha_\sigma}$, the regularized solution $\hat{u}_\alpha$ corresponding to the chosen value of the parameter α_σ, and $\hat{u}$ is a solution of the basic problem.

Proof: We assume that $\alpha = \alpha_\sigma$ and also make use of the extremal property of regularized solutions. Elementary calculations yield

$$\tilde{\rho}^2(\alpha_\sigma) + \alpha_\sigma \tilde{\gamma}^2(\alpha_\sigma) \le ||A_h\hat{u} - f_\delta||^2 + \alpha_\sigma \nu_L^2$$
$$\le (\hat{\mu}_A + h\nu_L + \delta)^2 + \alpha_\sigma \nu_L^2.$$

Using Eq. (17), we then have

$$(h\tilde{\gamma} + \hat{\mu}_A + \delta)^2 + \alpha_\sigma \tilde{\gamma}^2 \le (\hat{\mu}_A + h\nu_L + \delta)^2 + \alpha_\sigma \nu_L^2,$$

where we assumed for brevity that $\tilde{\gamma} = \tilde{\gamma}(\alpha_\sigma)$. Subtracting the quantity $(\mu_A + \delta)^2$ from both sides of the last estimate, we obtain

$$(h\tilde{\gamma} + \hat{\mu}_A + \delta)^2 + \alpha_\sigma \tilde{\gamma}^2 \le (\hat{\mu}_A + h\nu_L + \delta)^2 + \alpha_\sigma \nu_L^2.$$

We divide both sides of this inequality by $h^2 + \alpha_\sigma$ and complete the square, thus obtaining

$$\tilde{\gamma} + \sqrt{\frac{h(\hat{\mu}_A+\delta)}{h^2+\alpha_\sigma}}^{\,2} \le \nu_L + \sqrt{\frac{h(\hat{\mu}_A+\delta)}{h^2+\alpha_\sigma}}^{\,2}$$

This yields $\tilde{\gamma} \le \nu_L$; that is,

$$||Lu_\sigma - g|| \le \nu_L. \tag{21}$$

Using Eq. (21), we derive from Eq. (17)

$$||A_h u_\sigma - f_\varepsilon|| \le h\nu_L + \hat{\mu}_A + \delta,$$

and therefore

$$||Au_\sigma - f|| \le \mu_A + (\hat{\mu}_A - \mu_A) + 2(h\nu_L + \delta). \tag{22}$$

As in proving Theorem 3 on the convergence of regularized solutions, we derive from (21), (22) that the relation (20) is satisfied. □

Remark. The application of generalized criterion $\tilde{\rho}$ is free from a priori knowledge of any quantitative characteristics of the desired solution $\hat{u}$ of the basic problem. For $h = \delta$ generalized criterion $\tilde{\rho}$ coincides with the usual criterion $\tilde{\rho}$. In the case where it is known a priori that the inconsistency measure $\mu_A = 0$, it is appropriate to assume that $\hat{\mu}_A = 0$. If the element $g = 0$ and the set $D = D_{AL}$, we easily see that $\mu_L = 0$. If, in addition, the set $U_A = \{\hat{u}\}$, then $\nu_L = ||L\hat{u}||$. In this case the conditions of Lemma 19 reduce to the requirement that the solution of the basic problem $\hat{u} \in N_L$, the kernel of the operator L. The inequality (18) is satisfied a priori if $h \ll \delta = 2\delta_0$.

4. Similarly, we construct now a generalized criterion $\tilde{\rho}$ free from the a priori knowledge of quantitative characteristics of the desired solution. As in Subsection 10.2, we assume here that $\mu_A = 0$, $\nu_L > 0$. Also, we assume that the approximation condition (16) is satisfied.

In order to determine the appropriate values of the regularization parameter, we consider the equation

$$\tilde{\phi}^{1/2}(\alpha) = 2(h\tilde{\gamma}(\alpha) + \delta). \tag{23}$$

We show that under certain conditions Eq. (23) has a unique solution $\bar{\alpha}_\sigma$, $0 < \bar{\alpha}_\sigma < \infty$, where $\sigma = (h,\delta)$ (due to strict monotonicity of the functions $\tilde{\phi}$ and $\tilde{\gamma}$).

Lemma 20. Let $\mu_A = 0 < \nu_A$, $\mu_L < \nu_L$. If h and δ satisfy the inequality

$$\sqrt{2}\,(h\nu_L + \delta) < 2(h\mu_L + \delta), \tag{24}$$

for sufficiently small σ Eq. (23) has a unique root $\bar{\alpha}_\sigma$, $0 < \bar{\alpha}_\sigma < \infty$.

Proof: Using the approximation condition (16), we have

$$\tilde{\mu}_A = \inf_{u \in D} ||A_h u - f_\varepsilon|| \le ||A_h \hat{u} - f_\delta|| \le h\nu_L + \delta, \tag{25}$$

where $\hat{u}$ is the solution of the basic problem.

From the properties of the function $\tilde{\phi}^{1/2}$ proved earlier it follows that the values of this function exhaust a priori the interval $(\tilde{\mu}_A, \tilde{\nu}_A)$ for $\alpha \in (0,\infty)$. From Eqs. (24) and (25) it follows that the graph of the function $\tilde{\phi}^{1/2}$ intersects the line $\phi = 2(h\mu_L + \delta)$ for sufficiently small $\alpha > 0$ (see Figure 3).

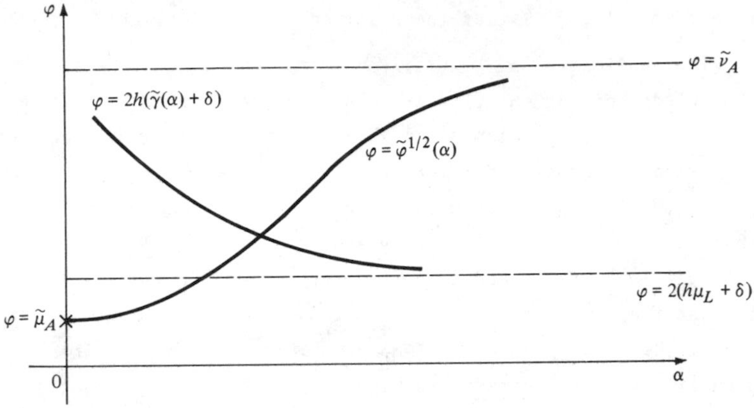

Figure 3

Since the inequality $\tilde{\gamma}(\alpha) \geq \mu_L$ is satisfied for any $\alpha > 0$, we have:

$$2(h\tilde{\gamma}(\alpha) + \delta) \geq 2(h\mu_L + \delta);$$

that is, the graph of the right-hand side of Eq. (23) is above the line $\phi = 2(h\mu_L + \delta)$. At the same time, as we know,

$$\lim_{\alpha\to\infty} \tilde{\gamma}(\alpha) = \mu_L,$$

which, together with the strict monotonicity of the continuous functions $\tilde{\phi}^{1/2}$ and $\tilde{\gamma}$, leads to the situation shown in Figure 3. We can clearly see that Eq. (23) has a unique root greater than zero. □

In the sequel we assume that the conditions of Lemma 20 are satisfied.

We call the choice of the regularization parameter α as the root $\bar{\alpha}_\sigma$ of Eq. (23) *generalized criterion* $\tilde{\phi}$. The following theorem proves the fact that the set of solutions $\hat{U}_A = \{u \in U_A: ||Lu-g|| \leq \sqrt{2}\,\nu_A\}$ of the equation $Au = f$ are approximated by the solution defined by generalized criterion $\tilde{\phi}$.

Theorem 29. Let the conditions of Lemma 20 be satisfied. Also, let the regularization parameter $\bar{\alpha}_\sigma$ be defined by generalized criterion $\tilde{\phi}$ (23). For any elements $u \in H$ and $g \in G$ we have

$$\lim_{\sigma\to 0} \inf_{\bar{u}\in\hat{U}_A} \{||A\bar{u}_\sigma - f|| + |(L\bar{u}_\sigma - L\bar{u},\ g)| + |(\bar{u}_\sigma - \bar{u},\ u)|\} = 0 \tag{26}$$

where the elements $\bar{u}_\sigma = u_{\bar{\alpha}_\sigma}$.

Proof: Using the extremal property of regularized solutions and the condition of the choice of the parameter (23), we have

$$\begin{aligned} 4(h_{\tilde{\gamma}}(\bar{\alpha}_\sigma)+\delta)^2 &= ||A_h\bar{u}_\sigma - f_\sigma||^2 + \bar{\alpha}_\sigma||L\bar{u}_\sigma - g||^2 \\ &\le ||A_h\hat{u} - f_\delta||^2 + \bar{\alpha}_\sigma||L\hat{u} - g||^2 \\ &\le (h\nu_L+\delta)^2 + \bar{\alpha}_\sigma\nu_L^2. \end{aligned} \tag{27}$$

Since $\tilde{\gamma}(\alpha) \ge \mu_L = \inf\limits_{u\in D}||Lu-g||$ for all $\alpha > 0$, we obtain, using the condition (24), that

$$2(h\nu_L+\delta)^2 \le 4(h\mu_L+\delta)^2 \le 4(h\tilde{\gamma}(\bar{\alpha}_\sigma) + \delta)^2. \tag{28}$$

From Eqs. (27), (28) it readily follows that

$$\nu_L^2 \ge \frac{(h\nu_L+\delta)^2}{\bar{\alpha}_\sigma}.$$

Using the last estimate, we have, from Eq. (27),

$$||L\bar{u}_\sigma - g||^2 \le \nu_L^2 + \nu_L^2 = 2\nu_L^2;$$

that is,

$$||L\bar{u}_\sigma - g|| \le \sqrt{2}\,\nu_L. \tag{29}$$

On the other hand, using also Eq. (29), we have

$$\begin{aligned} ||A_h\bar{u}_\sigma - f_\delta|| &\le \tilde{\phi}^{1/2}(\bar{\alpha}_\sigma) \le 2(h\tilde{\gamma}(\bar{\alpha}_\sigma) + \delta) \\ &\le 2(\sqrt{2}\,h\nu_L + \delta). \end{aligned}$$

Hence, using the approximation condition (16), we have

$$\begin{aligned} ||A\bar{u}_\sigma - f|| &\le ||A_h\bar{u}_\sigma - f_\varepsilon|| + h||L\bar{u}_\sigma - g|| + \delta \\ &\le \sqrt{2}\,h\nu_L + \delta + 2(\sqrt{2}\,h\nu_L + \delta). \end{aligned} \tag{30}$$

The estimates (29), (30) are similar to the estimates (8), (9) obtained earlier. Arguing in the same way as in proving Theorem 26, we can see that the relation (26) is satisfied. □

Remark. It is not hard to see that under the conditions of Theorem 29 the remark made after Theorem 26 holds in toto, as well as the following theorem, which is an analog of Theorem 27.

Theorem 30. Let the conditions of Theorem 29 be satisfied. Also, let the operator B be strongly subordinate to the operators A and L. Then

$$\lim_{\sigma \to 0} \inf_{\bar{u} \in \hat{U}_A} ||B\bar{u}_\sigma - B\bar{u}||_V = 0. \tag{31}$$

If the set $\hat{U}_A = \{\hat{u}\}$, we have from Eq. (31):

$$\lim_{\sigma \to 0} ||B\bar{u}_\sigma - B\hat{u}||_V = 0.$$

The condition (24) of Lemma 20 is satisfied a priori if $h \ll \delta$.

Chapter 3

Regular Methods for Solving Linear and Nonlinear Ill-Posed Problems

Section 11. Regularity of Approximation Methods

1. The analysis of the algorithms investigated so far for solving the basic problem enables us to formulate a very general principle of constructing approximate solutions.

Let $\Delta = \{A, f, g, L, D, \ldots\}$ be an s-vector of exact information of the basic problem. Also, let $\Delta_\sigma = \{A_h, f_\delta, g_\tau, L_t, \tilde{D}, \ldots\}$ be a vector of approximate information with similar components of the same dimension, where $\sigma = (h, \delta, \tau, t, \ldots)$ is an *accuracy vector* defining the accuracy of specification of approximate information.

Each mapping R_σ defined on any approximate information of Δ_δ, whose values are nonempty sets $U_\sigma = R_\sigma\{\Delta_\sigma\} \subset D$ is referred to as an *approximation method for solving the basic problem* (we note that the operators A and L may well be nonlinear). The approximation method R_σ is said to be *regular* if the following relations are satisfied:

$$\overline{\lim_{\sigma\to 0}} \sup_{u\in U_\sigma} ||Au-f||_F \leq \mu_A, \quad \overline{\lim_{\sigma\to 0}} \sup_{u\in U_\sigma} ||Lu-g||_G \leq \nu_L, \tag{1}$$

where $\sigma \to 0$ if and only if $h, \delta, \tau, \ldots$ tend to $0+$.

We define the *error* associated with the approximation method R_σ by

$$\Delta(U_\sigma, \hat{u}) = \sup_{u\in U_\sigma} |u - \hat{u}|, \tag{2}$$

where $\hat{u}$ is the solution of the basic problem and $|u| = (||Au||_F^2 + ||Lu||_G^2)^{1/2}$ $\forall u \in D$.

The approximation method R_σ is said to be *convergent (stable)* if

$$\lim_{\sigma\to 0} \Delta(U_\sigma, \hat{u}) = 0, \tag{3}$$

where $\Delta(U_\sigma,\hat{u})$ is the error of this method for any solution of the basic problem.

The following theorem yields the convergence criterion for the approximation method R_σ.

Theorem 31. A necessary and sufficient condition for the approximation method R_σ for solving the basic problem to be convergent is that it be regular.

Proof: Let R_σ be convergent. We shall show that it is regular.

It is seen that

$$\sup_{u\in U_\sigma} ||Au-f|| \le \sup_{u\ U_\sigma} ||Au-A\hat{u}|| + \mu_A \le \Delta(U_\sigma,\hat{u}) + \mu_A,$$

$$\sup_{u\in U_\sigma} ||Lu-g|| \le \sup_{u\ U_\sigma} ||Lu-L\hat{u}|| + \nu_L \le \Delta(U_\sigma,\hat{u}) + \nu_L.$$

Using (2), we derive from the above the relation (1). Therefore, the method R_σ is regular.

We next show that any regular method R_σ is convergent.

Assume the contrary. Then, for some $\varepsilon > 0$ there is at least one sequence $\{u_n\}$, $u_n \in U_{\sigma_n}$, where $\sigma_n = (h_n, \sigma_n, \tau_n, t_n,\ldots) \to 0$ as $n \to \infty$, such that

$$|u_n-\hat{u}| \ge \varepsilon > 0. \tag{4}$$

Since the method R_σ is regular, the following relations hold in conjunction with (1):

$$\overline{\lim_{n\to\infty}} ||Au_n-f|| \le \mu_A, \quad \overline{\lim_{n\to\infty}} ||Lu_n-g|| \le \nu_L. \tag{5}$$

Employing considerations similar to those given in proving Theorem 3 on convergence of regularized solutions to the solution of the basic problem, we establish that (4) yields

$$\lim_{n\to\infty} |u_n-\hat{u}| = 0,$$

which is not possible because of (3). The contradiction obtained proves that the method R_σ is convergent. □

It is not hard to see that the conditions (1) for the approximation method R_σ to be regular are equivalent to the following assertion: there exist nonnegative functions $\beta_1(\sigma)$ and $\beta_2(\sigma)$ such that

$$\lim_{\sigma\to 0} \beta_1(\sigma) = \lim_{\sigma\to 0} \beta_2(\sigma) = 0$$

and

$$||Au-f|| \leq \mu_A + \beta_1(\sigma), \quad ||Lu-g|| \leq \nu_L + \beta_2(\sigma), \quad \forall u \in U_\sigma. \tag{6}$$

2. The solution of the basic problem can be defined uniquely as the element $\hat{u} \in U_A$ for which

$$\nu_L = \inf_{u \in U_A} ||Lu-g|| = ||L\hat{u}-g||.$$

Regular approximation methods yield solutions converging to $\hat{u}$. However, it is sometimes necessary to define, in some sense, approximations to the entire set U_A or to some nonempty subset of U_A. We define the sets $\hat{U}_K$ as follows:

$$\hat{U}_K = \{u \in U_A: ||Lu-g|| \leq K\nu_L\},$$

where $K \geq 1$ is a constant. It is easily seen that for $K \geq 1$ the sets $\hat{U}_K \neq \emptyset$ since $\hat{u} \in \hat{U}_K$. On the other hand, for $K = 1$ we have $\hat{U}_{K-1} = \{\hat{u}\}$; that is, U consists of a single element and, furthermore, for $K = \infty$ the set $\hat{U}_\infty$ obviously coincides with the entire set U_A. In addition, $\hat{U}_{K_1} \geq \hat{U}_{K_2}$ if $k_1 \geq k_2 \geq 1$; that is, the sets $\hat{U}_K$ expand as $k \to \infty$.

An approximation method R_σ is called *K-regular* if for some $K \geq 1$ the relations

$$\overline{\lim_{\sigma\to 0}} \sup_{u \in U_\sigma} ||Au-f|| \leq \mu_A, \quad \overline{\lim_{\sigma\to 0}} \sup_{u \in U_\sigma} ||Lu-g|| \leq K\nu_L. \tag{7}$$

are satisfied.

We note that for $K = 1$ any K-regular approximation method R_σ is regular.

We shall now define the *accuracy* of the K-regular method R_σ, using

$$\Delta_{fgh}(U_\sigma,\hat{U}_K) = \sup_{u \in U_\sigma} \inf_{v \in \hat{U}_K} \{|(Au-Av, f)| + |(Lu-Lv, g)| + |(u-v, h)|\}, \tag{8}$$

depending on fixed elements $f \in F$, $g \in G$, $h \in H$. We call the K-regular method *weakly convergent* if

$$\lim_{\sigma\to 0} \Delta_{fgh}(U_\sigma,\hat{U}_K) = 0, \quad \forall f,g,h \tag{9}$$

from the corresponding spaces F, G and H.

The following theorem holds.

Theorem 32. Each K-regular method R_σ is weakly convergent.

Proof: It follows from (5) that for some $\sigma_0 > 0$

$$\sup_{u \in U_\sigma} |u| \leq \text{const}, \quad \sigma: \ 0 < \sigma \leq \sigma_0,$$

where, for instance, $\sigma_1 \leq \sigma_2$ designates that $h_1 \leq h_2$, $\sigma_1 \leq \delta_2, \ldots$. The completeness relationship also implies that

$$\sup_{u \in U_\sigma} ||u||_H \leq \text{const}, \quad \sigma: \ 0 < \sigma \leq \sigma_0. \tag{10}$$

We now choose arbitrary elements $u_\sigma \in U_\sigma$. It follows from (9), (10) that the family $\{u_\sigma\}$ is weakly compact in H, the family $\{Au_\sigma\}$ is weakly compact in F, and the family $\{Lu_\sigma\}$ is weakly compact in G. We distinguish any subfamily $\{u_{\sigma'}\} \subseteq \{u_\sigma\}$ such that

$$u_{\sigma'} \xrightarrow{\text{weakly}} u_0 \ \text{(in H)}, \quad Au_{\sigma'} \xrightarrow{\text{weakly}} f_0 \ \text{(in F)},$$
$$Lu_{\sigma'} \xrightarrow{\text{weakly}} g_0 \ \text{(in G)}$$

as $\sigma' \to 0$. From the joint weak closure of the operators A and L and the relations proved above it follows that $u_0 \in D$, $Au_0 = f_0$, $Lu_0 = g_0$. Using the weak lower semi-continuity of the norm in a Hilbert space as well as the relation (6), we establish:

$$\begin{aligned} ||Au_0 - f|| &\leq \varliminf_{\sigma' \to 0} ||Au_{\sigma'} - f|| \leq \varlimsup_{\sigma' \to 0} ||Au_{\sigma'} - f|| \leq \mu_A, \\ ||Lu_0 - g|| &\leq \varliminf_{\sigma' \to 0} ||Lu_{\sigma'} - g|| \leq \varlimsup_{\sigma' \to 0} ||Lu_{\sigma'} - g|| \leq K\nu_L. \end{aligned} \tag{11}$$

The first relation in (11) yields $u_0 \in U_A$. Hence

$$\lim_{\sigma' \to 0} ||Au_{\sigma'} - f|| = \mu_A = ||Au_\sigma - f||_F. \tag{12}$$

The second relation in (11) yields $u_0 \in \hat{U}_K$.

What we have proved above and the arbitrariness of the family $u_{\sigma'}$ imply that

$$\lim_{\sigma \to 0} \inf_{u \in U_K} \{|(Au_\sigma - Au), f)| + |(Lu_\sigma - Lu, g)| + |(u_\sigma - u, h)|\} = 0$$

for $\forall f \in F$, $g \in G$, $h \in H$.

Exploiting the arbitrariness of the family u_σ, we obtain from this the required relation (7). □

Remark 1. It follows from (12) and the weak convergence of

$$Au_{\sigma'} \xrightarrow{\text{weakly}} Au_0$$

that

$$\lim_{\sigma'\to 0} ||Au_{\sigma'}-Au_0|| = 0.$$

For the whole family $\{u_\sigma\}$ we have

$$\lim_{\sigma\to 0} \inf_{u_0\in \hat{U}_A} ||Au_\sigma-Au_0|| = 0.$$

Because of the arbitrary choice of the family u_σ, we have in fact

$$\lim_{\sigma\to 0} \sup_{u\in U_\sigma} \inf_{\bar{u}\in\hat{U}_K} ||Au_\sigma-A\ddot{u}|| = 0.$$

Assuming that

$$\Delta_{gh}(U_\sigma,\hat{U}_K) = \sup_{u\in U_\sigma} \inf_{v\in\hat{U}_K} \{||Au-Av|| + |(Lu-Lv, g)| + |(u-v, h)|\},$$

$$\forall g \in G,\ h \in H,$$

we derive from the preceding arguments that for each K-regular approximation method R_σ the following limit relation holds:

$$\lim_{\sigma\to 0} \Delta_{gh}(U_\sigma,\hat{U}_K) = 0, \quad \forall g \in G, \quad h \in H$$

(*strengthened* weak convergence).

In analogy with Theorem 31, we can prove the following.

Theorem 32'. A necessary and sufficient condition for the approximation method R_σ to be K-regular is that it be weakly convergent in strengthened form.

Remark 2. We now define the special approximation method R_K so that $R_K\{\Delta\} = \hat{U}_K$. By the definition of the set $\hat{U}_K$ for each $u \in \hat{U}_K$ we have the relations

$$||Au-f|| = \mu_A, \quad ||Lu-g|| \le K\nu_L,$$

and therefore

$$\lim_{K\to 1+} \sup_{u\in\hat{U}_K} ||Au-f|| \le \mu_A, \quad \overline{\lim_{K\to 1+}} \sup_{u\in\hat{U}_K} ||Lu-g|| \le \nu_L,$$

that is, the relations (1) are satisfied and, therefore, the method R_K is regular.

It is natural to call an approximation method R_K a *shrinking-regions method* since the sets $\hat{U}_{K_1} \subseteq U_{K_2}$ for any $K_1 \le K_2$.

3. The method of shrinking regions admits a generalization. In fact, consider some K-regular method R_σ, $K > 1$. We construct the approximation method $R_{\sigma,K}$ so that $R_{\sigma,K}\{\Delta_\sigma, K\} = U_{\sigma,K} = U_\sigma \cap \hat{U}_K$. It is easily seen that for the method $R_{\sigma,K}$ we have the relations

$$\overline{\lim_{\sigma\to 0, K\to 1}} \sup_{u\in U_{\varepsilon,K}} ||Au-f|| \le \mu_A, \qquad \overline{\lim_{\sigma\to 0, K\to 1}} \sup_{u\in U_{\sigma,K}} ||Lu-g|| \le \nu_L.$$

Therefore, the method $R_{\sigma,K}$, which we also refer to as *a method of shrinking regions*, is regular. We have thus proved the following assertion.

Theorem 33. For each K-regular approximation method for solving the basic problem one can construct at least one (using, for example, the method of shrinking regions) regular approximation method.

4. The investigation above implies:

(a) the Tikhonov regularization method is regular;

(b) the methods based on criteria ρ, γ and ϕ for choosing the regularization parameter are regular;

(c) the residual method and the method of quasisolutions (exact information) are regular;

(d) the methods based on criteria of the $\tilde{\rho}$ type for choosing the regularization parameter are regular;

(e) the methods based on criteria of the $\tilde{\phi}$ type for choosing the regularization parameter are K-regular for $K = \sqrt{2}$.

In the sequel we shall consider some other regular and K-regular approximation methods for solving the basic problem and their generalizations.

Section 12. The Theory of Accuracy of Regular Methods.

1. In addition to the problem of constructing regular approximation methods for solving the basic problem, the problem of estimating the accuracy of these methods is very essential.

Let $u \in D$ be any element. Also, let

$$\mu_A(u) = ||Au-f||_F.$$

It is seen that $\mu_A(u) \ge \mu_A$ for any $u \in D$. We estimate the deviation $||Au-Au_0||_F$, where $u \in D$ is any element, and $u_0 \in U_A = \{u \in D: ||Au_0-f|| \le \mu_A\}$.

We have

$$||Au-Au_0|| = ||Au-f-(Au_0-f)||^2 = \mu_A^2(u)$$

$$- 2(Au-f,Au_0-f) + \mu_A^2 = \mu_A^2(u) - \mu_A^2 - 2(Au_0-f, Au-Au_0).$$

By Theorem 5 we have

$$(Au_0-f, Au-Au_0) \geq 0 \quad \forall u \in D,$$

which yields the inequality

$$||Au-Au_0||^2 \leq \mu_A^2(u) - \mu_A^2. \tag{1}$$

We note that if $D = D_{AL}$, then $(Au_0-f, Au-Au_0) = 0$ and, furthermore, exact equality holds:

$$||Au-Au_0||^2 = \mu_A^2(u) - \mu_A^2, \tag{2}$$

thereby proving the following lemma.

Lemma 21. For any $u \in D$ and $u_0 \in U_A$ the estimate (1) holds. If $D = D_{AL}$ the equality (2) holds.

We next consider some K-regular approximation method R_σ for solving the basic problem.

Since for any $u \in U_\sigma \subset D$

$$\mu_A \leq ||Au-f||,$$

we have the relation

$$\lim_{\sigma\to 0} \sup_{u\in U_\sigma} ||Au-f|| = \mu_A,$$

which implies the existence of the nonnegative function ε_1 such that $\lim_{\sigma\to 0} \varepsilon_1(\sigma) = 0$ and in addition,

$$\sup_{u\in U} ||Au-f|| \leq \mu_A + \varepsilon_1(\sigma).$$

Thus, we have

$$\mu_A < \sup_{u\in U_\sigma} ||Au-f|| \leq \mu_A + \varepsilon_1(\sigma).$$

Applying Lemma 21, we obtain for $u \in U_\sigma$ and $u_0 \in U_A$ that

$$\sup_{u\in U_\sigma, u_0\in U_A} ||Au-Au_0||^2 \leq (\mu_A + \varepsilon_1(\sigma))^2 - \mu_A^2.$$

Due to the arbitrariness of the elements $u \in U_\sigma$ and $u_0 \in U_A$, we arrive at the following assertion.

Lemma 22. For any K-regular approximation method R_σ the following estimate holds:

$$\Delta_A(U_\sigma,U_A) = \sup_{u\in U_\sigma, u_0\in U_A} ||Au-Au_0||^2 \leq [(\mu_A+\varepsilon_1(\sigma))^2-\mu_A^2]^{1/2}. \quad (3)$$

Therefore

$$\lim_{\sigma\to 0} \Delta_A(U_\sigma,U_A) = 0. \quad (4)$$

Remark. Since $\hat{U}_K \subseteq U_A$ for any $1 \leq K < \infty$, it is seen that a relation similar to (4) will be satisfied:

$$\lim_{\sigma\to 0} \sup_{u\in U_\sigma, u_0\in \hat{U}_K} ||Au-Au_0|| = 0, \quad (5)$$

under the general assumptions about the information of the basic problem (in particular, the operators A and L).

Unfortunately, it is not possible to obtain a similar estimate for the deviation of values of the operator L, since, in general, the "approximate" solutions U_σ need not be contained in the set U_A. Hence we are content to have here a weaker estimate.

In fact, the second condition (6) in Section 11 implies the existence of a nonnegative function ε_2 such that

$$\sup_{u\in U_\sigma} ||Lu-g|| \leq K\nu_L + \varepsilon_2(\sigma), \quad \lim_{\sigma\to 0} \varepsilon_2(\sigma) = 0. \quad (6)$$

Hence for any $u \in U_\sigma$ and $u_0 \in \hat{U}_K$ we have

$$\sup_{u\in U_\sigma, u_0\in \hat{U}_K} ||Lu-Lu_0|| \leq \sup_{u\in U_\sigma} ||Lu-g|| + \sup_{u_0\in \hat{U}_K} ||Lu_0-g||$$
$$\leq 2K\nu_L + \varepsilon_2(\sigma), \quad (7)$$

and therefore only

$$\overline{\lim_{\sigma\to 0}} \sup_{u\in U_\sigma, u_0\in \hat{U}_K} ||Lu-Lu_0|| \leq 2K\nu_L. \quad (8)$$

However, the following assertion holds.

Let $\nu_L = 0$. For any K-regular approximation method R_σ we have the relation

$$\lim_{\sigma\to 0} \sup_{u\in U_\sigma, u_0\in \hat{U}_K} |u-u_0| = 0. \quad (9)$$

Indeed, in this case $\hat{U}_K = \{\hat{u}\}$ for any $K \geq 1$. It remains only to apply the estimates (5) and (8).

2. In the general case $(\nu_L > 0)$ a relation of the type (9) does not, in general, hold for K-regular methods. In order to obtain more meaningful results, we "weaken" the degree of accuracy of approximation solutions.

We consider a linear operator B defined on $D_{AL} \subset H$ into a Banach space V. We assume that the operator B satisfies the following conditions:

(a) The condition of B-completeness, that is,

$$||Bu||_V \leq \gamma_B|u|, \quad \forall u \in D, \quad \gamma_B = \text{const};$$

$$N_A \subseteq N_B, \quad N_A = \{u \in D_{AL}: Au = 0\}, \quad N_B = \{u \in D_{AL}: Bu = 0\}$$

where N_A, N_B are the *null spaces* of the operators A and B. (This condition can be satisfied a priori if $N_A = \{0\}$.)

(b) For any sequence $\{u_n\} \subset D_{AL}$, $n = 1,2,\ldots$, such that

$$u_n \xrightarrow{\text{weakly}} u_0 \quad (\text{in } H), \quad Au_n \xrightarrow{\text{weakly}} f_0 \quad (\text{in } F),$$
$$Lu_n \xrightarrow{\text{weakly}} g_0 \quad (\text{in } G),$$

the limit relation is satisfied:

$$\lim_{n\to\infty} ||Bu_n - Bu_0||_V = 0.$$

We call condition (b) of the operator B the *strengthened subordination* of the operator B to the operators A and L.

By the B-*error of a* K-*regular approximation method* R_σ we shall mean:

$$\Delta_B(U_\sigma, \hat{U}_K) = \sup_{u\in U_\sigma, u_0\in\hat{U}_K} ||Bu - Bu_0||_V.$$

Next, we define the *estimation function* ω_B by

$$\omega_B(B,R) = \sup_u ||Bu||_V \quad \text{for } \{u \in D_{AL}: ||Au|| \leq \varepsilon, \ ||Lu|| \leq R\},$$

where $\varepsilon < \infty$ and $R < \infty$ are constant.

The following theorem shows that the B-error of an approximation method can easily be expressed in terms of the estimation function.

Theorem 34. Let a K-regular approximation method R_σ be given. Also, let the relations (3) and (7) be satisfied. Then we have

$$\Delta_B(U_\sigma,\hat{U}_K) \leq \omega_B\left(\sqrt{(\mu_A+\varepsilon_2)^2-\mu_A^2},\ 2K\nu_L + \varepsilon_2\right). \tag{10}$$

Proof: The proof of (10) follows immediately from the definition of the estimation function.

3. Later we shall give some applications of the estimate (10) for estimating the accuracy of various K-regular methods. Now we discuss some simple properties of the estimation function.

Let

$$M_{\varepsilon,R} = \{u \in D_{AL}: ||Au|| \leq \varepsilon,\ ||Lu|| \leq R\}.$$

Furthermore, we assume that the operators A and L are jointly closed on the whole set D_{AL} (but not on the set D, as was assumed before).

Because of the assumptions made on the operator B and the joint closure of the operators A and L in the set D_{AL} (which are assumed to be satisfied in this section) we have the following lemma.

Lemma 23. For any $\varepsilon \geq 0$, $R \geq 0$ the function $\omega_B(\varepsilon,R)$ is defined. Moreover, there exists at least one element $h \in M_{\varepsilon,R}$ for which

$$\omega_B(\varepsilon,R) = ||Bh||_V.$$

Proof: For any $u \in M_{\varepsilon,R}$, due to the B-completeness condition, we have

$$||Bu||_V \leq \gamma_B|u| \leq \gamma_B(\varepsilon^2 + R^2)^{1/2} < +\infty,$$

that is, the function $\omega_B(\varepsilon,R)$ is actually defined for all $\varepsilon \geq 0$, $R \geq 0$.

Let $u_n \in M_{\varepsilon,R}$, $n = 1,2,\ldots$ be a maximizing sequence such that

$$\omega_B(\varepsilon,R) - \frac{1}{n} \leq ||Bu_n||_V \leq \omega_B(\varepsilon,R). \tag{11}$$

It is not hard to see that the sequences $\{u_n\}$, $\{Au_n\}$, and $\{Lu_n\}$ are bounded and, therefore are weakly compact in H, F and G, respectively. We may assume without loss of generality that they are weakly convergent in the respective spaces:

$$u_n \xrightarrow{\text{weakly}} h, \quad Au_n \xrightarrow{\text{weakly}} f, \quad Lu_n \xrightarrow{\text{weakly}} g.$$

Using the joint weak closure of the operators A and L, we conclude from the foregoing relations that

$$h \in D_{AL}, \quad Ah = f, \quad Lu = g.$$

The properties of a weakly convergent sequence imply that

$$||Ah|| \le \underline{\lim}_{n\to\infty} ||Au_n|| \le \varepsilon, \quad ||Lh|| \le \underline{\lim}_{n\to\infty} ||Lu_n|| \le R;$$

that is, $h \in M_{\varepsilon,R}$.

Using the condition of strengthened subordination of the operator B and taking the limit in Eq. (11), we conclude that the supremum is attained at the element h. □

Lemma 24. The estimation function $\omega_B(\varepsilon,R)$ possesses the following properties:

(a) $\omega_B(\varepsilon,R) \ge 0$;

(b) $\omega_B(\lambda\varepsilon,\lambda R) = \lambda\omega_B(\varepsilon,R), \quad \forall\lambda \ge 0$;

(c) $\omega_B(\varepsilon,R)$ is an increasing function of each variable;

(d) $\lim_{\varepsilon\to 0} \omega_B(\varepsilon,R) = 0, \quad \forall R < \infty$.

Proof: The properties (a), (b) and (c) are obvious. Let us prove (d). By Lemma 23, there are elements $h_\varepsilon \in M_{\varepsilon,R}$ for which

$$\omega_B(\varepsilon,R) = ||Bh_\varepsilon||_V.$$

We then have

$$\lim_{\varepsilon\to 0} ||Ah_\varepsilon|| = 0, \quad \overline{\lim}_{\varepsilon\to 0} ||Lh_\varepsilon|| \le R.$$

The above relations imply that the families of elements $\{h_\varepsilon\}$ and $\{Lh_\varepsilon\}$ are bounded and weakly compact in H and G, respectively. We distinguish a subfamily $\{h_{\varepsilon'}\} \subseteq \{h_\varepsilon\}$ such that

$$h_{\varepsilon'} \xrightarrow{\text{weakly}} h_0 \ \ (\text{in } H), \quad Ah_{\varepsilon'} \xrightarrow{\text{weakly}} 0 \ \ (\text{in } F),$$
$$Lh_\varepsilon \xrightarrow{\text{weakly}} g_0 \ \ (\text{in } G)$$

as $\varepsilon' \to 0$. Due to the joint weak closure of the operators A and L we have

$$h_0 \in D_{AL}, \quad Ah_0 = 0, \quad Lh_0 = g_0;$$

that is, h_0 belongs to the null space of the operator A and therefore it belongs to the null space of the operator B under our assumption. However, by the same assumption, the operator B is strongly subordinate to the operators A and L. Therefore we have

$$0 = \lim_{\varepsilon'\to 0} ||Bh_{\varepsilon'} - Bh_0||_V = \lim_{\varepsilon'\to 0} \omega_B(\varepsilon',R).$$

From the arbitrariness of the subfamily h_ε, it follows that

$$\lim_{\varepsilon\to 0} \omega_B(\varepsilon,R) = 0,$$

which was to be proved. □

Let

$$\Delta_{AB}(U_\sigma,\hat{U}_K) = \Delta_A(U_\sigma,\hat{U}_K) + \Delta_B(U_\sigma,\hat{U}_K).$$

The next theorem is an immediate corollary of Lemmas 22 and 24 and Theorem 34.

Theorem 35. Let a K-regular method R_σ be given. Then the error

$$\Delta_{AB}(U_\sigma,\hat{U}_K) \leq ((\mu_A+\varepsilon_1)^2-\mu_A^2)^{1/2} + \omega_B(((\mu_A+\varepsilon_1)^2-\mu_A^2)^{1/2},2K\nu_L+\varepsilon_2) \qquad (12)$$

approaches 0 as $\sigma \to 0$.

Remark 1. If the measure of consistency $\mu_A > 0$, then $((\mu_A+\varepsilon_1)^2-\mu_A^2)^{1/2} = O(\sqrt{\varepsilon_1})$; if $\mu_A = 0$, then $((\mu_A+\varepsilon_1)^2-\mu_A^2)^{1/2} = \varepsilon_1$. Therefore, for consistent problems $(\mu_A = 0)$ the order of accuracy of any K-regular method R_σ is essentially higher than that for inconsistent problems $(\mu_A > 0)$.

Section 13. The Computation of the Estimation Function

1. In order to effectively construct estimates of the B-error of K-regular methods, it is necessary to know various techniques for computing the estimation function $\omega_B(\varepsilon,R)$. Later we shall consider exact methods for computing estimation functions and, in addition, we construct exact majorizing functions with respect to the order of the quantities ε and R with the estimation function.

Let the space H be separable, and let the set D_{AL} be dense in H. Also, we assume that there exists a system of elements $u_i \in D_{AL}$, $i = 1,2,\ldots$, which is dense in H and A-orthogonal; that is, the scalar products $(Au_i,Au_j) = \sigma_{ij}$, where σ_{ij} is the Kronecker delta. Furthermore, let the quadratic forms $||Au||^2$, $||Lu||^2$ and $||Bu||^2$ (we assume that the space V is also Hilbert) be written as

$$||Au||^2 = \sum_{i=1}^{\infty} \xi_i^2, \quad ||Lu||^2 = \sum_{i=1}^{\infty} g_i\xi_i^2, \quad ||Bu||^2 = \sum_{i=1}^{\infty} \lambda_i\xi_i^2,$$

where $\xi_i = (Au_i,Au_i)$, $i = 1,2,\ldots$. In this case, we say that the quadratic forms $||Au||^2$, $||Lu||^2$ and $||Bu||^2$ are *spectrally similar*. Concerning g_i and λ_i we assume that there exist some natural numbers m_1 and m_2 such that

$$g_{i+1} > g_i > 0, \quad \forall i \geq m_1, \quad \lambda_{i+1} > \lambda_i > 0, \quad \forall i \geq m_2$$

and

$$\lim_{i\to\infty} g_i = \lim_{i\to\infty} \lambda_i = \infty.$$

Under the foregoing assumptions, the function

$$\omega^2(\varepsilon,R) = \max_{\xi} \sum_{i=1}^{\infty} \beta_i \xi_i^2, \tag{1}$$

where the vector $\xi_3 = (\xi_1,\xi_2,\ldots)$ satisfies the conditions

$$\sum_{i=m_0}^{\infty} \xi_i^2 \leq \varepsilon^2, \quad \sum_{i=1}^{\infty} g_i \xi_i^2 \leq R^2. \tag{2}$$

Thus, computing the estimation function $\omega_B(\varepsilon,R)$ is reduced to solving the infinite-dimensional quadratic programming (1), (2).

Following the method developed by V. K. Ivanov and T. I. Koroljuk [31], we substitute $\xi_i^2/\varepsilon^2 = \mu_i$. Then, we have

$$\frac{\omega^2(\varepsilon,R)}{\varepsilon^2} = \max_{\mu} \sum_{i=1}^{\infty} \lambda_i \mu_i^2, \tag{3}$$

where the vector $\mu = (\mu_1,\mu_2,\ldots)$ satisfies the condition

$$\sum_{i=1}^{\infty} \mu_i \leq 1, \quad \sum_{i=1}^{\infty} g_i \mu_i \leq \frac{R^2}{\varepsilon^2}, \quad \mu_i \geq 0 \tag{4}$$

Let $g = g(\lambda) \geq 0$, $\lambda \geq 0$, be an increasing convex function and let $g_i = g(\lambda_i)$, $i = 1,2,\ldots$. We consider the points $M_i = (\lambda_i, g_i)$ in the (λ,g) plane. Also, we denote by Γ the polygonal line composed of lines connecting the points M_i in sequence.

Lemma 25. The polygonal line Γ is an increasing convex function.

Proof: Let us consider the points M_i, M_{i+1} and M_{i+2}, $i \geq m_2$. Obviously, it suffices to show that the point M'_{i+1} (see Figure 4) lies above the point M_{i+1} on the segment $M_i M_{i+2}$. Let the ordinate of the point M'_{i+1} be equal to g'_{i+1}. It is easy to show that

$$g'_{i+1} = \frac{g_{i+2}-g_i}{\lambda_{i+2}-\lambda_i} (\lambda_{i+1}-\lambda_i) + g_i.$$

Then

$$g'_{i+1}-g_i = \frac{g_{i+2}-g_i}{\lambda_{i+2}-\lambda_i}(\lambda_{i+1}-\lambda_i) + g_i - g_{i+1}$$

$$= \frac{\lambda_{i+1}-\lambda_i}{\lambda_{i+2}-\lambda_i} g_{i+2} + \frac{\lambda_{i+2}-\lambda_{i+1}}{\lambda_{i+2}-\lambda_i} g_i - g_{i+1}$$

$$= \alpha g(\lambda_{i+2}) + (1-\alpha)g(\lambda_i) - g(\alpha\lambda_{i+2}+(1-\alpha)\lambda_i)$$

where

$$\alpha = \frac{\lambda_{i+1} - \lambda_i}{\lambda_{i+2} - \lambda_i} > 0.$$

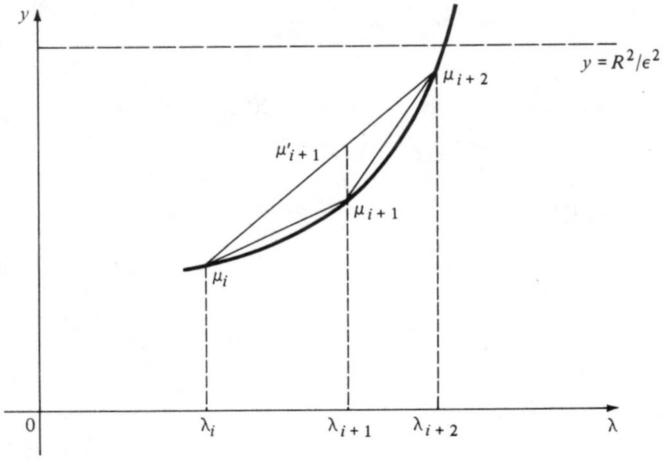

Figure 4

Using the convexity of the function $g(\lambda)$, $\lambda \geq 0$, we have $g'_{i+1} \geq g_i$, which was to be proved. □

Next, under the same conditions (convexity of the function $g(\lambda)$), let

$$Q = \{(\lambda,g): \lambda = \sum_{i=1}^{\infty} \mu_i\lambda_i, \quad g = \sum_{i=1}^{\infty} \mu_i g_i\}.$$

It is seen that $\Gamma \subseteq Q$.

The following lemma holds.

Lemma 26. The region $Q = \bigcup_{i=2}^{\infty} Q_i$, where Q_i are triangles $M_1M_iM_{i+1}$, $i \geq 2$, each pair of which, Q_i and Q_{i+1}, intersect only on the segment M_iM_{i+1}.

Proof: The proof of the lemma is elementary.

Now we can give the following geometric interpretation of the problem (3), (4): among the points $M = (\lambda, g)$ of the region Q satisfying the condition $g \leq R^2/\varepsilon^2$ (Figure 4), find the point with the maximal abscissa.

The next theorem is a corollary of Lemmas 25 and 26 and the remark made above.

Theorem 36. (V. K. Ivanov and T. I, Koroljuk [31]). Let ε be so small that $g_{m_1} \leq R^2/\varepsilon^2$ and let $g(\lambda)$, $\lambda \geq 0$ be an increasing convex function such that $g_i = g(\lambda_i)$, $i = 1,2,\ldots$. We denote by m the smallest index for which $g_m \leq R^2/\varepsilon^2$. Then we have

$$\omega_B^2(\varepsilon,R) = \varepsilon^2\lambda_m \tag{5}$$

if

$$g_m = R^2/\varepsilon^2,$$

and we have

$$\omega^2(\varepsilon,R) = \frac{R^2(\lambda_{m+1}-\lambda_m)+\varepsilon^2(\lambda_m g_{m+1}-\lambda_{m+1}g_m)}{g_{m+1}-g_m} \tag{6}$$

if

$$g_m < R^2/\varepsilon^2.$$

Proof: The polygonal line Γ bounds the region Q from the right and below. By the hypothesis of the theorem, the required index m exists. If $g_m = R^2/\varepsilon^2$ the line $g = R^2/\varepsilon^2$ intersects the polygonal line Γ at the point (λ_m, g_m). Therefore, the function $\omega_B^2(\varepsilon,R)$ can be estimated in conjunction with (5). If $g_m < R^2/\varepsilon^2$, $\omega_B^2(\varepsilon,R)/\varepsilon^2$ is equal to the abscissa of the intersection point of the lines $g = R^2/\varepsilon^2$ and Γ. This condition yields (6).

Remark. We always have

$$\lambda_m\varepsilon^2 \leq \omega_B^2(\varepsilon,R) < \varepsilon^2\lambda_{m+1};$$

therefore, (5) is asymptotic as $m \to \infty$, that is, $\varepsilon \to 0$.

2. We now apply the above technique for computing the estimation function given a triple of operators A, L and B.

We note that if

$$g(\lambda) = \lambda^{1+2\zeta}$$

where $\zeta > 0$, we can easily see that

$$\omega(\varepsilon,R) \leq \varepsilon^{\frac{2\zeta}{1+2\zeta}} R^{\frac{1}{1+2\zeta}} . \tag{7}$$

In fact, solving the equation

$$g(\lambda_{m+1}) = R^2/\varepsilon^2,$$

we obtain

$$\lambda_m = \varepsilon^{-\frac{2\zeta}{1+2\zeta}} R^{\frac{2}{1+2\zeta}} .$$

By Theorem 36,

$$\omega^2(\varepsilon,R) = \lambda_m\varepsilon^2 = \varepsilon^{\frac{4\zeta}{1+2\zeta}} R^{\frac{2}{1+2\zeta}} .$$

yielding (7).

As $\zeta \to \infty$, we obtain from (7) that

$$\omega(\varepsilon,R) = O(\varepsilon),$$

that is, the limit accuracy defined by the level of ε. It is interesting to note that as $\zeta \to \infty$ the dependence of the estimate for $\omega(\varepsilon,R)$ on R weakens. It is possible to interpret the condition $g(\lambda) = \lambda^{1+2\zeta}$ for large ζ as a strong condition for smoothness required of the estimated solution.

We give several examples of the application of the estimate (7).

<u>Example 1</u>. Let $H = F = G = V = L_2[-\pi,\pi]$. It is well known fact that each function $u \in L_2[-\pi,\pi]$ can be expressed in terms of a Fourier series

$$u(x) = \sum_{m=-\infty}^{\infty} \xi_m u_m(x), \qquad u_m(x) = \frac{1}{\sqrt{2\pi}} e^{imx}.$$

Let $A = E$ be the identity operator on $L_2[-\pi,\pi]$;

$$||Lu||^2 = \sum_{m=-\infty}^{\infty} m^{2\sigma}|\xi_m|^2, \qquad ||Bu||^2 = \sum_{m=-\infty}^{\infty} m^{2\mu}|\xi_m|^2, \quad 0 < \mu < \sigma.$$

Then, obviously, $g(\lambda) = \lambda^{1+2\zeta}$, where $\zeta = (\sigma-\mu)/2\mu$. Using (7), we have

$$\omega(\tau,R) \leq \tau^{1-\mu/\sigma}R^{\mu/\sigma}. \tag{8}$$

In particular, the estimate (8) holds if $Lu = d^\sigma u/dx^\sigma$, $Bu = d^\mu u/dx^\mu$ when σ, μ are positive integers, $\sigma > \mu$.

<u>Example 2</u>. As before, let $H = G = F = V = L_2[-1,1]$. Also, let

$$A = E, \quad Lu = \frac{d}{dx}\left[(1-x^2)\frac{du}{dx}\right]$$

and let $\{u_n : n = 0,1,\ldots\}$ be a system of Legendre polynomials normalized in $L_2[-1,1]$. Since the system $\{u_n\}$ is complete in $L_2[-1,1]$, each function $u \in L_2[-1,1]$ can be written as

$$u(x) = \sum_{n=0}^{\infty} \xi_n u_n(x).$$

Due to well-known properties of Legendre polynomials, we have

$$||Lu||^2_{L_2[-1,1]} = \int_{-1}^{1} \left[\frac{d}{dx}(1-x^2)\frac{du}{dx}\right]^2 dx = \sum_{i=0}^{\infty} n^2(n+1)^2\xi_n^2.$$

Let the operator B be such that

$$||Bu||^2_{L_2[-1,1]} = \sum_{n=0}^{\infty} n^{2\mu}(n+1)^{2\mu}\xi_n^2, \qquad 0 < \mu < 1.$$

Then $g(\lambda) = \lambda^{1/\mu}$, $(1-\mu)/2\mu = \zeta$. Using (7), we have

$$\omega(\tau,R) \leq \tau^{1-\mu}R^{\mu}. \tag{9}$$

Eqs. (8) and (9) can be used for estimation of the accuracy of representation of observed data by the corresponding Fourier series.

Example 3. We now consider the problem of analytic extension of the function W which is harmonic in the circle $|z| < 1$ and square-integrable on the boundary $|z| = 1$.

Let $u_r(\phi) = W(re^{i\phi})$, $0 < r < 1$, and let $H = F = G = V = L_2[0,2\pi]$. We assume that $A = E$, $Lu_r(\phi) = u_1(\phi)$, $Bu_r(\phi) = u_\rho(\phi)$, $0 < r < \rho < 1$.

Putting

$$u_\tau(\phi) = \sum_{m=-\infty}^{\infty} \xi_m u_m(\phi), \qquad u_m(\phi) = \frac{1}{\sqrt{2\pi}} e^{im\phi},$$

we have

$$||Lu_\tau(\phi)||^2 = \sum_{i=-\infty}^{\infty} \frac{1}{\tau^{2m}} |\xi_m|^2, \quad ||Bu_\tau(\phi)|| = \sum_{m=-\infty}^{\infty} (\frac{\rho}{\tau})^{2m}|\xi_m|^2.$$

Making simple computations, we see that $g(\lambda) = \lambda^{1+2\zeta}$, where $\zeta = \sigma/2(1-\sigma)$, $\sigma = \ln g/\ln r$. We note that $0 < \sigma < 1$.

Using (7), we have

$$\omega(\tau,R) \leq \tau^{\sigma}R^{1-\sigma}, \tag{10}$$

if $||u_1(\phi)|| \leq R$.

A result similar to (10) was proved in [31] by means of other techniques.

Example 4. Let $G = H$ and $L = E$. We assume that $A^*A = g_1(B^*B)$, where $g_1(\lambda)$ is a convex increasing function of λ and the operators A and B are completely continuous.

Let $g_1(\lambda) = \lambda^{\frac{1+2\zeta}{2\zeta}}$ and let $\{e_i\}$ be a complete orthonormalized (in H) system of eigenelements of the operator A^*A; $A^*Ae_i = \mu_i e_i$, $\mu_i > 0$. We may then express each element of H as

$$u = \sum_{i=1}^{\infty} u_i e_i, \quad u_i = (u, e_i)_H,$$

so that

$$||Au||^2 = \sum_{i=1}^{\infty} \mu_i u_i^2, \quad ||Lu||^2 = \sum_{i=1}^{\infty} u_i^2, \quad ||Bu||^2 = \sum_{i=1}^{\infty} \mu_i^{\frac{2}{1+2}} u_i^2.$$

Assuming in this case that $\mu_i u_i^2 = \xi_i^2$, we obtain

$$||Au||^2 = \sum_{i=1}^{\infty} \xi_i^2, \quad ||Lu||^2 = \sum_{i=1}^{\infty} \mu_i^{-1}\xi_i^2, \quad ||Bu||^2 = \sum_{i=1}^{\infty} \mu_i^{-1/1+2} \xi_i^2,$$

that is, $\lambda_i = \mu_i^{-1/1+2\zeta}$, $g_i = \mu_i^{-1}$ and therefore, $g(\lambda) = \lambda^{1+2\zeta}$. Thus, we have proved (7) for our case.

Let us indicate when $g_1(\lambda) = \lambda^{\frac{1+2\zeta}{2\zeta}}$. Let $C: V \to F$ be a completely continuous linear operator and let $A = CB$. It is easy to verify that if $B = (C^*C)^\zeta$, $\zeta > 0$, the foregoing relation holds. It will hold as well for $\zeta = 1/2$ if $B = C^*$.

Example 5. We consider now the problem of solving the evolution equation

$$\frac{du}{dt} = Du, \quad u(0) = f, \quad 0 < t \leq T, \tag{11}$$

where D is self-adjoint nonnegative definite linear operator in H with unbounded discrete spectrum. We shall assume that the problem (11) has a solution and that $||u(T)||_H \leq R$. Let $Au = u(0)$, $Bu = u(t)$, $Lu = u(T)$. Also, let $\{d_i\}$ be a complete orthonormalized (in H) system of eigenelements of the operator D:

$$Dd_i = \mu_i d_i, \quad \mu_i > 0, \quad \lim_{i\to\infty} \mu_i = \infty.$$

Then, for each $u \in H$ we have

$$u = \sum_{i=1}^{\infty} \xi_i d_i, \quad \xi_i = (u,d_i)_H.$$

Let $u(0) = \Sigma_{i=1}^{\infty} \xi_i d_i$. We have

$$||Au||_H^2 = \sum_{i=1}^{\infty} \xi_i^2, \quad ||Bu||_H^2 = \sum_{i=1}^{\infty} \exp\{2\mu_i t\}\xi_i^2,$$

$$||Lu||_H^2 = \sum_{i=1}^{\infty} \exp(2\mu_i T)\xi_i^2,$$

implying that $\lambda_i = \exp(2\mu_i t)$, $g_i = \exp(2\mu_i T)$ and therefore $g(\lambda) = \lambda^{1+2\zeta}$, where $\zeta = (T-t)/2t$. Using (7), we have

$$||u(t)||_H \leq \omega(\tau,R) \leq \tau^{(1-(t/T))}R^{t/T}. \tag{12}$$

3. In applications, the exact methods of computing the estimating function $\omega(\varepsilon,R)$ as well as the techniques for estimating the function from above are essential. We discuss here one technique based on moment inequalities.

Let E_λ, $0 < \lambda < \infty$ be a decomposition of the identity in H and let

$$b_k = ||Au||_F^2 = \int_0^\infty \lambda^K d(E_\lambda u,u)_H, \quad b_\varepsilon = ||Lu||_G^2 = \int_0^\infty \lambda^\ell d(E_\lambda u,u)_H,$$

$$b_m = ||Bu||_V^2 = \int_0^\infty \lambda^m d(E_\lambda u,u)_H.$$

As is well known, for any positive p and q the moment inequality

$$b_m^{p+q} \leq b_{m+q}^p \; b_{m-p}^q$$

holds, where if we take $m+q = k$, $m-p = \ell$, we obtain

$$b_m \leq b_k^{(m-\ell)/(k-\ell)} b_\ell^{(k-m)/(k-\ell)}$$

and therefore

$$||Bu||_P^2 \leq ||Au||_F^{2(m-\ell)/(k-\ell)} ||Lu||_G^{2(k-m)/(k-\ell)},$$

yielding immediately the estimate

$$\omega(\tau,R) \leq \tau^{(m-\ell)/(k-\ell)} R^{(k-m)/(k-\ell)}.$$

In particular, if the conditions of Example 4 are satisfied, $\ell = 0$, $k = 1$, $m = 2\zeta/(1+2\zeta)$, and the estimate is the same as (7).

Example 6. Let $H = F = G = V = L_2(-\infty,\infty)$ and let $A = E$, $B = d^\mu/dx^\mu$, $L = d^\sigma/dx^\sigma$, $0 < \mu < \sigma$. Also, let $Fu = \hat{u}$ denote the Fourier transform of the function $u \in L_2[-\infty,\infty]$. Then

$$||Au||^2 = \int_{-\infty}^{\infty} |\hat{u}(\omega)|^2 d\omega, \quad ||Bu||^2 = \int_{-\infty}^{\infty} \omega^{2\mu}|\hat{u}(\omega)|^2 d\omega,$$

$$||Lu||^2 = \int_{\infty}^{\infty} \omega^{2\sigma}|\hat{u}(\omega)|^2 d\omega.$$

We have $k = 0$, $\ell = 2\sigma$, $m = 2\mu$ and therefore

$$\hat{\omega}(\tau,R) \leq \tau^{1-\mu/\sigma}R^{\mu/\sigma},$$

which coincides with (8). We may regard the operators L and B as fractional derivative operators.

4. We now define the function $\hat{\omega}(\varepsilon,R)$:

$$\hat{\omega}(\varepsilon,R) = \sup_u ||Bu||_V \quad \text{for } u \in D_{AL} := \{u: R^2||Au||^2 + \varepsilon^2||Lu||^2 \leq 2\varepsilon^2R^2\}, \tag{13}$$

and examine the relationship between the function $\hat{\omega}(\varepsilon,R)$ and the estimate $\omega(\varepsilon,R)$.

Lemma 27. For any $\varepsilon \geq 0$, $R \geq 0$

$$\omega(\varepsilon,R) \leq \hat{\omega}(\varepsilon,R) \leq \sqrt{2}\,\omega(\varepsilon,R). \tag{14}$$

Proof: Let

$$M_{\varepsilon,R} = \{u \in D_{AL}: R^2||Au||^2 + \varepsilon^2||Lu||^2 \leq 2\varepsilon^2R^2\}.$$

It is easily seen that

$$\hat{M}_{\varepsilon,R} \supseteq M_{\varepsilon,R},$$

which proves the inequality

$$\omega(\varepsilon,R) \leq \hat{\omega}(\varepsilon,R).$$

Next, we consider the set $M_{\sqrt{2}\varepsilon,\sqrt{2}R}$. It is easily seen that for each $\varepsilon > 0$, $R > 0$ this set contains a set $\hat{M}_{\varepsilon,R}$. Then, obviously,

$$\hat{\omega}(\varepsilon,R) \leq \omega(\sqrt{2}\varepsilon,\sqrt{2}R) = \sqrt{2}\,\omega(\varepsilon,R).$$

where we have exploited one of the properties of the function $\omega(\varepsilon,R)$, proved in Lemma 24. □

Thus the functions $\omega(\varepsilon,R)$ and $\hat{\omega}(\varepsilon,R)$ are of the same order with respect to ε and R; in this sense they are equivalent. The substitution of the estimate $\omega(\varepsilon,R)$ for $\hat{\omega}(\varepsilon,R)$ makes the estimation of accuracy of K-regular approximation methods of solving the basic problem slightly more coarse.

5. We now describe an effective technique for computing the function $\hat{\omega}(\varepsilon,R)$. As in Subsection 13.1, we assume that the quadratic forms $||Au||^2$, $||Lu||^2$ and $||Bu||^2$ are spectrally similar. In this case let

$$\lambda_{i+1} \geq \lambda_i \geq 0, \quad g_{i+1} \geq g_i \geq 0 \quad \text{for all} \quad i = 1,2,\ldots$$

and

$$\lim_{i\to\infty} \lambda_i = \lim_{i\to\infty} g_i = +\infty, \quad \lim_{i\to\infty} \lambda_i/g_i = 0.$$

We shall need the following

<u>Remark</u>. For any solution u_0 of the problem (12) we have

$$R^2||Au_0||^2 + \varepsilon^2||Lu_0||^2 = 2\varepsilon^2R^2.$$

In fact, if

$$\hat{\omega}(\varepsilon,R) = ||Bu_0||_V$$

and

$$R^2||Au_0||^2 + \varepsilon^2||Lu_0||^2 < 2\varepsilon^2R^2,$$

assuming that $u_\lambda = \lambda u_0$, $\lambda > 1$, we can choose a value of the parameter λ such that $u_\lambda \in \hat{M}_{\varepsilon,R}$. Obviously,

$$||Bu_\lambda|| = \lambda||Bu_0|| > ||Bu_0|| = \hat{\omega}(\varepsilon,R),$$

which contradicts the definition of the element u_0, thus proving our assertion.

It follows from the foregoing that

$$\hat{\omega}(\varepsilon,R) = \sup\{||Bu||_V : u \in D_{AL} \text{ and } R^2||Au||^2 + ||Lu||^2\varepsilon^2 = 2\varepsilon^2R^2\}. \tag{16}$$

<u>Theorem 37</u>. Let the quadratic forms $||Au||^2$, $||Lu||^2$ and $||Bu||^2$ be spectrally similar and let λ_i and g_i satisfy the requirements we have imposed on them. Then

$$\hat{\omega}^2(\varepsilon,R) = 2\varepsilon^2 R^2 \max_{i\geq 1} \frac{\lambda_i}{R^2+\varepsilon^2 g_i}. \tag{17}$$

Proof: Under the conditions of the theorem we have

$$R^2||Au||^2 + \varepsilon^2||Lu||^2 = \sum_{i=1}^{\infty} (R^2+\varepsilon^2 g_i)\xi_i^2,$$

$$||Bu||^2 = \sum_{i=1}^{\infty} \lambda_i \xi_i^2.$$

Letting

$$\mu_i^2 = \frac{(R^2+\varepsilon^2 g_i)\xi_i^2}{2\varepsilon^2 R^2}$$

we have

$$\frac{\hat{\omega}^2(\varepsilon,R)}{2\varepsilon^2 R^2} = \sup_{\mu} \sum_{i=1}^{\infty} \frac{\lambda_i}{R^2+\varepsilon^2 g_i}\mu_i^2,$$

where the vector $\mu = (\mu_1,\mu_2,\ldots)$ satisfies

$$\sum_{i=1}^{\infty} \mu_i^2 = 1,$$

thus implying (17). □

Remark. We note that the conditions of Theorem 37 are weaker than the conditions under which Theorem 36 was proved. The conditions (14) are a priori satisfied if there exists a nonnegative function g such that

$$\lim_{\lambda\to\infty} \frac{\lambda}{g(\lambda)} = 0 \tag{18}$$

and, in addition, $g_i = g(\lambda_i)$.

Theorem 38. Let the quadratic forms $||Au||^2$, $||Lu||^2$ and $||Bu||^2$ be spectrally similar and let λ_i and g_i be such that

$$\lambda_i \geq \lambda_i \geq 0, \quad g_{i+1} \geq g_i \geq 0, \quad \lim_{i\to\infty} \lambda_i = \lim_{i\to\infty} g_i = +\infty.$$

Then

$$\lim_{i\to\infty} \lambda_i/g_i = 0 \tag{19}$$

is a necessary and sufficient condition in order that

$$\lim_{\varepsilon\to 0} \omega(\varepsilon,R) = 0.$$

Proof: It follows from the inequalities (14) that we need to prove only that the condition (19) is necessary and sufficient for the relation

$$\lim_{\varepsilon\to 0} \hat{\omega}(\varepsilon,R) = 0$$

to be satisfied. If (19) is satisfied, Eq. (17) trivially yields the required relation.

Suppose then that (19) is not satisfied, We can find a subsequence of integers i_s, $s = 1,2,\ldots$, such that $\lambda_{i_s}/g_{i_s} \geq a > 0$, $s = 1,2,\ldots$, where a is an integer. Assuming $\mu_i = 0$ for $i \neq i_s$, $\mu_i = 1$ for $i = i_s$, we have from the above:

$$\frac{\hat{\omega}^2(\varepsilon,R)}{2\varepsilon^2R^2} \geq \frac{\lambda_{is}}{R^2+\varepsilon^2 y_{i_s}} \geq \frac{a}{R^2/g_{i_s}+\varepsilon^2}, \quad s = 1,2,\ldots$$

and therefore

$$\frac{\hat{\omega}^2(\varepsilon,R)}{2\varepsilon^2R^2} \geq \frac{a}{2};$$

that is, $\hat{\omega}(\varepsilon,R) \not\to 0$ as $\varepsilon \to 0$. □

We can easily generalize the method of computing $\hat{\omega}(\varepsilon,R)$. In fact, let E_λ, $\lambda > 0$, be the spectral decomposition of the identity in H such that the quadratic forms $||Au||^2$, $||Bu||^2$, $||Lu||^2$ are given by

$$||Au||^2 = \int_0^\infty d(E_\lambda u,u), \qquad ||Lu||^2 = \int_0^\infty g(\lambda)d(E_\lambda u,u),$$

$$||Bu||^2 = \int_0^\infty \lambda d(E_\lambda u,u),$$

where g is a nonnegative function satisfying the limit relation

$$\lim_{\lambda\to\infty} \frac{\lambda}{g(\lambda)} = 0. \tag{20}$$

Using the same technique as in computing the function $\hat{\omega}(\varepsilon,R)$, we have

$$\omega(\varepsilon,R) = 2\varepsilon^2R^2 \sup_{\lambda\geq 0} \frac{\lambda}{R^2+\varepsilon^2 g(\lambda)}, \tag{21}$$

In order that the relation

$$\lim_{\varepsilon\to 0} \omega(\varepsilon,R) = 0$$

be satisfied, it is necessary and sufficient that Eq. (20) be satisfied.

If the function g is continuously differentiable, we have only to solve a scalar equation for the determination of the extremal values of $\hat{\omega}$:

$$R^2 + g(\lambda)\varepsilon^2 = \lambda\varepsilon^2 g'(\lambda). \tag{22}$$

In particular, using Eq. (22), we easily obtain for the function $g(\lambda) = \lambda^{1+2\zeta}$, $\zeta > 0$:

$$\hat{\omega}(\varepsilon,R) = \left(\frac{2}{1+2\zeta}\right)^{1/2}(2\zeta)^{-\frac{\zeta}{1+2\zeta}}\varepsilon^{\frac{\zeta}{1+2\zeta}}R^{\frac{1}{1+2\zeta}} = O\left(\varepsilon^{\frac{\zeta}{1+2\zeta}}R^{\frac{1}{1+2\zeta}}\right)$$

(compare with the unimprovable estimate (7)).

Section 14. Examples of Regular Methods

1. Let the elements f, g and the operators A and L of the basic problem be given approximately; that is, we know $f_\delta \in F$, $g_\tau \in G$ and operators A_h and L_t such that

$$||f-f_\delta|| < \delta, \quad ||g-g_\tau|| < \tau, \quad ||Au-A_hu|| \leq h|u|,$$

$$||Lu-L_tu|| \leq t|u|, \quad \forall u \in D.$$

We consider the problem of constructing K-regular methods when the initial data is approximate. To begin with, we assume that $\delta = h = 0$, that is, there is no error in specifying f and the operator A.

Let

$$\tilde{\nu}_L = \inf_{u\in U_A} ||L_tu-g_\tau||_G.$$

By Theorem 1, there exists a unique element $\hat{u}_{t\tau} \in U_A$ such that

$$||L_t\hat{u}_{t\tau}-g_\tau|| = \tilde{\nu}_L.$$

Theorem 39. We have

$$\lim_{t,\tau\to 0} |\hat{u}_{t\tau}-\hat{u}| = 0,$$

where $\hat{u}$ is the solution of the basic problem.

Proof: It is seen that $\tilde{\nu}_L \leq ||L_t\hat{u}-g_\tau|| \leq \nu_L + t|\hat{u}| + \tau$. Therefore

$$||L\hat{u}_{t\tau}-g|| \leq \tilde{\nu}_L + h|\hat{u}_{t\tau}| + \tau \leq \nu_L + t(|\hat{u}_{t\tau}| + |\hat{u}|) + 2\tau.$$

On the other hand, since

$$||A\hat{u}_{t\tau}-f|| = \mu_A,$$

for sufficiently small t and τ the variables $|\hat{u}_{t\tau}|$ are uniformly bounded, that is, $|\hat{u}_{t\tau}| \leq$ constant (not depending on t and τ). Then

$$\lim_{t,\tau\to 0} ||Au_{t\tau}-f|| = \mu_A, \quad \overline{\lim_{t,\tau\to 0}} ||Lu_{t\tau}-g|| \leq \nu_L,$$

that is, the method constructed is regular. By Theorem 31, it is also convergent. □

A corollary of Theorem 39 is that if there is no error in specifying the element f and the operator A, we can construct the approximation method in the "classical" way, that is, by analogy with the formulation of the basic problem.

2. We now investigate the general case where h and δ are non-zero. We assume in addition that the characteristic of the basic problem is known only approximately, that is, (μ_A,ν_L). Thus, we are given $\hat{\mu}_A \geq \mu_A$ instead of μ_A and $\hat{\nu}_L \geq \nu_L$ instead of ν_L.

We have

$$\begin{aligned} ||A_h\hat{u}-f_\varepsilon|| &\leq \hat{\mu}_A + h|\hat{u}| + \delta, \\ ||L_t\hat{u}-g_\tau|| &\leq \hat{\nu}_L + t|\hat{u}| + \tau. \end{aligned} \tag{1}$$

Consider the sets U_σ, where $\sigma = (\delta,h,t,\tau,\hat{\mu}_A-\mu_A,\hat{\nu}_A-\nu_A)$. Notice that $u \in U_\sigma$ for each σ so that the sets U_σ are nonempty. We thereby obtain an approximation method R_σ for solving the basic problem.

Theorem 40. The method R_σ constructed above is regular (therefore it is convergent). For sufficiently small σ,

$$\Delta_B(U_\sigma,\hat{u}) \leq \omega_B(((\mu_A+\varepsilon_1)^2-\mu_A^2)^{1/2},\ 2\nu_L+\varepsilon_2), \tag{2}$$

where

$$\varepsilon_1 = \hat{\mu}_A-\mu_A+2h\tilde{C}+2\varepsilon, \quad \varepsilon_2 = \hat{\nu}_A-\nu_A+2t\tilde{C}+2\delta,$$

$$\tilde{C} = 2(||f|| + ||g|| + \hat{\mu}_A + \hat{\nu}_A + 2\delta + 2\tau).$$

Proof: For any $u \in U_\sigma$ we have

$$||Au-f|| \leq \hat{\mu}_A + 2h|u| + 2\delta, \quad ||Lu-g|| \leq \hat{\nu}_L + 2t|u| + 2\tau.$$

adding which we obtain

$$||Au-f|| + ||Lu-g|| \leq \hat{\mu}_A + \hat{\nu}_L + 2(h+t)|u| + 2\delta + 2\tau$$

and therefore

$$|u| \leq ||f|| + ||g|| + \hat{\mu}_A + \hat{\nu}_L + 2(h+t)|u| + 2\delta + 2\tau.$$

Let h and t be so small that $1-2(h+t) \geq 1/2$. Then

$$|u| \leq \tilde{C} \leq \text{constant}$$

uniformly for all $u \in U_\sigma$ and σ.

From the foregoing inequalities we obtain, for any $u \in U_\sigma$,

$$||Au-f|| \leq \mu_A + \varepsilon_1,$$
$$||Lu-g|| \leq \nu_L + \varepsilon_2, \tag{3}$$

where $\varepsilon_i(\sigma) \to 0$ as $\sigma \to 0$, $i = 1,2$.

Eq. (3) implies that the method is regular and therefore convergent. Using Theorem 34, we derive from (2) the required estimate (2). ◻

We shall call this approximation method R_σ "universal".

3. Next, we generalize the residual method, the basic idea of which was explained in Section 8 for the case with exact data.

We assume that in addition to δ and h we know also $\hat{\mu}_A$; $\hat{\mu}_A \geq \mu_A$, and the accuracy vector $\sigma = (\delta, h, t, \tau, \hat{\mu}_A - \mu_A)$.

Let

$$D_\sigma = \{u \in D\colon \ ||A_h u - f_\delta|| \leq \hat{\mu}_A + h|u| + \delta\}.$$

where, in conjunction with the first relation in (1), the set D_σ is non-empty. It makes sense to pose the following problem: Find the element $u_\sigma \in D_\sigma$ for which

$$||L_t u_\sigma - g_\tau|| = \inf_{u \in D_\sigma} ||L_t u - g_\tau|| = \tilde{\tilde{\nu}}_L. \tag{4}$$

It is not possible to prove that the problem (4) has a solution in the general case $(h \neq 0)$. Hence we do the following.

We define the sets

$$D_{\sigma,\varepsilon} = \{u \in D_\sigma\colon \ ||L_t u - g_\tau|| \leq \tilde{\tilde{\nu}}_L + \varepsilon\},$$

where $\varepsilon > 0$ is arbitrary; $0 < \varepsilon \leq \varepsilon_0 < +\infty$. We obtain thereby the approximation method $R_{\sigma,\varepsilon}$ which we shall call the *residual method*.

We have the following.

<u>Theorem 41</u>. The residual method $R_{\sigma,\varepsilon}$ is regular and therefore convergent. For sufficiently small σ,

$$\Delta_B(U_\sigma, \hat{u}) \leq \omega_B(((\mu_A+\varepsilon_1)^2 - \mu_A^2)^{1/2}, \ 2\nu_L + \varepsilon_2), \tag{5}$$

where

$$\varepsilon_1 = \hat{\mu}_A - \mu_A + 2h\tilde{C} + 2\delta, \quad \varepsilon_2 = t(\tilde{C}+|u|)+2\tau+\varepsilon,$$

$$\tilde{C} = 2[||f|| + ||g|| + \hat{\mu}_A + \nu_L + t|\hat{u}| + 2(\delta+\tau) + \varepsilon].$$

Proof: The variable $\tilde{\tilde{\nu}}_L$ obviously satisfies

$$\tilde{\tilde{\nu}}_L \le ||L_t\hat{u}-g_\tau|| \le \nu_L + t|\hat{u}| + \tau,$$

where $\hat{u}$ is the solution of the basic problem. Hence, for any $u \in D_{\sigma,\varepsilon}$ we have

$$||Au-f|| \le \hat{\mu}_A+2h|u|+2\delta, \quad ||Lu-g|| \le \nu_L+t(|u|+|\hat{u}|)+2\tau+\varepsilon.$$

and therefore

$$|u| \le ||f|| + ||g|| + \hat{\mu}_A + \nu_L + (2h+t)|u| + t|\hat{u}| + 2(\delta+\tau) + \varepsilon.$$

Let h and t be so small that $1-2(h+\tau) \ge 1/2$. Then

$$|u| \le \tilde{C} \le \text{constant}$$

uniformly for all $u \in D_{\sigma,\varepsilon}$ and σ,ε.

Thus, for all $u \in D_{\sigma,\varepsilon}$,

$$||Au-f|| \le \mu_A + \varepsilon_1, \quad ||Lu-g|| \le \nu_L + \varepsilon_2, \tag{6}$$

where, obviously, $\varepsilon_i = \varepsilon_i(\sigma) \to 0$ as $\sigma \to 0$, $\varepsilon \to 0$.

Eq. (6) implies that the residual method is regular. By Theorem 34 Eq. (5) holds as well. □

Remark. We emphasize the fact that the application of the residual method $R_{\sigma,\varepsilon}$ does not require that the variables t and τ, which characterize the approximation error of the operator L and the element g, be specified. Nor does it require that the variable ν_L or its estimate be known, making this method more valuable.

If the characteristic of the basic problem is $(0,\nu_L)$; that is, we consider the combined problem, it would be natural to assume that $\hat{\mu}_A = 0$.

4. We now generalize the method of quasi-solutions, the principal idea behind which was explained in Section 8.

We suppose that in addition to t and τ the quantity $\hat{\nu}_L$, $\nu_L \le \hat{\nu}_L$, and the accuracy vector $\sigma = (\delta,h,t,\tau,\hat{\nu}_L-\nu_L)$ are given.

Let

$$Q_\sigma = \{u \in D\colon ||L_tu-g_\tau|| \le \hat{\nu}_L + t|u| + \tau\}.$$

By the second relation in (1), the set Q_σ is nonempty. We consider the problem: find the elements $u_\sigma \in Q_\sigma$ for which

$$||A_h u_\sigma - f_\delta|| = \inf_{u \in Q_\sigma} ||A_h u - f_\delta|| = \tilde{\mu}_A.$$

As before, we define the sets $Q_{\sigma,\varepsilon} = \{u \in Q_\sigma: ||A_h u - f_\delta|| \leq \tilde{\mu}_A + \varepsilon\}$, where ε: $0 < \varepsilon \leq \varepsilon_0 < +\infty$ (for $t = 0$ let $\varepsilon = 0$). This defines an approximation method $K_{\sigma,\varepsilon}$ which we shall call the *method of quasi-solutions*.

Theorem 42. The method of quasi-solutions $K_{\sigma,\varepsilon}$ is regular (and therefore convergent). For sufficiently small σ the following estimation holds:

$$\Delta_B(U_\sigma, \hat{u}) \leq \omega_B(((\mu_A+\varepsilon_1)^2 - \mu_A^2)^{1/2}, (1+\sqrt{2})\nu_L + \varepsilon_2), \tag{7}$$

where

$$\varepsilon_1 = h\tilde{C} + h|\hat{u}| + 2\delta + \varepsilon, \quad \varepsilon_2 = \hat{\nu}_L - \nu_L + 2t\tilde{C} + 2\tau,$$

$$\tilde{C} = 2[||f|| + ||g|| + \mu_A + \hat{\nu}_L + h|\hat{u}| + 2(\delta+\tau)] + \varepsilon.$$

Proof: Since $\hat{u} \in Q_\sigma$, we can obtain the following upper bound:

$$\tilde{\mu}_A \leq ||A_h\hat{h} - f_\delta|| \leq \mu_A + h|\hat{u}| + \delta.$$

Therefore, for any $u \in Q_{\delta,\varepsilon}$ we have

$$||Au-f|| \leq \mu_A + h|u| + h|\hat{u}| + 2\delta + \varepsilon,$$

$$||Lu-g|| \leq \hat{\nu}_L + 2t|u| + 2\tau.$$

Arguing as before, assuming h and t so small that $1-(h+2t) \geq 1/2$, we obtain

$$|u| \leq \tilde{C} \leq \text{constant}$$

uniformly for all $u \in Q_{\sigma,\varepsilon}$ and σ,ε.

From the foregoing relations it then follows that for all $u \in Q_{\sigma,\varepsilon}$

$$||Au-f|| \leq \mu_A + \varepsilon_1, \quad ||Lu-g|| \leq \nu_L + \varepsilon_2,$$

implying that the method of quasi-solutions is regular. By Theorem 34, the estimate (7) holds as well. □

Remark. The application of the method of quasi-solutions does not require specification of the variables h and δ which characterize the accuracy of specification of the operator A, nor the element f. However, it does require definite information about the variable ν_L. Therefore, the residual method is close to the method of quasi-solutions. However, they differ essentially in the technique of a priori specification of variables necessary for computation.

5. The shortcoming of the methods above consists of the fact that they require that either the problem of finding common points of the two sets or the conditional extremal problem be solved. It is possible to solve both problems by means of the methods developed in the general theory of extremal problems. This shortcoming can be overcome under certain conditions. We shall formulate the method which is formally close to the regularization method but still differs from it essentially in that it does not require that the problem of choosing the regularization parameter be solved.

We shall now consider the combined problem; that is, the characteristic of the basic problem is $(0,\nu_L)$. We assume that the solution $\hat{u}$ of the basic problem is known to be contained in the sphere $S_C = \{u \in D: |u| \leq C\}$.

We define the quadratic functional

$$\phi[u] = ||A_h u - f_\delta||^2 + \frac{(hC+\delta)^2}{(\hat{\nu}_L + tC + \tau)^2} ||L_t u - g_\tau||^2, \quad u \in D.$$

Also, we consider the problem of finding elements $u \in D$ such that

$$\min_{u \in D} \phi[u] = \phi[u_\sigma]. \tag{8}$$

Using Theorem 2 on existence and uniqueness of regularized solutions for $\alpha = (hC = \delta)^2/(\hat{\nu}_L+tC+\tau)^2$, we can assert that the problem (7) always has a unique solution. We write this solution as u_σ, $\sigma = (h,t,\delta,\tau,\hat{\nu}_L-\nu_L)$. In this way we obtain the approximation method R_σ which we shall call the *deterministic Bayes method*.

Theorem 43. The deterministic Bayes method R_σ defined is $\sqrt{2}\,\nu_L$-regular. For sufficiently small σ we have

$$\Delta_B(u_\sigma,\hat{u}) \leq \Delta_B(\varepsilon_1, (\sqrt{2}+1)\nu_L + \varepsilon_2), \tag{9}$$

where

$$\varepsilon_1 = h\tilde{C} + \delta + \sqrt{2}(hC+\delta), \qquad \varepsilon_2 = \sqrt{2}(\hat{\nu}_L - \nu_L) + t\tilde{C} + \tau + \sqrt{2}(tC+\tau),$$

$$\tilde{C} = 2[||f|| + ||g|| + (\sqrt{2}+1)(\delta+\tau) + \sqrt{2}C(h+t) + \sqrt{2}\,\hat{\nu}_L).$$

Proof: Making use of the extremal property of the element u_σ, we obtain

$$||A_h u_\sigma - f_\delta||^2 + ||L_t u_\sigma - g_\tau||^2 \left(\frac{hC+\delta}{\nu_L + tC + \tau}\right)^2 \leq ||A_h \hat{u} - f_\delta||^2$$

$$+ \left(\frac{hC+\delta}{\hat{\nu}_L + tC + \tau}\right)^2 ||L_t \hat{u} - g_\tau||^2 \leq (hC+\delta)^2 + \left(\frac{hC+\delta}{\hat{\nu}_L + tC+\tau}\right)^2 (\hat{\nu}_L + tC + \tau)^2$$

$$= 2(hC + \delta)^2$$

yielding

$$||Au_\sigma - f|| \leq h|u_\sigma| + \delta + \sqrt{2}(hC+\delta), \quad ||Lu_\sigma - g|| \leq t|u_\sigma| + \tau + \sqrt{2}(\hat{\nu}_L + tC + \tau).$$

Provided $1-(h+t) \geq 1/2$, we have

$$|u|_\sigma \leq \tilde{C} \leq \text{constant}.$$

Hence, for sufficiently small h and t

$$||Au_\sigma - f|| \leq \varepsilon_1, \quad ||Lu_\sigma - g|| \leq \sqrt{2}\, \nu_L + \varepsilon_2.$$

yielding the estimate (9) and $\sqrt{2}\, \nu_L$-regularity. □

As we have noted, the advantage of this method is that obtaining approximate solutions u_σ is relatively simple. However, to apply this method, it is necessary to have sufficiently complete a priori information, in particular, whether the solution $\hat{u}$ of the basic problem belongs to the sphere S_C. If $h = t = 0$, that is, the operators A and L are given exactly, information of the kind mentioned is not necessary. Thus the application of this method becomes more feasible. For $h = t = 0$ and $g = g_\tau = 0$, the method becomes especially simple because the functional ϕ is given by

$$\phi[u] = ||Au - f_\delta||^2 + \frac{\delta^2}{\hat{\nu}_L^2} ||Lu||^2, \quad u \in D.$$

If the set $D = D_{AL}$ and, therefore, is linear, Eq. (9) can be effectively solved by solving the corresponding Euler equation:

$$A_h^*(A_h u - f_\delta) + \left(\frac{hC+\delta}{\hat{\nu}_A + tC + \tau}\right)^2 L_t^*(L_t u - g_\tau) = 0. \tag{10}$$

6. Now we consider the problem of *statistic regularization* of degenerate and ill-conditioned systems of linear algebraic equations, following the *a posteriori Bayes procedure*. We compare this procedure with the approach we have taken earlier.

Let the state of a system be characterized by a random vector u of dimension n and let, in addition, the measured data be characterized by a random vector f of dimension m, $m \geq n$. We assume that $p(u)$ is the a priori probability density of the state vector. Let $p(f|u)$ denote the conditional probability density of the vector f for given state u. According to the Bayes rule, the a posteriori probability density $p(u|f)$ satisfies

$$p(u|f) = \frac{p(f|u)p(u)}{p(f)}, \tag{11}$$

where $p(f)$ is the a priori probability density of the vector f.

Let f_i denote the ith component of f. We adopt the following basic hypothesis: the a priori probability density of the state vector u at the ith measurement is the a posteriori probability density of the vector u after the (i-1)th measurement. Using (11), we have

$$p_i(u|f_i) = \frac{p_i(f_i|u)p_{i-1}(u|f_{i-1})}{p(f_i)}, \qquad i = 1,2,\ldots,k,$$

where $p_i(u|f_i)$ is the a posteriori probability density of the vector u after i measurements, and $p_0(u|f_0) \equiv p(u)$.

It is easy to see that

$$p_k(u|f_k) = \left\{p(u) \prod_{i=1}^{k} p_i(f_i|u)\right\}\left\{\prod_{i=1}^{k} p(f_i)\right\}^{-1}.$$

It is natural to choose the state vector $\hat{u}_k$ corresponding to the k measurements by the following "maximum likelihood" principle:

$$p_k(\hat{u}_k|f_k) = \sup_u p_k(u|f_k). \tag{12}$$

If

$$p(u) = (\text{constant})\exp\{-(u-u_0)^T C(u-u_0)\},$$

$$p_i(f_i|u) \equiv p(f_i|u) = (\text{constant})\exp\{-(Au-f_i)^T \overline{P}(Au-f_i)\},$$

where C and $\overline{P}$ are positive definite matrices of order $n \times n$ and $m \times m$, respectively, u_0 is a vector of dimension n, and A is an arbitrary matrix of dimension $m \times n$, then Eq. (12) yields

$$\left(\frac{1}{k} C + A^T\overline{P}A\right)\hat{u}_k = A^T\overline{P}\hat{f}_k + \frac{1}{k} Cu_0,$$

$$\hat{f}_k = \frac{1}{k} \sum_{i=1}^{k} f_i. \tag{13}$$

In Eq. (10) let $h = t = \tau = \hat{\nu}_A - \nu_A = 0$. Eq. (10) becomes

$$A^*(Au-f_\varepsilon) + \frac{\delta^2}{\nu_A^2} L^*(Lu-y) = 0. \tag{14}$$

We shall make some assumptions. Let $H = G = R_n$, where $R_n = \{u: u = (u_1, u_2, \ldots, u_n)\}$ with Euclidean norm; i.e.,

$$||u||_{R_n} = (u_1^2 + u_2^2 + \ldots + u_n^2)^{1/2}.$$

We assume that A is a rectangular $m \times n$ matrix, that is to say, the corresponding operator acts on R_n into R_m. For F we take the space R_m equipped with the norm $||f||_p$ defined by

$$||f||_p = (Pf, f)_{R_m},$$

where P is a positive definite matrix of order $m \times m$. As is easily seen, the operator A^* is given by

$$A^* + A^T P.$$

Define $g = Lu_0$. Denote the matrix $\nu_L^{-2} L^* L$ by C and assume that it is positive definite.

Eq. (14) assumes the form

$$(\delta^2 Cu + A^T PA)u = A^T Pf_\delta + \delta^2 Cu_0. \tag{15}$$

If $\overline{P} = \frac{1}{\sigma^2} P$, where $\sigma > 0$, characterizes the accuracy of f_i measurements, Eq. (13) assumes the form

$$(\frac{\sigma^2}{k} C + A^T PA)\hat{u}_k = \frac{\sigma^2}{k} Cu_0 + A^T P\hat{f}_k. \tag{16}$$

It is not hard to see that the quantity σ^2/K characterizes the accuracy of the vector $\hat{f}_k$ obtained on the basis of the full experiment consisting of k subsequent measurements of the vector.

We have thereby established a complete analogy between the deterministic Bayes method described above and the *a posteriori Bayes regularization* reduced to the solution of Eqs. (13).

In Eq. (16), let $\alpha = \sigma^2/K$. Then Eq. (16) assumes the form

$$(\alpha C + A^T PA)\hat{u}_k = A^T P\hat{f}_k + \alpha Cu_0$$

and therefore

$$\hat{u}_k = \alpha(\alpha C + A^T PA)^{-1} Cu_0 + (\alpha C + A^T PA)^{-1} A^T P\hat{f}_k.$$

If the accuracy of the experiment increases or the number of measurements of the vector f increases indefinitely (that is, $k \to \infty$), we see that $\alpha \to 0$. It is natural to calculate the limits of the matrices

$$\alpha(\alpha C + A^T PA)^{-1} C \quad \text{and} \quad (\alpha C + A^T PA)^{-1} A^T P$$

as $\alpha \to 0$.

To simplify the solution of this problem we assume that $P = E_m$, that is, P is the identity. Also, let

$$R_\alpha = (\alpha C + A^T A)^{-1} A^T, \qquad Z_\alpha = \alpha(\alpha C + A^T A)^{-1} C.$$

We write the matrix C as the product $C = K^T K$, where K is a square nonsingular matrix of order $n \times n$. Then

$$R_\alpha = K^{-1}(\alpha E + N^T N)^{-1} N, \quad Z_\alpha = \alpha K^{-1}(\alpha E + N^T N)^{-1} K,$$

where the matrix $N = AK^{-1}$ is of the same type as A.

Since the matrix $N^T N$ is nonnegative definite and symmetric there exists an orthogonal matrix Q such that $N^T N = Q^T \Lambda Q$, where $\Lambda = \operatorname{diag}\{s_1^2, s_2^2, \ldots, s_m^2\}$ is composed of eigenvalues of the matrix $N^T N$ in decreasing order: $s_1^2 \geq s_2^2 \geq \ldots \geq s_m^2$.

It is easy to see that

$$Z_\alpha = \alpha K^{-1} R^T (\alpha E + \Lambda)^{-1} RK$$

and therefore

$$\lim_{\alpha \to 0} Z_\alpha = K^{-1} R^T \hat{E}_{n-r} RK,$$

where the matrix $\hat{E}_{n-r} = \operatorname{diag}\{\underbrace{0, 0, \ldots, 0}_{r}, \underbrace{1, 1, \ldots, 1}_{n-r}\}$, and r denotes the rank of the matrix $N^T N$ (or N).

We note that for $r = n$ (that is, if the matrix N has full rank), the matrix $\hat{E}_0$ is zero and, therefore,

$$\lim_{\alpha \to 0} Z_\alpha = 0 \quad (r = n).$$

In order to determine $\lim_{\alpha \to 0} R_\alpha$, we note that

$$\lim_{\alpha \to 0} (\alpha E + N^T N)^{-1} N^T = N^+,$$

where N^+ is the Moore-Penrose pseudoinverse matrix of N (in Section 24 this fact will follow from more general considerations). Hence

$$\lim_{\alpha\to 0} R_\alpha = K^{-1}N^+.$$

We have thus found both limits. Noting that $\hat{f}_k \to f$ (in probability) as $\alpha \to 0$ and using the limiting relations obtained above for R_α and Z_α, we have

$$\lim_{\alpha\to 0} \hat{u}_k = K^{-1}N^{-1}f + K^{-1}R^T E_{n-r} RKu_0$$

in probability.

If $r = n$, then

$$\lim_{\alpha\to 0} \hat{u}_k = K^{-1}N^+ f,$$

independent, obviously, of the vector u_0 characterizing the a priori state of the system under investigation. In this case

$$\hat{u}_k = K^{-1}N^+\hat{f}_k.$$

Section 15. The Principle of Residual Optimality for Approximate Solutions of Equations with Nonlinear Operators

1. Let U and F be two complete metric spaces. We consider the problem of obtaining an approximate solution of the equation

$$Au = f, \quad f \in F, \tag{1}$$

where $A: U \to F$, when $\{A, f, \mu_A = \inf_{u\in D_A} \rho_F(Au,f)\}$ are known only approximately. A solution of Eq. (1) is understood in the sense of "the method of least squares", that is to say, is an element $\bar{u} \in D_A$ for which

$$\inf_{u\in D_A} \rho_F(Au,f) = \rho_F(A\bar{u},f).$$

We call $\mu_A \geq 0$, as before, the measure of inconsistency of (1). In the case where (1) has a solution in the usual sense, $\mu_A = 0$. If A is not invertible, let

$$U_A = \{\bar{u} \in D_A: \rho_F(A\bar{u},f) = \mu_A\} \neq \emptyset.$$

Next, we construct a stable method of solving Eq. (1) satisfying minimal a priori requirements, namely:

(a) the algorithm for solution of (1) with approximate data does not use quantitive information about the solution;

(b) the stabilizing functional $||Lu-g||$ is to be chosen starting from the natural condition of approximation of the initial operator A (which is easy to write in all familiar cases) rather than a priori, as is done in the formulation of the basic problem.

We assume that instead of exact $\{A,f,\mu_A\}$ we have approximate $\{A_K,f_\delta,\hat{\mu}_A\}$ such that $f_\sigma \in F$ and $\rho_F(f,f_\sigma) \leq \delta$, $0 \leq \hat{\mu}_A-\mu_A \leq \Delta\mu$. Furthermore, the operators $A_h\colon U \to F$, $h > 0$, are defined on $D_{A_h} \supseteq D$, with the nonempty set $D \subseteq D_A$ and the set $U_D = D \cap U_A \neq \emptyset$.

Let $\zeta(h;u) \equiv \rho_F(A_hu,Au)$, $u \in D$. Also, let there be known the estimate of the quantity

$$\zeta(h;u) \leq \overline{\zeta}(h)\Omega(u), \quad u \in D, \tag{2}$$

where the function $\overline{\zeta}(h)$, $\lim_{h\to 0} \overline{\zeta}(h) = 0$ characterizes the order of the error of approximation of the operator A and Ω is an *estimating* functional. Obtaining estimates of type (2) is a classical problem in theory of approximation methods and therefore can be obtained by the usual analysis. We call the condition (2) an *approximation condition* and assume it to be satisfied.

Example. On the set $D \subseteq D_A$ let there be given a family of "projection" operators P_h, $h > 0$; that is, operators defined on D for any $h > 0$, with range in D satisfying the condition: $\lim_{n\to 0} \rho_u(u,P_hu) = 0$, $u \in D$. We assume that the estimate of the convergence rate is also known:

$$\rho_u(u,P_hu) \leq \overline{\zeta}_0(h)\Omega(u), \quad u \in D.$$

where $\overline{\xi}_o(h)$ is a function of the type $\overline{\xi}(h)$ and $\Omega(u)$ is the estimating functional. We suppose that the operator A satisfies a *Lipschitz condition* on D with constant K; that is, for any $u,v \in D$

$$\rho_F(Au,Av) \leq K\rho_u(u,v).$$

We define an *approximating family of operators* by letting $A_h = AP_h$ on D. We then have

$$\rho_F(A_hu,Au) = K\rho_u(P_hu,u) \leq K\overline{\xi}_0(h)\Omega(u), \quad u \in D;$$

that is, the approximation condition (2) is satisfied for $\overline{\zeta}(h) \equiv K\overline{\xi}_0(h)$. We note that estimates similar to that for the quantity $\rho(u,P_hu)$, $u \in D$ are widely known in the theory of approximations for various classes of spaces and sets.

We define the set of *formal* solutions of the problem (1) as the set $U_\sigma = \{u \in D\colon \hat{\rho}_F(A_hu,f_\delta) \leq \xi(h)\Omega(u) + \hat{\mu}_A - \delta\}$ where the vector $\sigma = (h,\delta,\Delta\mu)$ characterizes the accuracy of approximate A_h, f_δ, $\hat{\mu}_A$ for the problem. Using the triangle inequality and the approximation condition (2), for any element $\bar{u} \in U_D$ we obtain

$$\rho_F(A_h\bar{u}, f_\delta) \leq \xi(h)\Omega(u) + \hat{\mu}_A + \delta.$$

Therefore $U_\sigma \supseteq U_D$ and thereby is nonempty.

In order to construct stable approximations, we suggest the following *optimality principle:* Choose such $u_\sigma \in U_\sigma$ for which

$$m_\sigma = \inf_{u \in U_\sigma} \Omega(u) = \Omega(u_\sigma).$$

It is easily seen that this principle leads to choosing as an approximate solution of the problem (1) such elements of the set of formal solutions U_σ which have a minimal admissible residual. Hence, it is reasonable to call this principle the *principle of residual optimality.*

2. For approximate realization of the principle of residual optimality we do not need to seek elements u_σ to solve the problem exactly. Thus we assume that the sequence $\varepsilon_n \to 0$ is defined as $n \to \infty$ and the elements $u_\zeta \in U_\sigma$ are defined somehow such that

$$\Omega(u_\zeta) \leq m_\sigma + \varepsilon_n, \quad n = 1,2,... \tag{3}$$

where the vector $\zeta = (h, \delta, \Delta\mu, \varepsilon_n)$ characterizes the accuracy of solution of the optimization problem (3).

The following theorem holds.

Theorem 44. Let the operator A be continuous on the set D closed in U. Also, for any $C \geq \mu_A$ let the set $S_C = \{u \in D\colon \rho_F(Au, f) \leq C, \Omega(u) \leq C\}$ be compact in the space U. Then u_ζ converges:

$$\lim_{\zeta \to 0} \rho_u(u_\zeta, U_D) = 0. \tag{4}$$

Proof: It is seen that $m_\sigma \leq \Omega(\bar{u})$ for any $\bar{u} \in U_D$. Therefore

$$\Omega(u_\zeta) \leq \Omega(\bar{u}) + \varepsilon_n, \quad n = 1,2,...,$$

$$\begin{aligned} \rho_F(Au_\zeta, f) &\leq \rho_F(A_h u_\zeta, f_\delta) + \rho_F(A_h u_\zeta, Au_\zeta) \\ &\quad + \rho_F(f, f_\delta) \leq \xi(h)\Omega(u_\zeta) + \hat{\mu}_A + \delta + \xi(h)\Omega(u_\zeta) + \delta \\ &\leq 2\xi(h)(\Omega(\bar{u}) + \varepsilon_n) + 2\delta + \hat{\mu}_A \end{aligned} \tag{5}$$

and

$$\overline{\lim_{\to 0}}\, \Omega(u_\zeta) \leq \Omega(\bar{u}), \quad \forall \bar{u} \in U_D, \tag{6}$$

$$\lim \rho_F(Au_\zeta, f) = \mu_A. \tag{7}$$

Eqs. (6) and (7) and the conditions of the theorem imply that the family $\{u_\zeta\}$ is compact in U. Let $u_m = u_{\zeta_m}$, $m = 1,2,\ldots$ be any convergent sequence as $m \to \infty$ and $\lim_{m\to\infty} u_m = u_0$. Due to the choice of $u_m \in U_\sigma$ and the closure of D we have $u_0 \in D$. Assuming $\zeta = \zeta_m$ in Eq. (7) and taking into account the continuity of the operator A on D, we have $\rho_F(Au_0, f) \leq \mu_A$. This yields $u_0 \in U_A$ and therefore $u_0 \in D \cap U_A = U_D$, which fact as well as the compactness of the family u_ζ prove the theorem. □

<u>Remark 1</u>. If the problem (1) has a unique solution (that is, the set $U_D = \{\bar{u}\}$ consists of a single element), instead of (4) we have convergence in the usual sense:

$$\lim_{\to 0} \delta_u(u_\zeta, \bar{u}) = 0.$$

<u>Remark 2</u>. If $\mu_A = 0$, closure of D is not required. It is enough to assume that A is closed on D, as an examination of the proof of the theorem shows.

3. Let $h = \delta = \Delta\mu = 0$. The set U_Ω of solutions of the problem

$$m_\Omega = \inf_{u \in U_D} \Omega(u) = \Omega(\hat{u}), \quad \forall \hat{u} \in U_D \tag{8}$$

is said to be a set of *Ω-minimal solutions* of the problem (1). The following theorem is interesting.

<u>Theorem 45</u>. Let the conditions of Theorem 44 be satisfied. Furthermore, let the functional Ω be lower semi-continuous on D. The set of Ω-minimal solutions is nonempty.

<u>Proof</u>: Let $\varepsilon_n > 0$, $\varepsilon_n \to 0$, $n = 1,2,\ldots$ and let the elements $u_n \in U_D$ satisfy the relations

$$\Omega(u_n) \leq m_\Omega + \varepsilon_n, \quad n = 1,2,\ldots\,. \tag{9}$$

The sequence $\{u_n\}$ is obviously compact. We may assume without loss of generality that $\{u_n\}$ is convergent. Let $\hat{u} = \lim_{n\to\infty} u_n$. Since $u_n \in D$ and D is closed, $\hat{u} \in D$. Furthermore, due to the continuity of A on D and the choice of the sequence u_n we have: $\rho_F(A\hat{u}, f) \leq \mu_A$. Thus, we have constructed the element $\hat{u} \in U_D$. Taking advantage of the lower semi-continuity of the functional Ω on D and using (9), we obtain:

$$\Omega(\hat{u}) \le \underline{\lim}_{n\to\infty} \Omega(u_n) \le \overline{\lim}_{n\to\infty} \Omega(u_n) \le m_\Omega \tag{10}$$

implying that $\hat{u} \in U_\Omega$, namely, the set of Ω-minimal solutions of the problem (1). □

Remark 3. If Eq. (1) has a unique solution $\hat{u} \in U_D$, this solution will obviously be the Ω-minimal solution of the problem (1).

Remark 4. We can define the Ω-minimal solution of the problem (1) as unique also in the case where the spaces U and F are linear, the operator A is linear, the set D is closed and convex and the functional $\Omega(u)$ is strictly convex, at least on the set U_D (naturally, we assume that all the conditions of Theorem 45 are satisfied).

Under the conditions of Theorem 45, the assertion of Theorem 44 becomes more precise. In fact, we have

Theorem 46. Let the conditions of Theorem 45 be satisfied. Let the elements u_ζ be defined, as before, by Eq. (3). Then

$$\lim_{\zeta\to 0} \rho_U(u_\zeta, U_\Omega) = 0.$$

Proof: The relations (6) and (7) hold for any $u \in U_D$, including any $\hat{u} \in U_\Omega$:

$$\overline{\lim}_{\zeta\to 0} \rho_F(Au_\zeta, f) = \mu_A.$$

Due to the lower semi-continuity of the functional Ω on D, we have for the subsequence $\{u_m\} = \{u_{\zeta_m}\}$ chosen in Theorem 44 and for $u_0 = \lim_{m\to\infty} u_m$:

$$\Omega(u_0) \le \underline{\lim}_{n\to\infty} \Omega(u_m). \tag{11}$$

Eqs. (6) and (11) yield

$$\Omega(u_0) \le \Omega(\hat{u}) = m_\Omega,$$

which together with the fact proved in Theorem 44 that $u_0 \in U_D$ leads us to the conclusion that $u_0 \in U_\Omega$, the set of Ω-minimal solutions. This as well as the compactness of the family $\{u_\zeta\}$ proves our assertion. □

Remark 5. In proving Theorem 46 we actually showed that under the conditions of this theorem a stronger limiting relation holds:

$$\lim_{\zeta\to 0} \max\{\rho_u(u_\zeta, U_\Omega),\ |\Omega(u_\zeta) - m_\Omega|,\ |\rho_F(Au_\zeta, f) - \mu_A|\} = 0. \tag{12}$$

4. Next, we assume that the spaces U and F are reflexive Banach spaces.

Theorem 47. Let the set D be weakly closed, let the operator A be weakly continuous (that is, for each sequence $\{u_n\} \subset D$ with $u_n \xrightarrow{\text{weakly}} u_0$ (in U), the sequence $Au_n \xrightarrow{\text{weakly}} Au_0$ (in F)), let the functional Ω be weakly lower semi-continuous on D and let each set $W_C = \{u \in D: ||Au|| \leq C, ||Lu|| \leq C\}$ be bounded in U (C = const). Then the set of Ω-minimal solutions U_Ω is nonempty.

Proof: Let $\varepsilon_n \to 0$ and let $u_n \in U_D$ be such that

$$\Omega(u_n) \leq m_\Omega + \varepsilon_n, \quad n = 1,2,\ldots . \tag{13}$$

It is easy to see that $\{u_n\} \subseteq W_C$ for

$$C = \max\{m_\Omega + \sup_n \varepsilon_n, \ \mu_A = ||f||_F\}.$$

Since the space U is reflexive, W_C is weakly compact. We assume without loss of generality that the sequence $u_n \xrightarrow{\text{weakly}} u_0 \in D$ (in U); we have used here the weak closure of the set D. Due to the weak continuity of the operator A we have $Au_n \xrightarrow{\text{weakly}} Au_0$. Using the lower semi-continuity of the norm, we derive that

$$||Au_0 - f||_F \leq \underline{\lim}_{n\to\infty} ||Au_n - f||_F \leq \mu_A,$$

yielding $u_0 \in U_D$. Using the weak semi-continuity of the functional Ω on D as well as the relation (13), we obtain:

$$\Omega(u_0) \leq \underline{\lim}_{n\to\infty} \Omega(u_n) \leq m_\Omega,$$

implying $u_0 \in U_\Omega$. □

The conditions of Theorem 47 are satisfied if, for example, the set D is closed and convex, the operator A is linear and continuous, and the functional Ω is lower semi-continuous and convex on D.

We proceed to proving the following assertion.

Theorem 48. Let the conditions of Theorem 47 be satisfied. For the elements $u_\zeta \in U_\sigma = \{u \in D: ||A_h u - f_\delta||_F \leq \xi(h)\Omega(u) + \mu_A + \delta\}$ satisfying the condition $\Omega(u_\zeta) \leq m_\sigma + \varepsilon_n$, $\varepsilon_n \to 0+$, $n \to \infty$, we have

$$\liminf_{\zeta\to 0} \inf_{u\in U_\Omega} \{|u^*(u-u_\zeta)| + |f^*(Au-Au_\zeta)| + |m_\Omega - \Omega(u_\zeta)| + |\,||Au_\zeta - f||_F - \mu_A|\} = 0, \tag{14}$$

where u^* and f^* are arbitrary linear functionals in U and F, respectively.

Proof: The relations (6) and (7) hold in this case.

Assuming that $\bar{u} = \hat{u} \in U_\Omega$, we obtain

$$\overline{\lim_{\zeta\to 0}}\, \Omega(u_\zeta) \le m_\Omega, \tag{15}$$

$$\overline{\lim_{\zeta\to 0}}\, ||Au_\zeta - f||_F \le \mu_A. \tag{16}$$

which imply that the family $\{u_\zeta\}$ is weakly compact in U.

Let $u_m = u_{\zeta_m}$, $m = 1,2,\ldots$ define a weakly convergent subsequence of this family: $u_m \xrightarrow{\text{weakly}} u_0$ (in U). Due to the weak closure of the set D, $u_0 \in D$. Using the weak continuity of the operator A and the lower semi-continuity of the norm in a Banach space, we have $Au_n \xrightarrow{\text{weakly}} Au_0$ (in F) and, therefore,

$$||Au_0 - f||_F \le \underline{\lim_{m\to\infty}}\, ||Au_m - f||_F \le \overline{\lim_{n\to\infty}}\, ||Au_m - f||_F \le \mu_A, \tag{17}$$

yielding $u_0 \in U_D$; that is, it is the solution of the problem (1). Moreover, it is seen that

$$\lim_{m\to\infty} ||Au_m - f||_F = \mu_A. \tag{18}$$

Further, due to the weak lower semi-continuity of the functional Ω on D, we have:

$$\Omega(u_0) \le \underline{\lim_{m\to\infty}}\, \Omega(u_m) \le \overline{\lim_{m\to\infty}}\, \Omega(u_m) \le m_\Omega;$$

that is, $u_0 \in U_\Omega$. Hence

$$\lim_{m\to\infty} \Omega(u_m) = m_\Omega. \tag{19}$$

Therefore, any weakly convergent sequence $\{u_m\}$ of the family $\{u_\zeta\}$ converges in U to one of the solutions $u_0 \in U_\Omega$ and, in addition, $Au_m \xrightarrow{\text{weakly}} Au_0$ (in F). Finally, the relations (18) and (19) hold. The assertion of the theorem follows from the weak compactness of the family $\{u_\zeta\}$ and its properties proved above. □

Remark 6. If $\mu_A = 0$, Theorem 48 holds without the basic assumption that the space F is reflexive. In addition, we may take F to be simply a normed space.

<u>Remark 7</u>. Let the space F satisfy the *Efimov-Stechkin conditions*; that is, weak convergence $f_n \xrightarrow{\text{weakly}} f_0$ (in F) and convergence of the norms: $||f_n||_F \longrightarrow ||f_0||_F$ imply strong convergence: $||f_n - f_0||_F \longrightarrow 0$. In this case the assertion of Theorem 48 admits some strengthening, namely:

$$\lim_{\zeta\to 0} \{|m_\Omega - \Omega(u_\zeta)| + \inf_{u\in U_\Omega} (|u^*(u-u_\zeta)| + ||Au - Au_\zeta||_F\} = 0$$

<u>Remark 8</u>. Let the conditions of Theorem 48 be satisfied. Also, let $\Omega(u) = ||Lu-g||_G^2$, where L is an operator acting from U into a reflexive Banach space G and having domain $D_L \supseteq D$. If the operator L is weakly continuous on D, we can easily show that the functional Ω is weakly lower semi-continuous on D. Following the arguments used in proving Theorem 48, we prove that

$$\lim_{\zeta\to 0} \{\inf_{u\in U_\Omega} (|u^*(u-u_\zeta)| = |f^*(Au-Au_\zeta)| + |g^*(Lu-Lu_\zeta)|$$
$$+ |\,||Au_\zeta - f|| - \mu_A| + |\,||Lu_\zeta - g||_G - m_\Omega|\} = 0.$$

where g^* is any linear functional on G.

If the spaces F and G are Efimov-Stechkin spaces, we can make the foregoing relation more precise. In fact, we have

$$\lim_{\zeta\to 0} \{\inf_{u\in U_\Omega} (|u^*(u-u_\zeta)| + ||Au-Au_\zeta||_F + ||Lu-Lu_\zeta||_G\} = 0. \tag{20}$$

We note that the Efimov-Stechkin conditions are satisfied, for example, for any Hilbert space as well as for Banach spaces of the type L_p, $p > 1$.

5. Our results so far do not indicate whether the approximations are strongly convergent in the main space U. The next theorem gives sufficient conditions for this convergence.

<u>Theorem 49</u>. Let the set D be weakly closed, let the operators A and L be weakly continuous on D and let $\Omega(u) \equiv ||Lu-g||_G^2$, $u \in D$. If the spaces F and G satisfy the Efimov-Stechkin conditions and if, in addition, for any $u,v \in D$ the condition

$$\gamma(||u-v||_U) \le ||Au-Av||_F + ||Lu-Lv||_G \tag{21}$$

(the completeness condition) is satisfied, where γ, $\gamma(0) = 0$, is a continuous strictly increasing function on the positive axis, the relation

$$\lim_{\zeta\to 0} \inf_{u\in U_\Omega} (||u-u_\zeta||_U + ||Au-Au_\zeta||_F + ||Lu-Lu_\zeta||_G) = 0. \tag{22}$$

holds for the approximations u_ζ defined in Theorem 48.

Proof: Using (21), it is easy to show that $u_\zeta \in W_C$ for some C. Therefore, Eq. (20) holds. Taking $u = \hat{u}$, $v = u_\zeta$ in Eq. (21) and also using the properties of the function γ, we see that Eq. (22) is satisfied. □

Remark 9. If the operators A and L are linear and the spaces F and G are strictly convex, we may replace (21) by the simpler condition:

$$\gamma(||u||_U) \leq ||Au||_F + ||Lu||_G, \quad u \in D.$$

Furthermore, if the set D is convex, we can easily show that the Ω-minimal (here $\Omega(u) = ||Lu-g||_G^2$, $u \in D$) solution $\hat{u}$ is defined uniquely. In this case it is possible to replace weak continuity of the operators A and L by continuity only (using the well-known Banach-Saks theorem).

These results are also valid when the operator A (and the operator L) is unbounded. In this case it is possible to replace weak continuity of these operators by weak closure. The operator A is said to be weakly closed if

$$u_n \in D_A, \quad u_n \xrightarrow{\text{weakly}} u_0, \quad Au_n \xrightarrow{\text{weakly}} f_0, \quad n \to \infty,$$

imply $u_0 \in D_A$, $Au_0 = f_0$. We should replace the conditions of weak lower semi-continuity of the functional Ω and weak closure of the set D by the condition of weak lower semi-closure of this functional. By weak lower semi-closed functionals we mean the functionals Ω for which the facts that

$$u_n \in D, \quad u_n \xrightarrow{\text{weakly}} u_0, \quad \Omega(u_n) \leq \Omega_0 < +\infty,$$

imply that $u_0 \in D$, $\Omega(u_0) \leq \underline{\lim}_{n\to\infty} \Omega(u_n)$. We note that the functional $\Omega(u) = ||Lu-g||_G^2$, $u \in D$, is weakly lower semi-closed on D if the operator L is weakly closed on D.

6. In the Banach space U let the projections P_m, $m = 1,2,\ldots$ be defined. Also, for elements of a subset $D \subset U$ let the following relation hold:

$$||u-P_m u||_U \leq \tau_m ||Lu||_G, \quad m = 1,2,\ldots,$$

where L is a closed linear operator defined on D and, in addition $r_m \to 0$ as $m \to \infty$. We assume that the operator $A\colon U \to F$ is linear and bounded, $||A|| \leq K$. Letting $A_h := AP_m$, $h = 1/m$, $m = 1,2,\ldots$, we easily see that for any $u \in D$,

$$||A_h u-Au||_F \le K||u-P_m u||_U \le Kr_m||Lu||_G;$$

that is, the approximation condition (2) is satisfied for $\xi(h) = Kr_m$, $\Omega(u) = ||Lu||_G$, $u \in D$.

We now consider the case where $U = L_2[a,b]$ is the space of square-integrable functions on the interval $[a,b]$, and the functions in D are sufficiently "smooth" as defined below. On the interval $[a,b]$ we take a mesh of nodes:

$$a = x_1 < x_2 < \dots < x_n = b$$

and put

$$P_m u = \frac{u(x_{i+1})-u(x_i)}{x_{i+1}-x_i}(x-x_i) + u(x_i), \quad x_i \le x \le x_{i+1}, \quad i = 1,2,\dots,m-1.$$

It is easy to show that for a sufficiently smooth function $u = u(x)$ we have

$$||u-P_m u||_{L_2} \le \frac{h^2}{2}\,\left|\left|\frac{d^2u}{dx^2}\right|\right|_{L_2}, \qquad h = (x_{i+1}-x_i).$$

Obviously, it suffices to assume that u has a generalized L_2 second derivative in the Sobolev sense [80].

Let the operator A be an integral operator:

$$(Au)(x) = \int_a^b k(x,\xi)u(\xi)d\xi, \quad a \le x \le b, \tag{22}$$

where the function k is such that $\int_a^b \int_a^b k^2(x,\xi)dxd\xi \le K^2$.

For all $u \in D$, that is, the set of twice differentiable (in the sense indicated) functions, we have

$$||Au-AP_m u||_{L_2} \le (Kh^2/2)||Lu||_{L_2}, \qquad u \in D,$$

where the operator $L = d^2/dx^2$. Therefore, $\Omega(u) \equiv ||Lu||_{L_2}$, $u \in D$, and $\xi(h) = Kh^2/2$. If the integral equation $Au = f$ with the operator A in Eq. (22) has a unique solution $\bar{u} \in D$, the methods suggested above are applicable and, furthermore, the approximate solutions converge, as can easily be seen, in the metric of the space $W_2^{(2)}[a,b]$.

In fact, the conditions guarantee convergence of approximations in the norm

$$|u| = (||Au||^2_{L_2} + ||Lu||^2_{L_2})^{1/2},$$

which is equivalent to the norm of the space $W_2^{(2)}[a,b]$ given by

$$||u||_{W_2^{(2)}} = (||u||_{L_2}^2 + ||Lu||_{L_2}^2)^{1/2}.$$

Section 16. The Regularization Method for Nonlinear Equations

1. In the basic problem we shall allow the operators A and L to be nonlinear. As before, we assume that the set

$$U_A = \{u \in D\colon\ ||Au-f||_F = \mu_A = \inf_{u \in D} ||Au-f||_F\} \neq \emptyset.$$

We define the set

$$\hat{U} = \{\hat{u} \in U_A\colon\ ||L\hat{u}-g||_G = \nu_L = \inf_{u \in U_A} ||Lu-g||_G\},$$

which we shall regard to be the set of solutions of the basic problem in the nonlinear case.

Next, we assume that the operators A and L are jointly weakly closed on D; that is to say, for any sequence $u_n \in D$, $n = 1,2,\ldots$, such that

$$u_n \xrightarrow{\text{weakly}} u_0 \quad (\text{in } H), \qquad Au_n \xrightarrow{\text{weakly}} f_0 \quad (\text{in } F),$$
$$Lu_n \xrightarrow{\text{weakly}} g_0 \quad (\text{in } G),$$

it follows that $u_0 \in D$, $Au_0 = f_0$, $Lu_0 = g_0$.

We note that in the linear case joint weak closure of the operators A and L was a direct consequence of their joint closure. If $L = E$, we should, obviously, require the operator A to be weakly closed. This requirement is satisfied a priori if the operator A transforms each bounded sequence in H into a bounded sequence in F.

Let the condition $W_C = \{u \in D\colon\ ||Au|| \leq C,\ ||Lu|| \leq C\} \neq \emptyset$ for some $C > 0$ be an analog of the completeness condition. We assume that the set W_C is bounded (therefore, weakly compact) in H.

Now we define the parametric functional

$$\Phi_\alpha[u] = \Phi_\alpha[u;A,f;L,g] = ||Au-f||_F^{\sigma_1} + ||Lu-g||_G^{\sigma_2}, \qquad u \in D,$$

where $\sigma_1,\sigma_2 \geq 1$ and $\alpha > 0$ is the regularization parameter. Also, we pose the problem of finding the element $u_\alpha \in D$ minimizing the functional Φ_α:

$$m_\alpha = \inf_{u \in D} \Phi_\alpha[u] = \Phi_\alpha[u_\alpha]. \tag{1}$$

Theorem 50. Under the assumptions above, the problem (1) has at least one solution for any jointly weakly closed operators A and L and for elements $f \in F$, $g \in G$.

Proof: Let a minimizing sequence $\{u_n\} \subset D$ be such that

$$m_\alpha \le \Phi_\alpha[u_n] \le m_\alpha + \frac{1}{n}, \qquad n = 1,2,\dots . \tag{2}$$

Then it is seen that

$$||Au_n|| \le ||f|| + (m_\alpha+1)^{1/\sigma_1}, \qquad ||Lu_n|| \le ||g|| + \left(\frac{m_\alpha+1}{\alpha}\right)^{1/\sigma_2}. \tag{3}$$

Assuming $C = \max\{||f|| + (m_\alpha+1)^{1/\sigma_1}, ||g|| + (\frac{m_\alpha+1}{\alpha})^{1/\sigma_2}$, we see that $\{u_n\} \subset W_C$; therefore, it is weakly compact. We assume without loss of generality that it is also weakly convergent:

$$u_n \xrightarrow{\text{weakly}} u_0 \quad (\text{in } H), \qquad Au_n \xrightarrow{\text{weakly}} f_0 \quad (\text{in } F),$$
$$Lu_n \xrightarrow{\text{weakly}} g_0 \quad (\text{in } G). \tag{4}$$

The property of joint weak closure implies that

$$u_0 \in D, \quad Au_0 = f, \quad Lu_0 = g_0. \tag{5}$$

We shall show that the element u_0 is a solution of the problem (2). Making use of the weak lower semi-continuity of a norm in a Hilbert space and using (2), (4) and (5), we have

$$m_\alpha \le \Phi_\alpha[u_0] \le \underline{\lim}_{n\to\infty} \Phi_\alpha[u_n] \le \overline{\lim}_{n\to\infty} \Phi_\alpha[u_n] \le m_\alpha;$$

that is, $\Phi_\alpha[u_0] = m_\alpha$. □

Let $U_\alpha = \{u \in D\colon \Phi_\alpha[u] = m_\alpha\}$. Theorem 50 shows that for any $\alpha > 0$ the mapping $U_\alpha = R_\alpha\{A,f;L,g\}$ as a function of $\{A,f;L,g\}$ is well-defined.

Let $U_\alpha = R_\alpha\{A,f;L,g\}$. We shall show that the set U_α as $\alpha \to 0$ converges (in a sense to be defined) to the set of exact solutions $\hat{U}$. However, first we prove the following theorem.

Theorem 51. Under the assumption of joint weak closure of the operators A and L on D and the weak compactness of the sets W_C in H, the set $\hat{U}$ is nonempty.

Proof: Suppose $\bar{u}_\alpha \equiv U_\alpha$. We have

$$\Phi_\alpha[\bar{u}_\alpha] \leq \Phi_\alpha[\bar{u}] \leq ||A\bar{u}_\alpha - f||^{\sigma_1} + \alpha||L\bar{u}-g||^{\sigma_2}$$

for any $u \in U_A$, yielding the inequalities

$$||L\bar{u}_\alpha - g|| \leq ||L\bar{u}-g||, \qquad ||A\bar{u}_\alpha - f|| \leq (\mu_A^{\sigma_1} + \alpha\nu_L^{\sigma_2})^{1/\sigma_1} \tag{6}$$

which hold for any $\bar{u} \in U_A$, $\bar{u}_\alpha \in U_\alpha$.

The inequalities (6) imply that the families $\{\bar{u}_\alpha\}$, $\{A\bar{u}_\alpha\}$ and $\{L\bar{u}_\alpha\}$ are weakly compact in H, F, and G, respectively. Let $\{\bar{u}_{\alpha'}\}$ be a subfamily such that

$$\bar{u}_{\alpha'} \xrightarrow{\text{weakly}} \hat{u}_0 \ (\text{in } H), \qquad A\bar{u}_{\alpha'} \xrightarrow{\text{weakly}} \hat{f} \ (\text{in } F),$$
$$L\bar{u}_{\alpha'} \xrightarrow{\text{weakly}} \hat{g} \ (\text{in } G)$$

as $\alpha' \to 0$. Then $\hat{u} \in D$, $A\hat{u} = \hat{f}$, $L\hat{u} = \hat{g}$.

Putting $\alpha = \alpha'$ in (6) and using the properties of weakly convergent sequences, we have

$$||L\hat{u}-g|| \leq \nu_L, \qquad ||A\hat{u}-f|| \leq \mu_A, \tag{8}$$

yielding $\hat{u} \in \hat{U}$, thus completing the proof of the theorem. □

Remark 1. In contrast to the case where the operators A and L are linear, the set $\hat{U}$ may consist of more than one element. If the set $U_A = \{\hat{u}\}$, it is obvious that $\hat{U} = \{\hat{u}\}$; that is, it consists of a single element.

2. We now analyze in more detail the proof of Theorem 50. Putting $\bar{u} = \hat{u}$ in (6) and using (7), we obtain:

$$\lim_{\alpha'\to 0} ||L\bar{u}_{\alpha'} - g|| = ||L\hat{u}-g||, \qquad \lim_{\alpha'\to 0} ||A\bar{u}_{\alpha'} - f|| = ||A\hat{u}-f||,$$

which together with (7) implies the strong convergence

$$\lim_{\alpha'\to 0} ||Au_{\alpha'} - A\hat{u}|| = \lim_{\alpha'\to 0} ||Lu_{\alpha'} - L\hat{u}|| = 0 \tag{9}$$

for some element $\hat{u} \in \hat{U}$.

As the measure of proximity of the regularized solutions U_α to the set of exact solutions $\hat{U}$, we define the quantity

$$\beta(\alpha) = \sup_{u\in U_\alpha} \inf_{\hat{u}\in\hat{U}} \{||Au-A\hat{u}||_F + ||Lu-L\hat{u}||_G .$$

The following theorem holds.

Theorem 52. With the properties of the operators A and L defined above, the set U_α approximates the set $\hat{U}$ in the sense that

$$\lim_{\alpha\to 0} \beta(\alpha) = 0.$$

Proof: Let $\varepsilon > 0$ be given. We define the set

$$O_\varepsilon[\hat{u}] = \{u \in D: \inf_{\hat{u}\in\hat{U}} \{||Au-A\hat{u}||_F + ||Lu-L\hat{u}||_G\} < \varepsilon\}.$$

It suffices to show that we can find $\underline{\alpha} = \underline{\alpha}(\varepsilon) > 0$ such that the inclusion $U_\alpha \subset O_\varepsilon[\hat{U}]$ holds for all α, $0 < \alpha \leq \underline{\alpha}$. We assume the converse is true. Then there exists an $\varepsilon_0 > 0$ and a sequence $\{\bar{u}_n\} = \{\bar{u}_{\alpha_n}\} \subset U_{\alpha_n}$ $(\alpha_n \to 0,\ n \to \infty)$ such that

$$\inf_{\hat{u}\in\hat{U}} \{||A\bar{u}_n-A\hat{u}||_F + ||L\bar{u}_n-L\hat{u}||_G\} \geq \varepsilon_0 > 0.$$

Repeating the main points of the proof of Theorem 51, we prove the existence of a subsequence $\{\bar{u}_{n_s}\} \subseteq \{\bar{u}_n\}$ where $s \to \infty$, and an element $\hat{u} \in \hat{U}$ for which relations similar to (9) hold:

$$\lim_{s\to\infty} (||A\bar{u}_{n_s}-A\hat{u}||_F + ||L\bar{u}_{n_s}-L\hat{u}||_G) = 0.$$

We have thus arrived at a contradiction of the choice of the sequence $\{\bar{u}_{n_s}\}$. Then, for sufficiently small values α the sets $U_\alpha \subset O_\varepsilon[\hat{U}]$, $\varepsilon > 0$, which was to be proved. □

Remark. If the set $\hat{U} = \{\hat{u}\}$, then, obviously, the function

$$\beta(\alpha) = \sup_{u\in U_\alpha} \{||Au-A\hat{u}||_F + ||Lu-L\hat{u}||_G\}.$$

Corollary. Let an operavor $B: H \to V$, where V is a Banach space, be defined on D, such that a condition similar to the B-completeness condition holds:

$$\gamma(||Bu-Bv||_V) \leq ||Au-Av||_F + ||Lu-Lv||_G, \quad u,v \in D, \tag{10}$$

where γ, $\gamma(0) = 0$, is a strictly increasing function continuous at 0, defined on the positive semi-axis.

If the conditions of Theorem 51 are satisfied, we have

$$\lim_{\alpha\to 0} \sup_{u\in U_\alpha} \inf_{\hat{u}\in\hat{U}} ||Bu-B\hat{u}||_V = 0.$$

In particular, if $B = E$, the identity operator in H, a similar relation holds. It is not hard to see in this case that condition (10) is sufficient for weak compactness of the sets W_C.

3. Let there be given, instead of the exact $\{A,f;L,g\}$, the approximation $\{A_h,f_\delta;L_t,g_\tau\}$ where

$$f_\delta \in F, \quad ||f_\delta-f|| \leq \delta, \qquad g_\tau \in G, \quad ||g_\tau-g|| \leq \tau,$$

and the operators A_h and L_t, defined for all $h \geq 0$, $t \geq 0$ on D, satisfy the condition of mutual weak closure as well as the approximation condition

$$||A_hu-Au|| \leq h|u|, \qquad ||L_tu-Lu|| \leq h|u|, \qquad \forall u \in D,$$

where, as before, $|u| = (||Au||^2 + ||Lu||^2)^{1/2}$.

Let $\tilde{W}_M = \{u \in D: ||A_hu|| \leq M, ||L_tu|| \leq M\}$. Using the approximation condition, we conclude that for any $u \in \tilde{W}_M$:

$$||Au|| \leq h|u| + M, \qquad ||Lu|| \leq t|u| + M,$$

and therefore

$$|u| \leq ||Au|| + ||Lu|| \leq (h+t)|u| + 2M;$$

that is, for sufficiently small h and t,

$$|u| < \frac{1}{1-(h+t)} 2M \leq C_1 = \text{const}.$$

Letting $C = \max\{hC_1+M, tC_1+M\}$, we obtain the inclusion $\tilde{W}_M \subseteq {}_C$, which implies that the set $\tilde{W}_M$ is bounded (hence weakly compact) in the space H. It follows from this remark and Theorem 50 that the sets $\tilde{U}_\alpha = R_\alpha[A_h,f_\delta,\ldots]$ are defined (and nonempty) for all $\alpha > 0$ if $\sigma = (\delta,h,t,\tau)$ is sufficiently small.

As the measure of proximity of the sets $\tilde{U}_\alpha$ and $\hat{U}$ we define the quantity

$$\beta_{AL}(\tilde{U}_\alpha,\hat{U}) = \sup_{u\in\tilde{U}_\alpha} \inf_{\hat{u}\in\hat{U}} \{||Au-A\hat{u}||_F + ||Lu-L\hat{u}||_G\}.$$

In the general case, as can easily be illustrated by examples,

$$\lim_{\alpha\to 0,\sigma\to 0} \beta_{AL}(\tilde{U}_\alpha,\hat{U}) \nrightarrow 0;$$

However, the following theorem holds.

Theorem 53. We can find $\alpha = \alpha(\delta,h)$ such that

$$\lim_{\sigma\to 0} \beta_{AL}(U_\sigma,\hat{U}) = 0,$$

where the set $U_\sigma \equiv \tilde{U}_{\alpha(\delta,h)}$.

Proof: For any $\tilde{u}_\alpha \in \tilde{U}_\alpha$ and $\hat{u} \in \hat{U}$ we have the inequalities

$$||A_h\tilde{u}_\alpha - f_\delta||^{\sigma_1} + ||L_t\tilde{u}_\alpha - g_\tau||^{\sigma_2} \le ||A_h\hat{u} - f_\delta||^{\sigma_1} + \alpha||L_t\hat{u} - g_\tau||^{\sigma_2}$$

$$\le (\mu_A + h|\hat{u}| + \delta)^{\sigma_1} + \alpha||L_t\hat{u} - g_\tau||^{\sigma_2} \le (||A\tilde{u}_\alpha - f|| + h|\hat{u}| + \delta)^{\sigma_1}$$

$$+ \alpha||L_t\hat{u} - g_\tau||^{\sigma_2} \le (||A_h\tilde{u}_\alpha - f_\delta|| + h|\tilde{u}_\alpha| + h|\hat{u}| + 2\delta)^{\sigma_1}$$

$$+ \alpha(\nu_L + t|\hat{u}| + \tau)^{\sigma_2},$$

yielding

$$||A\tilde{u}_\alpha - f|| \le ||A_h\tilde{u}_\alpha - f_\delta|| + h|\tilde{u}_\alpha| + \delta$$

$$\le [(\mu_A + h|\hat{u}| + \delta)^{\sigma_1} + \alpha(\nu_L + t|\hat{u}| + \tau)^{\sigma_2}]^{1/\sigma_1} + h|\tilde{u}_\alpha| + \varepsilon,$$

$$||L\tilde{u}_\alpha - g|| \le ||L_t\tilde{u}_\alpha - g_\tau|| + t|\tilde{u}_\alpha| + \tau$$

$$\le \left[\frac{(||A_h\tilde{u}_\alpha - f_\delta|| + h|\tilde{u}_\alpha| + h|\hat{}| + 2\delta)^{\sigma_1} - ||A_h\tilde{u}_\alpha - f_\delta||^{\sigma_2}}{} + (\nu_L + t|\hat{u}| + \tau)^{\sigma_2}\right]^{1/\sigma_2}.$$

Since the inequality

$$1 + a \le (1 + a^{1/\sigma})^\sigma$$

holds for any $a \ge 0$ and $\sigma \ge 1$, we obtain

$$\left(x^{\sigma_1} + y^{\sigma_2}\right)^{1/\sigma_i} \le x^{\sigma_1/\sigma_i} + y^{\sigma_2/\sigma_i}.$$

for any $x \ge 0$, $y \ge 0$ and σ_i, $i = 1,2$.

Using the last inequality and the preceding estimates, we have

$$||A\tilde{u}_\alpha - f|| \le \mu_A + h|\hat{u}| + 2\delta + \alpha^{1/\sigma_1}(\nu_L + t|\hat{u}| + \delta)^{\sigma_2/\sigma_1} + h|\tilde{u}_\alpha| + \delta,$$

$$||L\tilde{u}_\alpha - g|| \le \nu_L + t|\hat{u}| + t|\tilde{u}_\alpha| + 2\tau$$

$$+ \left[\frac{(||A_h\tilde{u}_\alpha - f_\delta|| + h|\tilde{u}_\alpha| + h|\hat{u}| + 2\delta)^{\sigma_1} - ||A_h\tilde{u}_\alpha - f_\delta||^{\sigma_1}}{\alpha}\right]^{1/\sigma_2}.$$

Noting that for any $x \ge 0$, $y \ge 0$ and $\sigma \ge 1$ the inequality

$$(x+y)^\sigma \le x^\sigma + \sigma(x+y)^{\sigma-1}y,$$

holds, we obtain

$$(||A_h\tilde{u}_\alpha - f_\delta|| + h|\tilde{u}_\alpha| + h|\hat{u}| + 2\delta)^{\sigma_1} - ||A_h\tilde{u}_\alpha - f_\delta||^{\sigma_1}$$

$$\le \sigma_1(h|\tilde{u}_\alpha| + h|\hat{u}| + 2\delta)(||A_h\tilde{u}_\alpha - f_\delta|| + h|\tilde{u}_\alpha| + h|\hat{u}| + 2\delta)^{\sigma_1 - 1}$$

$$\le \sigma_1(h|\tilde{u}_\alpha| + h|\hat{u}| + 2\delta)\Big\{\Big[(\mu_A + h|\hat{u}| + \delta)^{\sigma_1} + \alpha(\nu_L + t|\hat{u}| + \tau)^{\sigma_2}\Big]^{1/\sigma_1}$$

$$+ h|\tilde{u}_\alpha| + h|\hat{u}| + 2\delta\Big\}^{\sigma_1 - 1}.$$

We now choose a positive function $\alpha = \alpha(\delta,h) \to 0$ so that the relation

$$\frac{\delta+h}{\alpha(\delta,h)} = o(\delta+h). \tag{11}$$

is satisfied.

From the foregoing estimate it follows that

$$||L\tilde{u}_\alpha - g|| \le \nu_L + o(\delta+h)|\tilde{u}_\alpha| + o(\sigma),$$

where $o(\sigma) \to 0$ as σ tends to zero and does not depend on $|\tilde{u}_\alpha|$. Since

$$||A\tilde{u}_\alpha - f|| \le \mu_A + h|\tilde{u}_\alpha| + o(\sigma),$$

we have the inequality

$$|\tilde{u}_\alpha| \le C = \text{const},$$

where the constant C does not depend on the choice of $\tilde{u}_\alpha \in \tilde{U}_\alpha$ and σ. Therefore

$$\overline{\lim_{\sigma\to 0}} \sup_{u\in U_\sigma} ||Au-f|| \le \mu_A, \quad \overline{\lim_{\sigma\to 0}} \sup_{u\in U_\sigma} ||Lu-g|| \le \nu_L. \tag{12}$$

where $U_\sigma = \tilde{U}_{\alpha(\delta,h)}$ and $\alpha = \alpha(\delta,h)$ satisfies (11).

Using arguments similar to those used in proving Theorem 52, we derive from (12) that

$$\lim_{\sigma\to 0} \beta_{AL}(U_\sigma,\hat{U}) = 0. \qquad \square$$

Remark 1. If $\mu_A = 0$, we can choose the function $\alpha = \alpha(\delta,h)$ such that

$$\frac{\delta+h}{\sqrt{\alpha}} = O(\delta + h).$$

This fact can be established if we analyze the proof of Theorem 53.

Remark 2. If the operator B satisfies (10), the conditions of Theorem 53 imply the limit relations

$$\lim_{\sigma\to 0} \sup_{u\in U_\sigma} \inf_{\hat{u}\in\hat{U}} ||Bu-B\hat{u}||_V = 0.$$

4. As in the linear case, the approximation method R_σ satisfying the conditions (12) is said to be *regular*. The method R_σ is said to be β_{AL}-*convergent* if the relation

$$\lim_{\sigma\to 0} \beta_{AL}(U_\sigma,\hat{U}) = 0$$

is satisfied, where the function $\beta_{AL}(U_\sigma,\hat{U})$ is given by

$$\beta_{AL}(U_\sigma,\hat{U}) = \sup_{u\in U_\sigma} \inf_{\hat{u}\in\hat{U}} \{||Au-A\hat{u}||_F + ||Lu-L\hat{u}||_G\}.$$

Theorem 54. In order that the approximation method R_σ to be β_{AL}-convergent, it is necessary and sufficient that it be regular.

Proof: We prove sufficiency in the same way as we proved β_{AL}-convergence of regularized solutions in the previous theorem. We prove here only the necessity. Indeed, let the method R_σ be β_{AL}-convergent and let $u \in U_\sigma$. Using the triangle inequality we have for an arbitrary $\hat{u} \in \hat{U}_\sigma$:

$$||Au-f|| \le \mu_A + ||Au-A\hat{u}||,$$

$$||Lu-g|| \le \nu_L + ||Lu-L\hat{u}||.$$

Since the left-hand sides of these inequalities do not depend on $\hat{u}$, it is seen that

$$||Au-f|| \le \mu_A + \inf_{\hat{u}\in\hat{U}} \{||Au-A\hat{u}||_F +|||Lu-L\hat{u}||_G\},$$

$$||Lu-g|| \le \nu_L + \inf_{\hat{u}\in\hat{U}} \{||Au-A\hat{u}||_F + ||Lu-L\hat{u}||_G\}.$$

Hence

$$\sup_{u \in U_\sigma} ||Au-f|| \leq \mu_A + \beta_{AL}(U_\sigma,\hat{U}),$$

$$\sup_{u \in U_\sigma} ||Lu-g|| \leq \nu_L + \beta_{AL}(U_\sigma,\hat{U}),$$

which together with the β_{AL}-convergence of the method R_σ imply the regularity of this method. □

Corollary. For a suitable choice of the regularization parameter $\alpha = \alpha(\delta,h)$ the regularization method is regular. We note that the consistency of the regularization parameter with t and τ characterizing the accuracy of specification of the operator L and the element g is not necessary.

We shall indicate two more regular methods for solving the basic problem (the nonlinear case). We define the sets

$$U = \{u \in D: ||A_h u-f_\varepsilon|| \leq \hat{\mu}_A+h|u|+\delta,\ ||L_h-u-g_\tau|| < \hat{\nu}_L+t|u|+\tau\}.$$

where $\hat{\mu}_A \geq \mu_A$, $\hat{\nu}_A \geq \nu_A$ and $\sigma = (\delta,h,\tau,t,\Delta\mu,\Delta\nu)$. It is not hard to see that any element $\hat{u}$ belonging to the set of solutions $\hat{U}$ is contained in U_σ as well; that is to say, $\hat{U} \subseteq U_\sigma$. Therefore, for any σ the set U_σ is nonempty. It is easily seen that the method R_σ thus defined is regular.

Let a constant C be known for which

$$\sup_{\hat{u} \in \hat{U}} |\hat{u}| \leq C;$$

that is, the set $\hat{U} \subset S_C$. The existence of this constant can easily be proved. In fact, for any $\hat{u} \in \hat{U}$,

$$||A\hat{u}|| \leq ||A\hat{u}-f|| + ||f|| = \mu_A + ||f||,$$

$$||L\hat{u}|| \leq \nu_L + ||g||.$$

Therefore, let

$$C = \mu_A + \nu_L + ||f|| + ||g||.$$

We also define the sets

$$U_\sigma = \{u \in D: ||A_h u-f_\delta|| \leq \hat{\mu}_A+hC+\delta,\ \ ||L_t u-g_\tau|| \leq \hat{\nu}_L+tC+\tau\},$$

where $\sigma = (\delta,h,\tau,t,\Delta\mu,\Delta\nu)$.

It is easy to show that the sets $\hat{U} \subseteq U_\sigma$ for any σ, so that the sets U_σ are nonempty. The approximation method R_σ thus defined is regular.

It is possible to construct approximation methods of the residual type or quasi-solutions type in a similar way. However, we shall not do this here.

5. Instead, we return to the problem of an effective choice of a regularization parameter in the nonlinear case. We assume that $\mu_A = 0$, that is, the problem is consistent, and $\tau = t = 0$. Let $u_\infty \in D$ be such that the problem

$$\inf_{u \in D} ||Lu-g|| = ||Lu -g|| = \mu_L$$

has a solution.

Next, we assume that the element u_∞ is uniquely defined by the above condition. We can do without this condition, but the analysis becomes messy due to nonessential technicalities.

We consider the functional Φ_α for $\sigma_1 = \sigma_2 = 2$, that is, the same functional as in the linear case. Let $\tilde{\phi}(\alpha) = \Phi_\alpha[\tilde{u}_\alpha]$ where $\tilde{u}_\alpha$ are the pertinent solutions of the problem (1). We note that the value $\tilde{\phi}(\alpha)$ is *unique* for all $\alpha > 0$ regardless of possible nonuniqueness of solution of solution of the problem (1).

The following lemma holds.

Lemma 28. The function $\tilde{\phi}$ is continuous and nondecreasing for all $\alpha > 0$. If $\tilde{\mu}_A < \tilde{\nu}_A$, the values of $\tilde{\phi}(\alpha)$ exhaust the interval $(\tilde{\mu}_A^2, \tilde{\nu}_A^2)$, where $\tilde{\mu}_A = \inf_{u \in D} ||A_h u - f_\delta||_F$, $\tilde{\nu}_A = ||A_h u_\infty - f_\delta||$.

Proof: Let $\alpha \geq \alpha_0 > 0$. Since

$$||A\tilde{u}_\alpha - f_\delta||^2 + \alpha||L\tilde{u}_\alpha - g||^2 \leq ||A_h\hat{u} - f_\delta||^2 + \alpha||L\hat{u} - g||^2$$

$$\leq (\mu_A + h|\hat{u}| + \delta)^2 + \alpha\nu_L^2,$$

where $\tilde{u}_\alpha \in \tilde{U}_\alpha$ and $\hat{u} \in \hat{U}$, we have

$$\sup_{\alpha \geq \alpha_0} \sup_{u \in U_\alpha} ||L\tilde{u}_\alpha - g||^2 \leq \nu_L^2 + \frac{(\mu_A + hC + \delta)^2}{\alpha_0} = \mu_0^2 < +\infty,$$

where the constant C is such that the sphere $S_C \supset \hat{U}$.

Let α and β be such that $\alpha \geq \alpha_0$, $\beta \geq \alpha_0$. Then

$$\tilde{\phi}(\alpha) = ||A_h\tilde{u}_\alpha - f_\delta||^2 + \alpha||L\tilde{u}_\alpha - g||^2 \leq ||A\tilde{u}_\beta - f_\delta||^2 + \alpha||L\tilde{u}_\beta - g||^2,$$

$$\tilde{\phi}(\beta) = ||A_h\tilde{u}_\beta - f_\delta||^2 + \beta||L\tilde{u}_\beta - g||^2 \leq ||A\tilde{u}_\alpha - f_\delta||^2 + \beta||L\tilde{u}_\alpha - g||^2,$$

and therefore

$$\tilde{\phi}(\alpha) - \tilde{\phi}(\beta) \leq (\alpha-\beta)||L\tilde{u}_\beta - g||^2 \leq |\alpha-\beta|\mu_0^2,$$

$$\tilde{\phi}(\beta) - \tilde{\phi}(\alpha) \leq (\beta-\alpha)||L\tilde{u}_\alpha - g||^2 \leq |\alpha-\beta|\mu_0^2.$$

implying that the function $\tilde{\phi}$ is continuous (and even Lipschitz) as well as nondecreasing.

Choosing the element $u_\varepsilon \in D$, $\varepsilon > 0$, for which $||A_h u_\varepsilon - f_\delta|| < \tilde{\mu}_A + \varepsilon$, we obtain:

$$\tilde{\phi}(\alpha) \leq ||A_h u_\varepsilon - f_\delta||^2 + \alpha||Lu_\varepsilon - g||^2 < (\tilde{\mu}_A+\varepsilon)^2 + \alpha||Lu_\varepsilon - g||^2.$$

Due to the arbitrariness of the parameter α and the choice of sufficiently small $\varepsilon > 0$, we derive from the preceding inequality that

$$\lim_{\alpha\to 0+} \tilde{\phi}(\alpha) = \tilde{\mu}_A^2.$$

Next, we choose $\tilde{u}_\alpha \in \tilde{U}_\alpha$. For all $\alpha > 0$ the function $\tilde{\rho}(\alpha) = ||A_h\tilde{u}_\alpha - f_\delta||_F$ is defined. Since

$$||A_h\tilde{u}_\alpha - f_\delta||^2 + \alpha||L\tilde{u}_\alpha - g||^2 \leq ||A_h u_\infty - f_\delta||^2 + \alpha||Lu_\infty - g||^2$$
$$\leq ||A_h u_\infty - f_\delta||^2 + \alpha||L\tilde{u}_\alpha - g||^2,$$

we have

$$||A_h\tilde{u}_\alpha - f_\delta|| \leq \tilde{\nu}_A, \quad ||L\tilde{u}_\alpha - g|| \leq \mu_L + \frac{||A_h u_\infty - f_\delta||}{\sqrt{\alpha}},$$

yielding

$$||A_h\tilde{u}_\alpha - A_h u_\infty|| + ||L\tilde{u}_\alpha - Lu_\infty|| \to 0$$

as $\alpha \to +\infty$ and therefore,

$$\lim_{\alpha\to\infty} \tilde{\rho}(\alpha) = ||A_h u_\infty - f_\delta|| = \tilde{\nu}_A.$$

Since, obviously, $\tilde{\rho}(\alpha) \leq \tilde{\phi}(\alpha)$ for all α, the foregoing relation implies that the function $\tilde{\phi}$ assumes values arbitrarily close to $\tilde{\nu}_L$.

It follows from the continuity of $\tilde{\phi}$ that the values of this function exhaust the interval $(\tilde{\mu}_A, \tilde{\nu}_A)$. □

Let $\nu_A = ||Au_\infty - f||$. Since

$$\nu_A \leq \tilde{\nu}_A + h|u_\infty| + \delta, \quad \tilde{\nu}_A \leq \nu_A + h|u_\infty| + \delta,$$

the inequality

$$|\nu_A - \tilde{\nu}_A| \leq h|u_\infty| + \delta$$

holds. On the other hand, since

$$\tilde{\mu}_A \leq ||A_h\hat{u} - f_\delta|| \leq h|\hat{u}| + \delta \leq hC + \delta,$$

the condition $\tilde{\mu}_A < \tilde{\nu}_A$ is satisfied for sufficiently small h and δ if $\nu_L > 0$.

Lemma 29. Let $0 < \nu_L$ and let h and δ be so small that $\sqrt{2}(hC+\delta) < \nu_A - h(u_\infty) - \delta$. Then the equation

$$\tilde{\phi}(\alpha) = 2(hC + \delta)^2 \tag{13}$$

has at least one root $\alpha_\sigma > 0$, $\sigma = (h,\delta)$, where the constant C is such that $\hat{U} \subseteq S_C$.

Proof: The proof of Lemma 29 is immediate.

Let the parameter $\alpha_\sigma > 0$ be defined by Eq. (13). Let $U_\sigma \equiv \tilde{U}_\alpha$, thus defining some approximation method R_σ. We now investigate properties of this method.

For any $u_\sigma \in U_\sigma$ we have

$$||A_h u_\sigma - f_\delta||^2 + \alpha_\sigma ||Lu_\sigma - g||^2 = 2(hC+\delta)^2.$$

Hence

$$||Au_\sigma - f|| \leq ||A_h u_\sigma - f_\delta|| + h|u_\sigma| + \delta \leq \sqrt{2}(hC+\delta) + \delta + h|u_\sigma|.$$

On the other hand, we have

$$2(hC+\delta)^2 = ||A_h u_\sigma - f_\delta||^2 + \alpha_\sigma ||Lu_\sigma - g||^2 \leq ||A_h\hat{u} - f_\delta||^2 + \alpha_\sigma ||L\hat{u} - g||^2 \leq (hC+\delta)^2 + \alpha_\sigma \nu_L^2$$

and therefore

$$\alpha_\sigma \nu_L^2 \geq (hC+\delta)^2.$$

If $\nu_L > 0$, since

$$||Lu_\sigma - g||^2 \leq \frac{2(hC+\delta)^2}{\alpha_\sigma},$$

we obviously have

$$||Lu_\sigma - g|| \leq \sqrt{2}\,\nu_L, \quad \forall u_\sigma \in U_\sigma.$$

We have arrived at the following relations:

$$\lim_{\sigma\to 0} \sup_{u\in U_\sigma} ||Au-f|| = 0, \qquad \overline{\lim_{\sigma\to 0}} \sup_{u\in U_\sigma} ||Lu-g|| \le \sqrt{2}\, \nu_L. \tag{14}$$

Next, we show that Eqs. (14) yield an approximation property of the sets U_σ with respect to the set $\hat{U}_{\sqrt{2}\,\nu_L}$ defined as follows:

$$\hat{U}_{\sqrt{2}\,\nu_L} = \{\bar{u} \in U_A: \ ||Lu-g|| \le \sqrt{2}\, \nu_L\}.$$

In fact, we have

Theorem 55. Let $U_\sigma = \tilde{U}_{\alpha_\sigma}$, where the regularization parameter is defined by the criterion (13).

Then

$$\lim_{\sigma\to 0} \left\{ \sup_{u\in U_\sigma, \hat{u}\in\hat{U}_{\sqrt{2}\nu_L}} ||Au-A\hat{u}|| + \sup_{u\in U_\sigma} \inf_{\hat{u}\in\hat{U}_{\sqrt{2}\nu_L}} |(Lu-L\hat{u},g)| \right\} = 0$$

for any element $g \in G$, where $\sigma = (\delta,h)$.

Proof: It suffices to show that

$$\lim_{\sigma\to 0} \inf_{\hat{u}\in\hat{U}_{\sqrt{2}\nu_L}} |(Lu-L\hat{u},g)| = 0.$$

It follows from Eqs. (14) that for sufficiently small σ

$$\sup_\sigma \sup_{u\in U_\sigma} ||Au|| < +\infty, \qquad \sup_\sigma \sup_{u\in U_\sigma} ||Lu|| < +\infty$$

and, therefore, for some C the sets $U_\sigma \subset W_C$ for all sufficiently small σ. We separate the arbitrary elements $u_\sigma \in U_\sigma$. As we have seen, the family $\{u_\sigma\}$ is weakly compact in H. Thus we can pick a subsequence $\{u_{\sigma'}\} \subseteq \{u_\sigma\}$ such that

$$u_{\sigma'} \xrightarrow{\text{weakly}} \bar{u}_0 \ \ (\text{in } H), \qquad Au_{\sigma'} \xrightarrow{\text{weakly}} f_0 \ \ (\text{in } F),$$
$$Lu_{\sigma'} \xrightarrow{\text{weakly}} g_0 \ \ (\text{in } G).$$

Due to the joint weak closure of the operators A and L the relations given above yield

$$u_0 \in D, \quad Au_0 = f_0, \quad Lu_0 = g_0,$$

and therefore, the limit element $u_0 \in U_A$. From Eqs. (14) it also follows that $u_0 \in \hat{U}_{\sqrt{2}\,\nu_L}$. Obviously, this is enough to prove our assertion.

The theorem is proved.

Remark. In the general case the set U_σ constructed may consist of more than one element, which is different from the linear case, even if $\hat{U} = \{\hat{u}\}$.

6. Similar to the linear case, it is possible to construct the *deterministic Bayes method* for this case.

In fact, let $\mu_A = 0$, $\tilde{\nu}_A$, δ, h, t, τ be known and let a constant C be such that $\hat{U} \subseteq S_C$. Using this method, we define the functional

$$\Phi[u] \equiv ||A_h u - f_\delta||^2 + \left(\frac{hC+\delta}{\tilde{\nu}_A + tC + \tau}\right)^2 ||L_t u - g_\tau||^2, \quad u \in D.$$

Furthermore, we consider the minimization problem of defining elements $u_\sigma \in D$ for which

$$\inf_{u \in D} \Phi[u] = \Phi[u_\alpha].$$

Theorem 50 implies that this problem can have a solution (in contrast to the linear case, possibly a nonunique solution), so that the sets

$$U_\sigma = \{u_\sigma \in D: \ \Phi[u_\sigma] = \inf_{u \in D} \Phi[u]\} \neq \emptyset.$$

We have the following theorem on approximation of the set $U_{\sqrt{2}\,\nu_L}$ by means of elements of the sets U_σ, which is similar to Theorem 55.

Theorem 56. Let R_σ be the deterministic Bayes method defined above. Then

$$\lim_{\sigma \to 0} \left\{ \sup_{u \in U_\sigma} \ \sup_{\hat{u} \in \hat{U}_{\sqrt{2}\nu_L}} ||Au - A\hat{u}|| + \sup_{u \in U_\sigma} \ \inf_{\hat{u} \in \hat{U}_{\sqrt{2}\nu_L}} |(Lu - L\hat{u}, g)| \right\} = 0, \quad \forall g \in G,$$

where $\sigma = (\delta, h, t, \tau, \Delta\nu)$.

Proof: Let $u_\sigma \in U_\sigma$. It is then seen that for any $\hat{u} \in \hat{U}$

$$||A_h u_\sigma - f_\delta||^2 + \left(\frac{hC+\delta}{\tilde{\nu}_A + tC + \tau}\right)^2 ||L_t u_\sigma g_\tau||^2 \leq ||A_h \hat{u} - f_\delta||^2$$

$$+ \left(\frac{hC+\delta}{\tilde{\nu}_L + tC + \tau}\right)^2 ||L_t \hat{u} - g_\tau||^2 \leq (hC+\delta)^2 + (hC+\delta)^2 = 2(hC+\delta)^2.$$

Therefore

$$||Au_\sigma - f|| \leq ||A_h u_\sigma - f_\delta|| + L|u_\sigma| + \delta \leq \sqrt{2}(hC+\delta) + \delta + h|u_\sigma|,$$

$$||Lu_\sigma - g|| \leq ||L_t u_\sigma - g|| + t|u_\sigma| + \delta \leq \sqrt{2}(\tilde{\nu}_L + tC + \tau)^2 + \tau + t|u_\sigma|$$

for any $u_\sigma \in U_\sigma$ and $\hat{u} \in \hat{U}$, which further yield the relation (14). From now on, the proof of Theorem 56 follows the proof of Theorem 55 word-by-word. □

7. As in the linear case, the approximation method R_σ is said to be K-regular if the sets U_σ have the property:

$$\overline{\lim_{\sigma\to 0}} \sup_{u\in U_\sigma} ||Au-f|| \le \mu_A, \qquad \overline{\lim_{\sigma\to 0}} \sup_{u\in U_\sigma} ||Lu-g|| \le K\nu_L,$$

where $K \ge 1$. It is not hard to show that K-regular methods possess the following property.

<u>Theorem 57</u>. Let R_σ be a K-regular method. Then, if the set $\hat{U}_K = \{u \in U_A: ||Lu-g|| \le K\}$, for any elements $f \in F$, $g \in G$ and $u \in H$, we have the relation

$$\lim_{\sigma\to 0} \sup_{u_\sigma\in U_\sigma} \inf_{\hat{u}\in\hat{U}_K} \{|(u_\sigma-\hat{u},u| + |(Au_\sigma-A\hat{u},f)| + |(Lu_\sigma-L\hat{u},g)|\} = 0.$$

<u>Proof</u>: The proof of this theorem is similar to those of the two previous theorems. Therefore, it is omitted.

Chapter 4

The Problem of Computation and the General Theory of Splines

Section 17. The Problem of Computation and the Parameter Identification Problem

1. The stable computation of values of unbounded operators is one of the most important problems in computational mathematics. Let L be a linear operator with domain $D_L \subseteq U$ and range $Q_L \subseteq G$, where U and G are normed spaces and $||L|| = +\infty$; that is, there exists a sequence of elements $u_n \in D_L$, $||u_n|| = 1$, $n = 1,2,\dots$, such that $||Lu_n|| \to \infty$ as $n \to \infty$. Let $\hat{u} \in D_L$ and $\hat{g} = L\hat{u}$. We put $u_{n,\delta} = \hat{u} + \delta u_n$, where δ is an arbitrarily small number. Let $g_{n,\delta} = Lu_{n,\delta}$. Then

$$||g_{n,\delta}-\hat{g}||_G = \delta||L\bar{u}_n|| \to \infty \quad \text{as} \quad n \to \infty, \qquad \forall\delta > 0,$$

while $||u_{n,\delta}-\hat{u}|| = \delta$ may be arbitrarily small. Therefore, the problem of computing values of the operator in the case considered is unstable. Moreover, if we bear in mind arbitrary δ-approximations to the element $\hat{u}$ in U (that is, the elements $u_\delta \in U$ with $||u_\delta-\hat{u}|| \le \delta$), we can see that the values of the operator L may not even be defined on the elements u_δ; that is, $u_\delta \notin D_L$.

The problem is thus of effective construction of the elements $\hat{u}_\delta \in U$ on the basis of arbitrarily given δ-approximations to the element $\hat{u}$ satisfying the following relations:

1. $\hat{u}_\delta \in D_L$ for any $\delta > 0$;
2. $\lim\limits_{\delta\to 0} ||\hat{g}_\delta-\hat{g}||_G = 0; \quad \hat{g}_\delta = L\hat{u}_\delta.$

It is not hard to see that this problem is a particular case of the basic problem. Further we assume that $U = H$ and F are Hilbert spaces and the operator L is linear and closed; that is, when the relations

$$u_n \in D_L, \quad \lim_{n\to\infty} u_n = u_0, \quad \lim_{n\to\infty} Lu_n = g_0$$

are satisfied, we have $u_0 \in D_L$ and $Lu_0 = g_0$. Since in this case $A = E$ (identity) and, therefore, is closed, closure of L is sufficient for the condition of joint closure to be satisfied. Furthermore, since obviously

$$||u||_H^2 \leq ||u||_H^2 + ||Lu||_G^2, \quad \forall u \in D_L,$$

the completeness condition assumed in the basic problem is automatically satisfied.

Assuming $D = D_L$ and defining

$$\Phi_\alpha[u] = ||u-v||_H^2 + \alpha||Lu-g||_G^2, \qquad u \in D,$$

where $g \in G$ is a given element approximating the value $\hat{g}$ of the operator L on $\hat{u}$, $v \in H$ is any element, and $\alpha > 0$ is the regularization parameter.

We consider the following problem: Find an element $u_\alpha \in D$ such that

$$\Phi_\alpha[u_\alpha] = \inf_{u\in D} \Phi_\alpha[u]. \tag{1}$$

A corollary of the general theorems proved earlier is the fact that a single element $u_\alpha \in D$ satisfying (1) exists for each $\alpha > 0$ and any $v \in H$. Let $g = 0$. Thus the operator $\overline{S}_\alpha$ is well-defined on all of H, with range in D by $\overline{S}_\alpha v = u_\alpha$. Using the Euler equation, we can easily show that the operator $\overline{S}_\alpha$ is linear.

We shall show that the operator $\overline{S}_\alpha$ is bounded. Moreover, $||\overline{S}_\alpha|| \leq 1$. In fact, writing the Euler equation for the element u_α (for $\overline{g} = 0$), we have

$$(u_\alpha - v, h)_H + \alpha(Lu_\alpha, Lh)_G = 0, \quad \forall h \in D_L$$

where, putting $h = u_\alpha$, we obtain

$$||u_\alpha||_H^2 + \alpha||Lu_\alpha||_G^2 = (u_\alpha, v)_H \leq ||u_\alpha||_M ||v||_M, \quad \forall \alpha > 0.$$

Therefore, $||u_\alpha|| \leq ||v||$, $\forall \alpha > 0$. Since, by definition, $u_\alpha = \overline{S}_\alpha v$, then $||\overline{S}_\alpha v|| \leq ||v||$, $\forall v \in H$, and therefore $||\overline{S}_\alpha|| \leq 1$, i.e. $\overline{S}_\alpha$ is a contraction. It follows from the general regularization theorems that

$$\lim_{\alpha\to 0} ||u_\alpha - v|| = 0, \quad \forall v \in D_L.$$

$$\lim_{\alpha \to 0} ||u_\alpha - v|| = 0, \qquad \forall v \in D_L.$$

Therefore, the operator $\overline{S}_\alpha$ has the δ-property on elements of the set D_L.

If $\overline{D}_L = H$, there exists an operator L^* adjoint to L. Let $g \in D_{L^*}$. Then we easily derive from the Euler equation that (for any $g \subset D_{L^*}$)

$$u_\alpha = (E + \alpha L^* L)^{-1} v + \alpha(E + \alpha L^* L)^{-1} L^* g. \tag{2}$$

Letting $g = 0$, we have an explicit expression for $\overline{S}_\alpha$:

$$\overline{S}_\alpha = (E + \alpha L^* L)^{-1}.$$

Formula (2) shows that an algorithm R_α translating $\{v,g;L\}$ into the element $u_\alpha \in D_L$ is defined on $\{v,g;L\}$.

Let $v = \hat{u}$, that is, the element on which the value of the operator L is computed. Then we put

$$\hat{u}_\alpha = R_\alpha\{\Lambda\},$$

where $\Lambda = \{\hat{u},g;L\}$ is exact data for the problem. Let the element $\hat{u}$ be given approximately. In fact, we assume that only a δ-approximation to $\hat{u}$ is given, that is, an element $u_\delta \in H$, $||\hat{u} - u_\delta|| < \delta$. Let $\Lambda_\delta = \{u_\delta, g, L\}$ and

$$\tilde{u}_\alpha = R_\alpha\{\Lambda_\delta\}.$$

The function $\tilde{\rho}$ is defined by

$$\tilde{\rho}(\alpha) \equiv ||\tilde{u}_\alpha - u_\delta||_H, \qquad \alpha > 0.$$

Next, let $g = 0$. Then, obviously, $\mu_A = 0$, $\nu_L = ||L\hat{u}||$. We define the element u_∞ as follows. Let $N_L = \{u \in D_L\colon Lu = 0\}$ be the null space of the operator L. Then

$$\nu_A = \inf_{u \in N_L} ||u - \hat{u}||_H = ||u_\infty - \hat{u}||_H, \qquad u_\infty \in D_L$$

and

$$\tilde{\nu}_A = \inf_{u \in N_L} ||u - u_\delta||_H = ||\tilde{u}_\infty - u_\delta||_H, \qquad \tilde{u}_\infty \in N_L.$$

As was shown in Section 10, for criterion $\tilde{\rho}$ on the choice of a regularization parameter to be valid, it is sufficient that for all δ the inequality

$$\delta < \tilde{\nu}_A$$

be satisfied. It is not hard to see that $\tilde{\nu}_A = ||nP_{N^\perp}u_\delta||_H$, where $N^\perp$ is the orthogonal complement of the subspace N_L (in H). Therefore, the condition indicated above assumes the form

$$\delta < ||nP_{N^\perp}u_\delta||_H = \tilde{\nu}_L. \tag{3}$$

We now show that under the condition $\hat{u} \notin N_L$ the inequality (3) is satisfied if δ is sufficiently small.

In fact, in this case $\nu_A > 0$. We have

$$\nu_A \leq ||\tilde{u}_\infty - \hat{u}|| \leq \tilde{\nu}_A + \delta. \tag{4}$$

Similarly

$$\tilde{\nu}_A \leq \nu_A + \delta. \tag{5}$$

Therefore, $|\nu_A - \tilde{\nu}_A| \leq \delta$; for $\delta < \frac{1}{2}\nu_A$:

$$\delta < \frac{1}{2}(\nu_A - \tilde{\nu}_A) + \frac{1}{2}\tilde{\nu}_L \leq \frac{\delta}{2} + \frac{1}{2}\tilde{\nu}_L,$$

that is, $\delta < \tilde{\nu}_L$, thus proving that (3) is satisfied.

We have in fact proved

<u>Theorem 58</u>. Let $\hat{u} \notin N_L$. For $\delta < \frac{1}{2}\nu_A$ there exists a unique root α_δ of the equation

$$\tilde{\rho}(\alpha) = \delta,$$

and

$$\lim_{\delta\to 0} \{||\hat{u}_\delta - \hat{u}||_H + ||L\hat{u}_\delta - L\hat{u}||_G\} = 0,$$

where $\hat{u}_\delta = \hat{u}_{\alpha_\delta}$.

<u>Remark</u>. Obviously, $\hat{u}_\delta = \overline{S}_{\alpha_\delta} u_\delta$.

<u>Example</u>. We consider the problem of numerical differentiation. Let $H = G = L_2[a,b]$ and let the operator $L = d^n/dx^n$ be the operator of n-fold generalized differentiation in the Sobolev sense. If we know that $\hat{u} \in D_L$ and if, in addition, σ-approximations to $\hat{u}$ are given in $L_2[a,b]$:

$$||u_\delta - \hat{u}||_{L_2} = \left\{\int_a^b (u_\delta(x) - \hat{u}(x))^2 dx\right\}^{1/2} < \delta,$$

the Euler equation for determination of regularized solutions is seen to be

$$(-1)^n \alpha \frac{d^{2n}u}{dx^{2n}} + u = u_\delta, \qquad \left.\frac{d^{n+i}u}{dx^{n+i}}\right|_{x=a,x=b} = 0, \qquad i = 0,\ldots,n-1. \tag{6}$$

If values of the function $\hat{u}$ or its derivatives of order not higher than $2n$ are known a priori, we may include them in the boundary conditions.

The value of the parameter α_δ can be found from the equation

$$\int_a^b (\tilde{u}_\alpha(x)-u_\delta(x))^2 dx = \delta^2,$$

where $\tilde{u}_\alpha$ is the solution of Eq. (4).

We note that the set $N_L = \{u: u(x) = a_0+a_1x + \ldots + a_{n-1}x^{n-1}\}$, where $\{a_i\}$ are arbitrary; that is, consists of polynomials of degree not higher than $(n-1)$. The condition $\hat{u} \notin N_L$ implies that $\hat{u}$ is not a polynomial of degree higher than $(n-1)$. If $\hat{u} \in N_L$, we can obtain approximations to $\hat{u}$ if we take $\hat{u}_\delta \equiv \tilde{u}_\infty$, that is, by approximating the function u_δ by means of polynomials of degree not higher than $(n-1)$ in the mean-square sense. This follows from (4) and (5) for $\nu_A = 0$.

2. The family of operators $\{S_\alpha: \alpha > 0\}$ constructed in the preceding subsection has the following properties:

(1) The domain of S_α coincides with H;

(2) The domain of the operator L is the range of S_α;

(3) There exists a function $\alpha = \alpha(\delta)$ such that for any $\hat{u} \in D_L$

$$\lim_{\delta\to 0}\{||\hat{u}_\delta-\hat{u}||_H + ||L\hat{u}_\delta-L\hat{u}||_G\} = 0,$$

where

$$\hat{u}_\delta = S_{\alpha(\delta)}u_\delta, \qquad \forall u_\delta \in H: ||\hat{u}-u_\delta||_H < \delta.$$

We call the family of operators $\{S_\alpha\}$ satisfying conditions (1)-(3) an L-*smoothing* family.

It is not hard to see that the concept of the smoothing family of operators can easily be generalized to the more general case where the spaces U and G may be metric and the operator L is arbitrary. In this case we need to replace $||\cdot||_H$ and $||\cdot||_G$ by ρ_U and ρ_G, respectively, where ρ_U and ρ_G are distances in the spaces U and G.

Let us suppose that in the metric space U, in addition to the operator L mapping U into the metric space G, the operators L_i are given, $i = 1,2,\ldots,n$, with domains $D_i \subseteq U$ and range in G.

We say that the operator L *subordinates* the operators L_i,

$i = 1,\dots,n$, if the following conditions are satisfied:

(1) $D_i \supseteq D_L$, $i = 1,\dots,n$;

(2) it follows from the relations

$$\lim_{n\to\infty} \{\rho_U(u_n,\hat{u}) + \rho_G(Lu_n,L\hat{u})\} = 0, \quad u_n,\hat{u} \in D_L$$

that

$$\lim_{n\to\infty} \max_i \rho_G(L_i u_n, L_i\hat{u}) = 0.$$

We now prove a proposition:

Let there exist for the operator L at least one smoothing family $\{S_\alpha\}$. If the operator L subordinates the operators L_i, $i = 1,\dots,n$, then

$$\lim_{\delta\to 0} \{\rho_U(\hat{u}_\delta,\hat{u}) + \max_i \rho_G(L_i\hat{u}_\delta, L_i\hat{u})\} = 0, \quad \hat{u} \in D_L$$

where

$$\hat{u}_\delta = S_{\alpha(\delta)}u_\delta, \quad u_\delta \in U: \ \rho_U(u_\delta,\hat{u}) \le \delta.$$

Proof: Since the family $\{S_\alpha\}$ is L-smoothing, there exists a function $\alpha = \alpha(\delta)$ such that for the elements $\hat{u}_\delta = S_{\alpha(\delta)}u_\delta$ the following limit relation holds:

$$\lim_{\delta\to 0} \{\rho_U(\hat{u}_\delta,\hat{u}) + \rho_G(L\hat{u}_\delta,L\hat{u})\} = 0.$$

Using the condition of subordination of the operators L_i, $i = 1,\dots,n$, to the operator L, the result follows. □

The assertion that we have proved enables us to solve in a number of cases the problem of computing values of nonlinear operators, as well as "complicated" linear operators, if we choose the "simplest" operator L for which it is well known how to construct the smoothing family of operators.

Example. Define the operator $\bar{L}$ by

$$\bar{L}u \equiv u^{(m)} + f(x,u,u',\dots,u^{(m-1)}), \quad a \le x \le b,$$

where f is an arbitrary continuous function of its argument, and $D_{\bar{L}}$ coincides with the set of n-times differentiable (in the Sobolev sense) generalized functions $u \in L_2[a,b]$ together with its derivatives through the nth order. Let $L_1u = u$, $L_2u = u,\dots,L_mu \equiv u^{(m-1)}$. By the well-known Sobolev imbedding theorems the operator $Lu \equiv d^nu/dx^n$, $n \ge m$, subordinates the operators L_i. It is not hard to see that the operator L subordinates the operator $\overline{L}$ as well.

3. In applications, there are many problems involving the determination (identification) of parameters in differential equations on the basis of experimental information about the solution. Such problems are frequently called *identification* or *inverse* problems.

Example. Suppose we are given the equation

$$\frac{du}{dx} = au, \quad u(0) = u_0, \quad 0 \leq x \leq 1. \tag{7}$$

where $a > 0$ is the parameter to be determined. In order to determine a, solutions $\hat{u}$ must be measured on an interval $[0,1]$. Let the result of these measurements be a function $u_\delta \in L_2[0,1]$ such that

$$||u_\delta - \hat{u}||_{L_2} < \delta.$$

Using the measurements of u_δ, identify Eq. (7); that is, choose values of the parameter a which are most consistent with the experimental data, that is to say, the function u_δ.

Next, we formulate these inverse problems more generally and develop principles for solving them.

Let $A = \{a = (a_1, a_2, \ldots, a_n): a \in E_n\}$ be a nonempty set of vectors of dimension n of a Euclidean space E_n. Furthermore, let U and G be metric spaces and let $L_a[u]$ be an operator depending on $a \in A$ as a parameter and mapping U into G. We assume that the domain of the operators L_a does not depend on $a \in A$ and is a nonempty set $D \subset U$.

We assume that for some $\overline{a} \in A$ we have

$$L_{\overline{a}}\, \hat{u} = \hat{g}, \tag{8}$$

where $\hat{u} \in D$, $\hat{g} \in G$. Let $A_0 = \{a \in A: L_a[\hat{u}] = \hat{g}\}$. By hypothesis the set $A_0 \neq \emptyset$.

Let approximations in U and G be given respectively to $\hat{u}$ and $\hat{f}$: $u_\delta \in U$, $\rho_U(\hat{u}, u_\delta) < \delta$, $g_\delta \in G$, $\rho_G(g_\delta, \hat{g}) < \delta$. The problem consists of constructing (reconstructing) vectors a_δ from the admissible set A satisfying the approximation condition

$$\lim_{\delta \to 0} \rho_{E_n}(a_\delta, A_0) = 0$$

on the basis of the given u_δ and g_δ and the mathematical model $L_a u = g$.

We note here some difficulties in solving the problem.

More than one value of the parameter may satisfy (10), making this problem unstable even for exact specification of the elements $\hat{u}$ and $\hat{g}$. In this case, the problem of solving (8) can be well-posed for $\bar{a} \in A_0$.

2. The values $L_a u_\delta$; that is, the values of L_a on the approximations to the element $\hat{u}$, may not be defined. Therefore the determination of the parameter a directly from the equation $L_a u_\delta = f_\delta$ is impossible.

Next, we assume that model (8) is linear, that is,

$$L_a u \equiv \sum_{i=1}^{n} a_i L_i u, \tag{9}$$

where $L_i u$, $i = 1,\dots,n$, are operators on U into G, having domains $D_{L_i} \supseteq D$. It is easy to see that if the set A is convex and closed in E_m, the set A_0 will also be convex and closed. Therefore, a vector parameter $\bar{a} \in A_0$ is uniquely defined for which

$$\inf_{a \in A_0} ||a-a^*||_C^2 = ||a-a^*||_C^2,$$

where $a^* \in E_n$ denotes the test vector parameter and $||a||_C = (Ca,a)_{E_n}$, where C denotes the square positive definite matrix giving the a priori "loss" function while some vector parameter is being chosen from A.

We assume that there exists at least one subordinating operator L having domain $D_L = D$, and furthermore that there exists an L-smoothing family $\{S_\beta\colon \beta > 0\}$. Applying the regularization method, we define the functional

$$\Phi_\alpha[a] = \rho_G^\sigma(L_u[\hat{u}_\delta], f_\delta) + \alpha||a-a^*||_C^2, \quad a \in A,$$

where $\alpha > 0$ is the regularization parameter, $\hat{u}_\delta = S_{\beta(\delta)} u_\delta$, and $\sigma \geq 1$.

We now consider the problem of finding the vectors $a_\alpha \in A$ for which

$$\inf_{a \in A} \Phi_\alpha[a] = \Phi_\alpha[a_\alpha] = m_\alpha. \tag{10}$$

The following theorem holds.

<u>Theorem 59</u>. If the conditions formulated above are satisfied and if, in addition, the space G is normed, the problem (12) has a unique solution for all $\alpha > 0$.

Proof: By the conditions of the theorem, the functional

$$\rho_G^\sigma(L_a[\hat{u}_\delta],f_\delta) = ||L_a[\hat{u}_\delta]-f_\delta||_G^\sigma = \phi^\sigma(a)$$

is continuous and convex on A. The continuity property of A follows from the linearity of the model. In order to prove that the functional ϕ^σ is convex, we note that ϕ is convex and also that the inequality

$$(x+y)^\sigma \leq 2^{\sigma-1}(x^\sigma + y^\sigma)$$

holds for all $x \geq 0$, $y \geq 0$ and $\sigma \geq 1$. Using the last inequality, we easily prove that the functional ϕ^σ is convex.

This implies that for any a^1 and a^2 in A we have the inequality

$$\alpha\left|\left|\frac{a^1-a^2}{2}\right|\right|_C^2 \leq \frac{1}{2}\,\Phi_\alpha[a^1] + \frac{1}{2}\,\Phi_\alpha[a^2] - \Phi_\alpha\left[\frac{a^1+a^2}{2}\right]. \tag{11}$$

Next, we choose the sequence $\{a_s\}$ minimizing (10):

$$m_\alpha \leq \Phi_\alpha[a_s] \leq m_\alpha + \frac{1}{s}, \qquad s = 1,2,\ldots$$

Letting $a^1 = a_s$, $a^2 = a_{s+p}$ in Eq. (11), p being any natural number, we see that $\{a_s\}$ is Cauchy and therefore convergent. The closure and convexity of A as well as the continuity of $\Phi_\alpha[a]$ on A imply that $\lim_{s\to\infty} a_s = a_\alpha$ is an element of A and satisfies (12). □

Theorem 60. Let the conditions of Theorem 59 be satisfied. Also, let the regularization parameter $\alpha = \alpha(\delta)$ be chosen so that

$$\frac{\delta + \sum_{i=1}^{n} ||L_i\hat{u}_\delta - L_i\hat{u}||_F}{\alpha^{1/\sigma}} + \alpha \to 0 \quad \text{as} \quad \delta \to 0.$$

Then we have convergence:

$$\lim_{\delta\to 0} ||a_\delta - \bar{a}||_C = 0,$$

where $a_\delta = a_{\alpha(\delta)}$.

Proof: We have

$$\Phi_\alpha[a_\alpha] \leq \Phi_\alpha[\bar{a}], \tag{11}$$

that is,

$$||a_\alpha - a^*||_C^2 \leq \frac{||L_{\bar{a}}[\hat{u}_\delta]-f_\delta||_G^\sigma}{\alpha} + ||\bar{a}-a^*||_C^2 \leq \frac{(\delta + \sum_{i=1}^{m} \bar{a}\,||L_i[\hat{u}_\varepsilon]-L_i\hat{u}||_G^\sigma)}{} + ||\bar{a}+a^*||_C^2.$$

Letting here $\alpha = \alpha(\delta)$, as given by the theorem, we conclude that the family $\{a_\delta\} = \{a_{\alpha(\delta)}\}$ is bounded. Moreover,

$$\overline{\lim_{\delta\to 0}} ||a_\delta - a^*||_G \leq ||\bar{a} - a^*||_G. \tag{12}$$

Let $\{a_{\delta'}\}$ be a convergent subfamily: $\lim_{\delta'\to 0} a_{\delta'} = a_0$. It follows from (12) that $L_{a_0}[\hat{u}] = \hat{f}$, that is, $a_0 \in A_0$. It follows from (13) that $a_0 = \bar{a}$ independently of the choice of the subfamily $\{a_{\delta'}\}$. Therefore, we have the required relation: $\lim_{\delta\to 0} ||a_\delta - \bar{a}||_C = 0$. □

We note that we did not require the operators L_i, $i = 1,2,\ldots,m$, to be linear. Furthermore, a similar result holds in the case when the model is not linear. There is a considerable literature on the problem of parameter identification.

Section 18. Properties of Smoothing Families of Operators

1. As was shown in Section 17, the problem of computing values of the unbounded operator L given in a normed space U is unstable. Let $U = H$ be Hilbert and let $L: H \to G$ be a closed linear operator with domain D_L dense in H. Then, obviously, the adjoint L^* is well-defined. Let $\hat{u} \in D_L$ be an element and let $\tilde{u} \in H$ be such that $0 \leq ||\xi||_H < \delta$, where $\xi = \hat{u} - \tilde{u}$.

Using the solution of the variational problem

$$\inf_{u\in D_L} \Phi_\alpha[u] = \Phi_\alpha[\tilde{u}_\alpha], \quad \tilde{u}_\alpha \in D_L \quad (\Phi_\alpha[u] \equiv ||u-\tilde{u}||_H^2 + \alpha||Lu||_G^2) \tag{1}$$

one can construct a linear bounded operator $S_\alpha = (\alpha E + L^*L)^{-1}$ having smoothing properties. Indeed, as a specific method of choosing the parameter $\alpha = \alpha(\delta)$, let

$$\lim_{\delta\to 0} ||LS_\alpha\tilde{u} - L\hat{u}||_G = 0.$$

It is easy to see that for any element $\tilde{u} \in H$, $\tilde{u}_\alpha = S_\alpha\tilde{u} \in D_{L^*L}$; that is, those elements have a "reserve" of smoothness. This leads to the following problem.

Problem. Indicate all linear operators B for which

$$\lim_{\delta\to 0} ||BS_\alpha\tilde{u} - B\hat{u}|| = 0, \quad \hat{u} \in D_B, \quad D_B \subseteq D_{L^*L}$$

for given dependence of the parameter $\alpha = \alpha(\delta)$. Clarify the nature of this dependence on spectral properties of the operator B and the operator L.

Solving this problem is quite essential for practical applications of the method since the use of overly "strong" operators in order to construct smoothing families of operators leads to excessive "smoothing" of the initial information and furthermore, to the loss of the so-called "fine" structure of the solution.

2. Let E_λ, $0 \le \lambda_0 \le \lambda < +\infty$ be the spectral resolution of L^*L. Then for any $u \in D_L$

$$||Lu||^2 = \int_{\lambda_0}^{\lambda} \lambda(dE_\lambda u,u) < +\infty.$$

Let $\phi(\lambda)$, $\lambda_0 \le \lambda < +\infty$ be an arbitrary continuous positive function. We denote by L_ϕ any operator for which

$$||L_\phi u||^2 = \int_{\lambda_0}^{\lambda} \phi(\lambda)(dE_\lambda u,u) < +\infty, \qquad u \in D_{L_\phi} = D_\phi.$$

The following lemma holds.

Lemma 30. In order that $\tilde{u}_\alpha \in D_\phi$ for any $\tilde{u} \in H$, it is necessary and sufficient that for $\alpha > 0$

$$\sup_\lambda \frac{\phi(\lambda)}{(1+\alpha\lambda)^2} < +\infty. \tag{2}$$

Proof: Since

$$||L_\phi S_\alpha \tilde{u}||^2 = \int_{\lambda_0}^{\infty} \frac{\phi(\lambda)}{(1+\alpha\lambda)^2} (dE_\lambda \tilde{u},\tilde{u}),$$

we have

$$||L_\phi S_\alpha||^2 = \sup_\lambda \{\phi(\lambda)/(1+\alpha\lambda)^2\}.$$

For the operator $L_\phi S_\alpha$ to be bounded for $\alpha > 0$ it is necessary and sufficient that $\sup_\lambda \{\phi(\lambda)/(1+\alpha\lambda)^2\} < +\infty$.

From now on we assume that the condition of Lemma 30 is satisfied. Let

$$\Delta(\alpha,\delta;\hat{u}) = \sup_{\xi:||\xi||\le\delta} ||L_\phi\tilde{u}_\alpha - L_\phi\hat{u}||,$$

$$\Delta_0(\alpha;\hat{u}) = ||L_\phi S_\alpha\hat{u} - L_\phi\hat{u}||, \quad \Delta_1(\alpha,\delta) = \sup_{\xi:||\xi||\le\delta} ||L_\phi S_\alpha\tilde{u} - L_\phi S_\alpha\hat{u}||.$$

It is easy to verify that

$$\Delta(\alpha,\delta;\hat{u}) \le \Delta_0(\alpha;\hat{u}) + \Delta_1(\alpha,\varepsilon), \quad \Delta_1(\alpha,\varepsilon) \le \Delta_0(\alpha,\hat{u}) + \Delta(\alpha,\delta;\hat{u}) \tag{3}$$

Lemma 31. $\lim_{\alpha\to 0} \Delta_0(\alpha;\hat{u}) = 0, \quad \forall\hat{u} \in D_\phi$.

Indeed,

$$\Delta_0^2(\alpha;\hat{u}) = \int_{\lambda_0}^{\lambda} \frac{\alpha^2\lambda^2\phi(\lambda)}{(1+\alpha\lambda)^2}(dE_\lambda\hat{u},\hat{u}).$$

Since $\hat{u} \in D_\phi$, then

$$\int_{\lambda_0}^{\infty} \phi(\lambda)(dE_\lambda\hat{u},\hat{u}) < +\infty$$

and therefore

$$\lim_{N\to\infty} \int_N^{\infty} \phi(\lambda)(dE_\lambda\bar{u},\bar{u}) = 0.$$

We note that for any $N > \lambda_0$

$$\Delta_0^2(\alpha;\hat{u}) \le \max_{\lambda_0\le\lambda\le N} \frac{\alpha^2N^2}{(1+\alpha\lambda)^2} + \int_N^{\infty} \phi(\lambda)(dE_\lambda\hat{u},\hat{u}),$$

from which our assertion follows. □

The relations (3) and Lemma 31 yield the following.

Theorem 61. In order that

$$\lim_{\alpha,\delta\to 0} \Delta(\alpha,\delta;\hat{u}) = 0, \quad \forall\hat{u} \in D_\phi \tag{4}$$

it is necessary and sufficient that

$$\lim_{\alpha,\delta\to 0} \Delta_1^2(\alpha,\delta) = \lim_{\alpha,\delta\to 0} \sup_{\lambda} \frac{\phi(\lambda)\delta^2}{(1+\alpha\lambda)^2} = 0. \tag{5}$$

Remark. Let $\phi(\lambda) = \lambda^\sigma$, $0 < \sigma \le 2$ (this case is frequently encountered in applications). It is easy to compute that for $0 < \sigma \le 2$,

$$\sup_{\lambda} \frac{\lambda^\sigma}{(1+\alpha\lambda)^2} = \alpha^{-\sigma}(2-\sigma)^{2-\sigma}\frac{\sigma^2}{4}.$$

Condition (5) in this case is seen to be

$$\lim_{\alpha,\delta\to 0} \delta^2/\alpha^\sigma = 0.$$

In particular, for $\sigma = 1$ (that is, $L_\phi = L$), condition (5) is seen to be

$$\lim_{\alpha,\delta\to 0} \delta/\sqrt{\alpha} = 0.$$

It is easily seen that the condition

$$\lim_{\alpha,\delta\to 0} \frac{\delta}{\alpha} = 0$$

is sufficient for Eq. (4) to be satisfied for any admissible function ϕ.

Theorem 62. Let $\hat{u} \in H$, $||\tilde{u}-\hat{u}|| \le \delta$ and let

$$\lim_{\alpha,\delta\to 0} \sup_{\xi} ||L_\phi S_\alpha \tilde{u} - g|| = 0,$$

where $g \in H$. Then $\hat{u} \in D_\phi$ and $g = L_\phi \hat{u}$.

Therefore, we may evaluate the "smoothness" of the element g using the convergence of a regularizing algorithm.

Proof: We have $\tilde{u}_\alpha = S_\alpha \tilde{u}$, $||\tilde{u}_\alpha - \hat{u}|| \to 0$ as $\alpha,\delta \to 0$ for any $\hat{u} \in H$. Since $L_\phi \tilde{u}_\alpha \to g$, we obtain $\hat{u} \in D_\phi$, $L_\phi \hat{u} = g$ due to the closure of the operator L_ϕ. □

3. It is easy to see that the quantities $\Delta(\alpha,\delta;\hat{u})$ and $\Delta_0(\alpha;\hat{u})$ depend on the choice of the element $\hat{u} \in D_\phi$. Let $M \subset D_\phi$ and let

$$\Delta(\alpha,\delta,M) = \sup_{\hat{u}\in M} \Delta(\alpha,\delta;\hat{u}), \qquad \Delta_0(\alpha;M) = \sup_{\hat{u}\in M} \Delta_0(\alpha;\hat{u}).$$

We can easily show that

$$\Delta(\alpha,\delta;M) \le \Delta_0(\alpha;M) + \Delta_1(\alpha,\delta). \tag{6}$$

Theorem 63. If $\alpha = \alpha(\delta) > 0$ is such that

$$\lim_{\delta\to 0} \Delta(\alpha,\delta;M) = 0,$$

it is necessary that $\lim\limits_{\delta\to 0} \Delta_0(\alpha;M) = 0$. If $\lim\limits_{\alpha\to 0} \Delta_0(\alpha;M) = 0$ (the limit is unconditional), there exists $\alpha = \alpha(\delta) \to 0$, $\delta \to 0$, such that

$$\lim_{\delta\to 0} \Delta(\alpha,\delta;M) = 0.$$

Proof: Clearly

$$\Delta(\alpha,\delta;M) \ge \Delta(\alpha,0;M) = \Delta_0(\alpha;M), \tag{7}$$

proving the first assertion of the theorem.

In order to prove the second assertion, it suffices to choose $\alpha = \alpha(\delta) \to 0$ so that

$$\lim_{\delta,\alpha\to 0} \delta/\alpha = 0.$$

Then

$$\lim_{\delta\to 0} \Delta_0(\alpha;M) = \lim_{\delta\to 0} \Delta_1(\alpha,\delta) = 0$$

and, by virtue of (6),

$$\lim_{\delta\to 0} \Delta(\alpha,\delta;M) = 0.$$

Lemma 32. If $M = \{\hat{u} \in D_\phi : ||L_\phi u|| \leq R\}$ then $\Delta(\alpha,\delta;M) \geq R$.

We have

$$\Delta_0^2(\alpha,M) = \sup_{\hat{u}\in M} \int_{\lambda_0}^{\infty} \frac{\alpha^2\lambda^2}{(1+\alpha\lambda)^2}\, \phi(\lambda)(dE_\lambda \hat{u},\hat{u}) = \sup_{\lambda} \frac{\alpha^2\lambda^2}{(1+\alpha\lambda)^2} R^2 = R^2.$$

It remains only to use Eq. (7).

Theorem 63 and Lemma 32 underscore an essential difference between *uniform regularization* (on M) and *regularization at a point* (Theorem 61): to insure the convergence of $\Delta(\alpha,\delta;M)$ to zero it is necessary to take more narrow subsets from the domain of definition of the operator L_ϕ.

Theorem 64. Let $M = M_\psi = \{\hat{u} \in D_\psi : ||L_\psi u|| \leq R\}$ and let ψ be a positive continuous function satisfying the condition

$$\lim_{\lambda\to 0} \frac{\phi(\lambda)}{\psi(\lambda)} = 0. \tag{8}$$

Then

$$\lim_{\alpha\to 0} \Delta_0(\alpha;M_\psi) = 0.$$

Proof: We have

$$\Delta_0^2(\alpha,M_\psi) = \sup_{u\in M_\psi} \int_{\lambda_0}^{\infty} \frac{\alpha^2\lambda^2\phi(\lambda)\psi(\lambda)}{(1+\alpha\lambda)^2\psi(\lambda)}\,(dE_\lambda u,u)$$

$$\leq \inf_N \left\{\max_{\lambda\leq N} \frac{\alpha^2\lambda^2\phi(\lambda)}{(1+\alpha\lambda)^2\psi(\lambda)} + \max_{\lambda\geq N} \frac{\phi(\lambda)}{\psi(\lambda)}\right\} R^2,$$

which tends to zero as $\alpha \to 0$ in conjunction with the condition (8). □

It is not hard to see that the fact that the condition (8) is necessary in order that $\lim_{\alpha\to 0} \Delta_0(\alpha;M_\psi) = 0$.

Section 19. The Optimality of Smoothing Algorithms

1. Let $S_\alpha = (E + \alpha L^*L)^{-1}$, $\alpha > 0$, where L is a closed linear operator. We specify a linear operator B on H into V subordinated to the operator L. Then, obviously, for a specific choice of $\alpha = \alpha(\delta)$

$$\lim_{\delta\to 0} ||B_\alpha\tilde{u}-B\hat{u}||_V = 0, \quad \forall\hat{u} \in D_L,$$

where $B_\alpha = BS_\alpha$, $\tilde{u} \in H$, $||\tilde{u}-\hat{u}||_H < \delta$. Therefore, it is possible to use the smoothing family $\{S_\alpha\}$ in order to compute the value $\hat{v} = B\hat{u}$ when the element $\hat{u}$ is specified approximately. We note that the operator B_α is defined on all of H.

Let T be any operator defined on H into V. (We do not assume that the operator T is linear.)

Let

$$\omega_B(\delta,R;T) = \sup_{\substack{\xi\cdot||\xi||\leq\delta_0 \\ \hat{u}\in\hat{U}_R}} ||T\tilde{u}-B\hat{u}||_V,$$

where $\hat{U}_B = \{\hat{u} \in D_L\colon ||L\hat{u}||_G \leq R\}$, $\xi = \tilde{u}-\hat{u}$. Obviously, $\omega_B(\delta,R;T)$ characterizes the error in computation of the values of the operator B on the elements of the set $\hat{U}_R$ with the aid of the algorithm T.

Problem. Find an algorithm T_{opt} for which

$$\inf_T \omega_B(\delta,R;T) = \omega(\delta,R;T_{opt}). \tag{1}$$

The algorithm T_{opt} satisfying the relation (1) is said to be *optimal.* The quantity $\omega_B(\delta,R;T_{opt})$ is said to be the *optimal accuracy of computation of values of the operator* B *on the class* $\hat{U}_R$.

Next, we define the estimation function

$$\omega_B(\delta,R) = \sup_u ||Bu||_V \quad \text{on} \quad \{u \in D_L\colon ||u||_H \leq \delta,\ ||Lu||_G \leq R\}.$$

We now assume that the operator B is strongly subordinated to the operator $A = E$ and, in addition, that V is a Hilbert space. We have the following theorem on estimation from below of optimal accuracy of computation of values of the operator B.

Theorem 65.* Under the assumptions made above, for any algorithm T

$$\omega_B(\delta,R;T) \geq \omega_B(\delta,R). \tag{2}$$

*The proof of this assertion exploits a technique communicated to the author by A. G. Marchuk.

Proof: By Lemma 23, there exists an element $h \in D_L$ with $||h|| \leq \delta$, $||Lh|| \leq R$ for which

$$\omega_B(\delta,R) = ||Bh||_V.$$

It is also obvious that $(-h) \in D_L$ and

$$\omega_B(\delta,R) = ||B(-h)||_V.$$

By the definition of the quantity $\omega_B(\delta,R;T)$ we have

$$\omega_B(\delta,R;T) \geq \sup_{\substack{\xi\in\hat{U}_R:||\xi||\leq\delta,\\ \hat{u}\in\hat{U}_R}} ||T\tilde{u}-B\hat{u}||_V$$

$$\geq \max\left\{\sup_{\xi\in\hat{U}_R:||\xi||\leq\delta} ||Bh-T(h+\xi)||, \sup_{\xi\in U_R:||\xi||\leq\delta} ||B(-h)-T(-h+\xi)||\right\}$$

where the element h is as above. However,

$$\sup_{\xi\in\hat{U}_R:||\xi||\leq\delta} ||Bh-T(h+\xi)|| \geq ||Bh-T\theta||_V,$$

$$\sup_{\xi\in\hat{U}_R:||\xi||\leq\delta} ||B(-h)-T(-h+\xi)|| \geq ||B(-h)-T\theta||_V$$

where θ is the zero element of H. Consequently,

$$\omega_B(\delta,R;T) \geq \max\{||Bh-T\theta||, ||Bh+T\theta||\}.$$

Since

$$2Bh = Bh - T\theta + (Bh+T\theta),$$

using the triangle inequality we obtain

$$2||Bh||_V \leq ||Bh-T\theta||_V + ||Bh+T\theta||_V$$

$$\leq 2\max(||Bh-T\theta||_V, ||Bh+T\theta||_V).$$

Therefore

$$\omega_B(\delta,R;T) \geq ||Bh||_V = \omega_B(\delta,R), \quad \forall T.$$

Remark 1. It is not hard to see that Theorem 65 will hold under more general conditions. In fact, we may assume only that all the spaces are normed. It suffices to require only that the estimation function $\omega_B(\delta,R)$ be finite for any δ and R. In this case, it suffices to choose an admissible element h_ε for which $||Bh_\varepsilon||_V \geq \omega_B(\delta,R)-\varepsilon$, where $\varepsilon > 0$ is arbitrary. The main scheme of the proof remains unchanged.

Remark 2. If $A \neq E$ and $\mu_A = 0$ (that is, we consider the problem of computing values of the operator B on the solutions of the combined operator equation $Au = f$ with the approximately specified right side $\tilde{f} \in F$, $||f-\tilde{f}||_F < \delta$ and $\hat{U}_R = \{\hat{u} \in D_{AL}: A\hat{u} = f, ||L\hat{u}||_G \leq R\})$, the result obtained remains valid. We do not dwell on this problem because it is trivial.

2. It is rather difficult to construct and prove the existence of the optimal algorithm T_{opt} in the general case. However, under special conditions imposed on the operator B (the spectral likelihood of the operator L^*L) it is possible to prove that for some special choice of the parameter $\alpha = \alpha_{opt}$, $T_{opt} = B_{\alpha_{opt}}$, where the family of operators B_α has been defined above [82]. This leads us to the following problem.

Problem. Find the algorithm T_{qopt} for which

$$\omega_B(\delta,R;T_{qopt}) = (\omega_B(\delta,R)). \tag{3}$$

T_{qopt} is called *quasi-optimal* or *optimal with respect to order*. In this case, since the optimal algorithm is obviously defined by specifying the complete a priori quantitative information (that is, by the quantities δ and R), it is natural to seek quasi-optimal algorithms which require minimal a priori information.

It turns out that the solution of the problem (3), as we shall see later, is not unique.

Let the parameter $\alpha = \alpha(\delta) > 0$ and a smoothing family $\{S_\alpha\}$ be defined in accordance with the choice of the regularization parameter made by criterion $\tilde{\rho}$ (Theorem 58).

Let $\hat{u}_\delta = S_{\alpha(\delta)}\tilde{u}$. Then

$$||\hat{u}_\delta-\tilde{u}|| \leq \delta,$$

and therefore

$$||\hat{u}-\hat{u}_\delta|| \leq 2\delta. \tag{4}$$

On the other hand, due to the properties of regularized solutions, we have

$$||\hat{u}_\delta-\tilde{u}||^2 + \alpha(\delta)||L\hat{u}_\delta||^2 \leq ||\hat{u}-\tilde{u}||^2 + \alpha(\delta)||L\hat{u}||^2 \leq \delta^2$$
$$+ \alpha(\delta)||L\hat{u}||^2 = ||\hat{u}_\delta-\tilde{u}||^2 + \alpha(\delta)||L\hat{u}||^2;$$

that is,

$$||L\hat{u}_\delta|| \le ||L\hat{u}|| \le R.$$

Hence

$$||L\hat{u}_\delta - L\hat{u}|| \le 2R. \tag{5}$$

From the estimates (4), (5) and the definition of the estimation function $\omega_B(\delta,R)$ it follows that

$$||B\hat{u}_\delta - B\hat{u}|| \le 2\omega_B(\delta,R), \tag{6}$$

that is, the algorithm $\hat{B}_\delta = BS_{\alpha(\delta)}$ is optimal with respect to order.

It is not hard to see that the algorithm $\hat{B}_\delta$ is linear.

We now construct a smoothing algorithm S_δ using the *residual method*. Let $\hat{u}_\delta \in D_L$ with

$$||L\hat{u}_\delta|| = \inf_{u \in D_\sigma} ||Lu||, \quad D_\sigma = \{u \in D_L : ||u-\tilde{u}|| \le \delta\}.$$

The existence (also uniqueness in the case $N_L = \{0\}$) of the solution $\hat{u}_\delta$ of this problem follows from the general theorems proved in Chapter 3. Let $S_\delta^1\tilde{u} = \hat{u}_\delta$. Similar to the foregoing, we obtain

$$||L\hat{u}_\delta|| \le ||L\hat{u}|| \le R,$$

and therefore

$$||L\hat{u}_\delta - L\hat{u}|| \le 2R.$$

Similarly,

$$||\hat{u}_\delta - \hat{u}|| \le 2\delta$$

and therefore the estimate (6) holds again.

We now construct the smoothing algorithm S_δ^2 using the *method of quasisolutions*. Let $\hat{u}_\delta \in D_L$ with

$$\inf_u ||u-\tilde{u}|| = ||\hat{u}_\delta - \tilde{u}||, \quad u \in D_L : ||Lu|| \le R.$$

Since $||L\hat{u}_\delta|| \le R$,

$$||L\hat{u}_\delta - L\hat{u}|| \le 2R.$$

We show that $||\hat{u} - \hat{u}_\delta|| \le \delta$ and therefore

$$||B\hat{u}_\delta - B\hat{u}|| \le \omega_B(\delta,2R). \tag{7}$$

Indeed, the operator $S_\delta^2\tilde{u} = \hat{u}_\delta$ is simply projection onto the set $\hat{U}_R$. Since the set $\hat{U}_R$ is obviously convex, the projection operator of

$\hat{U}_R$ is a contracting operator. Hence

$$||\hat{u}-\hat{u}_\delta|| \leq ||\hat{u}-\tilde{u}|| \leq \delta,$$

as required.

The estimate (7) is proved.

We note that the realization of the algorithms $B_\delta^1 = BS_\delta^1$ and $B_\delta^2 = BS_\delta^2$ does not require that complete a priori information, namely, the quantities δ and R, be specified at the same time. In fact, in order for the algorithm B_δ^1 to be realized, it suffices to know δ; and in order for the algorithm B_δ^2 to be realized, it suffices to know R.

If we consider the smoothing algorithm S_δ^3, using complete information while defining the elements $\hat{u}_\delta \in D_L$ in conjunction with *the universal method*

$$\hat{u}_\delta \in D_L: \quad ||\hat{u}_\delta-\hat{u}|| \leq \delta, \quad ||L\hat{u}_\delta|| \leq R$$

then, since in this case

$$||\hat{u}-\hat{u}_\delta|| \leq 2\delta, \quad ||L\hat{u}-L\hat{u}_\delta|| \leq 2R,$$

we have

$$||B\hat{u}_\delta-B\hat{u}|| \leq 2\omega_B(\delta,R);$$

that is, the algorithm $B_\delta^3 = BS_\delta^3$ is also quasi-optimal.

Thus, the following theorem holds.

Theorem 66. The smoothing algorithms S_δ, S_δ^i, $i = 1,2,3$, define quasi-optimal algorithms for computation of values of the operator B. For the smoothing algorithm S_δ^3, complete a priori information is required, and for the smoothing algorithms S_δ, S_δ^1 and S_δ^2 only partial a priori information is necessary.

Remark 1. If the problem of solving the operator equation can be formulated as that of computing values of an unbounded operator, the assertions above will obviously hold for the new problem. We shall omit here the formulation of these results.

Remark 2. In Section 13 we described various cases of computing the estimation function $\omega(\delta,R)$. The results obtained can be used in order to estimate the accuracy of solution of the computational problem under various specifications of the operator B.

The assertion of Theorem 66 holds also for a smoothing algorithm constructed on the basis of the deterministic Bayes method; that is, minimization of the nonparameteric functional

$$||u-\tilde{u}||_H^2 + \frac{\delta^2}{R^2}||Lu||_G^2, \quad u \in D_L,$$

or the solution of the Euler equation

$$\frac{\delta_2}{R^2} L^*L + u = \tilde{u}$$

equivalent to the problem considered.

Section 20. The Differentiation Problem and Algorithms of Approximation of the Experimental Data

1. The Statement of the Problem. Let L be some closed linear differential operator defined on the class of functions of D_L specified on a finite or infinite interval $[a,b]$. We assume that $\overline{D}_L = L_2[a,b]$ (the bar denotes closure).

We denote by P the class of functions in D_L for which

$$\int_a^b (Lu)^2 dx < +\infty,$$

and by Q the class of functions for which

$$\int_a^b p(x)u^2(x)dx < +\infty,$$

where p is a given continuous positive function on $[a,b]$.

Let $D = P \cap Q$. We define the norm in D as follows:

$$||u||_D = ||p^{1/2}u||_{L_2} + ||Lu||_{L_2}.$$

We call the function f_δ a δ-*approximation* of a function $\hat{u} \in D$ if

$$\int_a^b p(x)[\hat{u}(x)-f_\delta(x)]^2 dx < \delta^2.$$

The basic problem consists of constructing, using given σ-approximations, functions of $u_\delta \in D$ such that the relation

$$||u_\delta(x)-\hat{u}(x)||_D \to 0, \quad \delta \to 0 \tag{1}$$

is satisfied.

In particular, when $Lu = d^n u/dx^n$, this reduces to the problem of stable differentiation of experimental data.

The algorithm for constructing approximations satisfying the condition (1) consists of solving the parametric variational problem

$$\min_{u \in D} \Phi_\alpha[u] = \Phi_\alpha[u_\alpha], \quad \Phi_\alpha[u] = \int_a^b p(u-f_\delta)^2 dx + \alpha \int_a^b (Lu)^2 dx, \tag{2}$$

where $\alpha > 0$ is a parameter.

The problem of existence of a solution of the problem (2) and of convergence of the solutions obtained in the sense of (1) to the function $\hat{u}$ has been considered in a general form. Now we focus our attention on effective techniques for constructing approximate solutions of the problem (2) and computing the basic numerical characteristics of the regularized solution u_α which are most frequently used for choosing the regularization parameter.

We consider the following numerical criteria:

$$\rho^2(\alpha) = \left[\int_a^b p(u_\alpha - f_\delta)^2 dx\right], \tag{3}$$

which is the square residual function on a regularized solution (it is used when the criterion of the choice of a regularization parameter by the condition $\rho(\alpha) = \delta$ is employed);

$$\gamma^2(\alpha) = \int_a^b (Lu_\alpha)^2 dx, \tag{4}$$

which is the numerical criterion used in choosing the regularization parameter by the condition $\gamma(\alpha) = R$, if it is known that $\int_a^b (L\hat{u})^2 dx \le R^2$;

$$\phi(\alpha) = \rho(\alpha) + \alpha\gamma(\alpha) = \Phi_\alpha[u_\alpha], \tag{5}$$

which is the numerical criterion used in choosing parameter values on the basis of values of the regularizing functional on regularized solution by the condition $\phi(\alpha) = \delta^2$; and

$$\theta(\alpha) = \int_a^b [L(u_\alpha - u_{\tau\alpha})]^2 dx, \quad \tau \simeq 1, \tag{6}$$

which is the numerical criterion used in choosing the so-called quasi-optimal values of the regularization parameter (see [91] and also Section 27).

Since the problem (2) is, as a rule, solved repeatedly for various values of the parameter α, for example, on the mesh $\alpha_{j+1} = \tau\alpha$, the effectiveness of the algorithm depends on both the effectiveness of solution of the problem and the number of operations necessary for computation of the basic criteria $\phi(\alpha)$, $\gamma(\alpha)$, $\phi(\alpha)$, and $\theta(\alpha)$. The algorithms given below are, in our opinion, effective in both senses and also convenient numerically.

We note that in order to compute $\theta(\alpha)$ we need to know the solution at the preceding step; that is, additional computer storage is required. Numerical solution of Eqs. (3)-(5) is considered in Section 26.

2. Decomposition of a Solution in Terms of Eigenfunctions of the Operator L.

2.1. The Analytic Basis of the Method. Let u_i, $i = 0,1,\ldots$, be a complete system of functions which is orthonormal in Q and satisfies the conditions

$$Lu_i = \lambda_i p^{1/2} u_i, \quad i = 0,1,\ldots,$$

where λ_i are constants.

Then the system $\{u_i\}$ is complete and orthogonal in D. We shall seek for a solution of the problem (2) the decomposition

$$u_\alpha(x) = \sum_{i=0}^{\infty} c_i^\alpha u_i(x). \tag{7}$$

Letting

$$\tilde{f}_i = \int_a^b p(x) f_\delta(x) u_i(x)\,dx, \quad i = 0,1,\ldots, \tag{8}$$

we obtain the following expressions for coefficients of (7):

$$c_i^\alpha = \frac{\tilde{f}_i}{1 + \alpha\lambda_i^2}, \quad i = 0,1,\ldots \tag{9}$$

Eqs. (5)-(6) are easily calculated. We have

$$\rho^2(\alpha) = \sum_{i=0}^{\infty} (c_i^\alpha - \tilde{f}_i)^2, \quad \gamma^2(\alpha) = \sum_{i=0}^{\infty} \frac{\lambda_i^2 \tilde{f}_i^2}{(1+\alpha\lambda_i^2)^2},$$
$$\phi(\alpha) = \rho^2(\alpha) + \alpha\gamma^2(\alpha), \quad \theta(\alpha) = \sum_{i=0}^{\infty} \lambda_i^2 (c_i^\alpha - c_i^{\tau\alpha})^2. \tag{10}$$

We compute the value of the operator L on a solution of the problem

(2) using the formula

$$Lu_\alpha(x) = \sum_{i=0}^{\infty} c_i^\alpha \lambda_i p^{1/2}(x) u_i(x).$$

2.2. Numerical Computation Technique. As a rule, the function f_δ is given approximately by its values at the nodes of some mesh

$$a \leq x_0 < x_1 < \ldots < x_{N+1} = b,$$

where $N \geq 1$ is an integer. In this connection, one should compute the coefficients $\tilde{f}_i$ using a quadrature formula (as a rule, of low order of accuracy). For numerical computation, one needs to replace infinite sums by finite sums in the formulas (7) and (10). The number of terms in the sums is to be given either a priori or during the computation of the coefficients $\tilde{f}_i$; for instance, by the criterion

$$\int_a^b p(x)\left[f_\delta(x) - \sum_{i=1}^{n} \tilde{f}_i u_i(x)\right]^2 dx = \int_a^b p(x) f_\delta^2(x)\,dx - \sum_{i=1}^{n} \tilde{f}_i^2 << \delta^2.$$

2.3. Specific Characteristics of the Algorithm. The computation of the coefficients $\tilde{f}_i$ in conjunction with the formulas (8) is the most time-consuming. The computation of $\tilde{f}_i$ in conjunction with the formulas (10) is elementary and does not depend on the choice of the system $\{u_i\}$. Therefore, the computation of the coefficients (9) and computation in conjunction with the formulas (10) can be displaced in the computer storage or split in time.

2.4. Examples

Example 1. Let $Lu = \frac{d}{dx}(1-x^2)\frac{du}{dx}$, $p(x) \equiv 1$, $a = -1$, $b = 1$. As the system $\{u_i\}$ we take a system of Legendre polynomials orthonormalized in $L_2[-1,1]$. Convergence in the sense (1) means that

$$\int_{-1}^{1} \left\{\frac{d}{dx}(1-x^2)\left[u_\delta'(x)-\hat{u}'(x)\right]\right\}^2 dx \rightarrow \int_{-1}^{1} \left[u_\delta(x)-\hat{u}(x)\right]^2 dx \rightarrow 0 \quad \text{as} \quad \delta \rightarrow 0. \tag{11}$$

It is easy to show that convergence in the sense (11) implies uniform convergence of u_δ to $\hat{u}$ together with its first derivatives on any interval $[-1+\eta, 1-\eta]$, $0 < \eta < 1$.

A similar result is obtained in [13] by means of another method.

Example 2. Let $Lu = d^2u/dx^2$, $p(x) \equiv 1$, $a = -\pi$, $b = \pi$. As the system $\{u_i\}$, we take a system of trigonometric functions;

$$u_0(x) \equiv \frac{1}{2\pi}, \quad u_{2i+1}(x) = \frac{1}{\pi} - \sin ix, \quad u_{2i}(x) = \frac{1}{\pi} \cos ix, \quad i = 1,2,\ldots .$$

Convergence in the sense (1) here means that

$$\int_{-\pi}^{\pi} (u''_\delta - \hat{u}')^2 dx + \int_{-\pi}^{\pi} (u_\delta - \hat{u})^2 dx \to 0 \quad \text{as} \quad \delta \to 0.$$

By standard imbedding theorems, the foregoing implies uniform convergence of the functions and their derivatives on the interval $[-\pi,\pi]$.

We note that this method can be regarded as a method of constructing regular analytic approximations to functions specified experimentally.

3. The Method of Finite-Difference Approximation to the Functional

3.1. The Essence of the Method. Let the functions f_δ be given on a discrete set of points $\{x_k\}$, $k = 0,1,\ldots,N-1$. We replace (2) by the finite-difference approximation:

$$\Phi_\alpha^h[\bar{u}] = \sum_{k=0}^{N-1} \mu_k \rho_k (u_k - f_k)^2 + \alpha \sum_{u=\ell}^{N-m-\ell} \nu_k (L_h \bar{u})_k^2, \tag{12}$$

where $\mu_k > 0$, $\nu_k > 0$ are the coefficients of the corresponding quadrature formulas, $p_k = p(x_k)$, $f_k = f_\delta(x_k)$, $\bar{u} = (u_0,u_1,\ldots,u_{N-1})$ is the required vector of solutions, $L_h\bar{u}$ is the matrix-vector approximation the function Lu:

$$(L_h\bar{u})_k = \sum_{j=k-\ell}^{k+m} \ell_{kj}^h u_j, \qquad i = \ell,\ell+1,\ldots,N-m-1.$$

Here, $\ell,m \geq 0$ are integers, $\ell+m < N-1$, and ℓ_{ij}^h are the coefficients of the difference scheme.

For example, if $Lu = du/dx$ we may put

$$(L_h\bar{u})_k = \frac{u_{k+1} - u_k}{h_k}, \qquad h_k = x_{k+1} - x_k.$$

Next, instead of the problem (2), we consider the finite-dimensional problem

$$\min_{\bar{u}} \Phi_\alpha^h[\bar{u}] = \Phi_\alpha^h[\bar{u}_\alpha], \tag{13}$$

whose solution $\bar{u}_\alpha$ is assumed to be the approximation to a solution of the problem (2); $L_h\bar{u}_\alpha$ is assumed to be the approximation to Lu_α.

3.2. Solution of the Problem (13). Since the quadratic form of (12) is, obviously, positive definite, a solution of the problem (13) is unique. This solution can be found either by solving the system of linear algebraic equations corresponding to the Euler equation for the functional (12) (we note that the matrix of this system has quasi-diagonal structure) or by minimizing directly the form (12) (using, for example, the method of relaxation or coordinatewise descent). In the latter case, it is natural to take a solution corresponding to its preceding value as an initial approximation for the parameter (since a solution of the problem (13) is obviously dependent on α).

If the parameter value considered is not "appropriate", the approximation to the solution $\bar{u}_\alpha$ can be defined quite roughly.

Approximate numerical values (3)-(6) can be found using the formulas

$$\rho^2(\alpha) \cong \sum_{k=0}^{N-1} \mu_k \rho_k (u_k^\alpha - f_k)^2, \qquad \gamma^2(\alpha) \cong \sum_{k=\ell}^{N-m-\ell} \nu_k (L_k \bar{u}_\alpha)_k^2,$$
$$\phi(\alpha) \cong \rho^2(\alpha) + \alpha\gamma^2(\alpha), \qquad \theta(\alpha) \cong \sum_{k=\ell}^{N-m-\ell} \nu_k [L_k(\bar{u}_\alpha - \bar{u}_{\tau\alpha})]_k^2. \tag{14}$$

3.3. Specific Characteristics of the Method. The method considered is quite universal and simple to use. It has an obvious shortcoming, namely the necessity of repeatedly solving a system of linear algebraic equations (having a special structure, but of high order). Furthermore, computation of the characteristics (14) requires, in general, longer computer time than the computation of analogous characteristics do in the previous method. It is also necessary to note that we obtain a pointwise solution only.

The method of finite-difference approximation of the Euler differential equation for the functional (2) has similar properties.

4. The Application of the Discrete Fourier Transform (DFT).

4.1. The Definition and the Main Property of the Discrete Fourier Transform. Let $a = 0$, $b = T$ and let values f_k be given at points $x_k = k\Delta x$, $k = 0,1,\ldots,N-1$, where N is an integer, and $\Delta x = T/N$. We define the complex-valued numbers F_m, $m = 0,1,\ldots,N-1$, as follows:

$$F_m = \sum_{k=0}^{N-1} f_k e^{-i\omega_m x_u}, \tag{15}$$

where $\omega_m = m\Delta\omega$, $m = 0,1,\ldots,N-1$, and $\Delta\omega = 2\pi/T$. The relations (15) define the DFT of the sequence $\{f_k\}$. The inverse to DFT (IDFT) is given by

$$f_k = \frac{1}{N}\sum_{m=0}^{N-1} F_m e^{i\omega_m x_k}. \tag{16}$$

In fact, if F_m is defined in accord with the formula (15),

$$\frac{1}{N}\sum_{m=0}^{N-1} F_m e^{i\omega_m x_k} = \frac{1}{N}\sum_{m=0}^{N-1}\left(\sum_{\ell=0}^{N-1} f_\ell e^{-i\omega_m x_\ell}\right)e^{i\omega_m x_k}$$

$$= \frac{1}{N}\sum_{\ell=0}^{N-1} f_e\left(\sum_{m=0}^{N-1} e^{\frac{2\pi i m(k-\ell)}{N}}\right) = f_k,$$

since

$$\sum_{m=0}^{N-1} e^{\frac{2\pi i m(k-\ell)}{N}} = \begin{cases} N, & k = \ell \bmod N, \\ 0, & k \neq \ell \bmod N. \end{cases} \tag{17}$$

Taking (17) into account, we obtain

$$\sum_{k=0}^{N-1} f_k^2 = \sum_{k=0}^{N-1}\left|\frac{1}{N}\sum_{m=0}^{N-1} F_m e^{i\omega_m x_k}\right|^2$$

$$= \frac{1}{N^\ell}\sum_{m,m=0}^{N-1} F_m F_{m'}\left(\sum_{k=0}^{N-1} e^{\frac{2\pi i(m-m')k}{N}}\right) = \frac{1}{N}\sum_{m=0}^{N-1}|F_m|^2;$$

that is, the basic equality

$$\sum_{k=0}^{N-1} f_k^2 = \frac{1}{N}\sum_{m=0}^{N-1}|F_m|^2, \tag{18}$$

holds if $\{F_m\}$ is the DFT of $\{f_k\}$.

4.2. Approximation and Transformation of the Functional (2). Let the operator L in (2) be defined by

$$Lu = a_0\frac{d^n u}{dx^n} + a_1\frac{d^{n-1}u}{dx^{n-1}} + \ldots + a_n = r\left(\frac{d}{dx}\right)u,$$

where $r(s) = a_0 s^n + a_1 s^{n-1} + \ldots + a_n$ is a polynomial of degree n.

We now approximate the functional (2), using

$$\Phi_\alpha[\bar{u}] = \Delta x\sum_{k=0}^{N-1}(u_k - f_k)^2 + \alpha\Delta x\sum_{k=0}^{N-1}(Lu|_{x=x_k})^2.$$

Let

$$U_m = \sum_{k=0}^{N-1} u_k e^{-i\omega_m x_m}, \quad m = 0,1,\ldots,N-1,$$

be the DFT of $\{u_k\}$, $k = 0,1,\ldots,N-1$. We define the function u by

$$u(x) = \frac{1}{N}\sum_{m=0}^{N-1} U_m e^{i\omega_m x}, \quad u(x_k) = u_k.$$

Since

$$\frac{d}{dx} e^{i\omega_m x} = i\omega_m e^{i\omega_m x}, \quad L e^{i\omega_m x} = r(i\omega_m) e^{i\omega_m x},$$

then, obviously,

$$Lu(x)\Big|_{x=x_k} = \frac{1}{N}\sum_{m=0}^{N-1} U_m r(i\omega_m) e^{i\omega_m x_k}$$

Therefore, $\{U_m r(i\omega_m)\}$ is the DFT for $Lu(x)\Big|_{x=x_k}$.

Using the basic property (18) and this result, we may write

$$\Phi_\alpha[\bar{u}] = \frac{\Delta x}{N}\sum_{m=0}^{N-1} \{|U_m - F_m|^2 + \alpha|U_m r(i\omega_m)|^2\} \equiv \Psi_\alpha[\bar{U}]. \tag{19}$$

4.3. <u>Finding an Approximate Solution</u>. It follows from (19) that the solution $\bar{U}_\alpha$ of the problem

$$\min_{\bar{U}} \Psi_\alpha[\bar{U}] = \Psi_\alpha[\bar{U}\ ] \tag{20}$$

is given by

$$U_m^\alpha = \frac{F_m}{1 + \alpha|r(i\omega_m)|^2}, \quad m = 0,1,\ldots,N-1. \tag{21}$$

Since the solution $\bar{u}_\alpha$ of the problem

$$\min_{\bar{u}} \Phi_\alpha[\bar{u}] = \Phi_\alpha[\bar{u}_\alpha]$$

is the IDFT of the solution of the problem (20), then

$$u_\alpha^k = \frac{1}{N}\sum_{m=0}^{N-1} U_m^\alpha e^{i\omega_m x_k}, \quad k = 0,1,\ldots,N-1. \tag{22}$$

Approximate numerical values can be computed, using

$$\rho^2(\alpha) \cong \frac{\Delta x}{N}\sum_{m=0}^{N-1} |U_m^\alpha - F_m|^2, \quad \gamma^2(\alpha) \cong \sum_{m=0}^{N-1} |U_m^\alpha r(i\omega_m)|^2,$$
$$\theta(\alpha) \cong \frac{\Delta x}{N}\sum_{m=0}^{N-1} |(U_m^\alpha - U_m^{\tau\alpha}) r(i\omega_m)|^2. \tag{23}$$

Approximate numerical values for $Lu_\alpha(x)\big|_{x=x_k}$ can be computed using

$$Lu_\alpha(x)\big|_{x=x_k} = \frac{1}{N}\sum_{m=0}^{N-1} U_m^\alpha r(i\omega_m)e^{i\omega_m x_k}.$$

4.4. The Specific Features of the Method. It is not hard to see that the characteristics (23) are computed not via the components of the true solution $\bar{u}_\alpha$ but via the components of the DFT of the solution. The problem of seeking for "appropriate" values of the parameter thereby becomes much easier. The reconstruction of the initial solution of the problem with the aid of the formulas (22) need be carried out only for the chosen values of the parameter. This essentially increases the efficiency of the method and makes it quite promising for future applications. In order to obtain the DFT and IDFT, it is possible to use the Fast Fourier Transform (FFT) [99] enabling one to make these transformations in an optimal number of operations. The shortcoming of the method is the necessity for specifying the data on a uniform mesh.

5. Approximation of Experimental Data by Piecewise Cubic Functions.

5.1. Interpolation. Let values $\{u_k\}$ be given on nodes of the mesh $\{x_k\}$:

$$a = x_0 < x_1 < x_2 < \dots < x_{n+1} = b.$$

We define the function u_h by

$$u_h(x) = u_k\left(\frac{\Delta u_k}{h_k} - \frac{s_k}{2}h_k - \frac{\Delta s_k}{6}h_k\right)(x-x_k)$$

$$+ \frac{s\alpha}{2}(x-x_k)^2 + \frac{\Delta u_k}{6h_k}(x-x_k)^3, \quad x_k \le x \le x_{k+1}, \tag{24}$$

$$k = 0,1,\dots,N,$$

where $\Delta u_k = u_{k+1} - u_k$, $h_k = \Delta x_k = x_{k+1} - x_k$ and the vector $\hat{s} = (s_1,s_2,\dots,s_N)$ is a solution of the system of algebraic equations:

$$C\hat{s} = B\hat{u}, \tag{25}$$

where $\hat{u} = (u_0,u_1,\dots,u_{N+1})$, the $N \times N$ matrix C is given by

$$C = \begin{bmatrix} \frac{1}{3}(h_0+h_1) & \frac{h_1}{6} & 0 & \cdots & 0 & 0 \\ \frac{h_1}{6} & \frac{1}{3}(h_1+h_2) & \frac{h_2}{6} & \cdots & 0 & 0 \\ \cdots & \cdots & \cdots & \cdots & \cdots & \cdots \\ \cdots & \cdots & \cdots & \cdots & \cdots & \cdots \\ \cdots & \cdots & \cdots & \cdots & \cdots & \cdots \\ 0 & 0 & 0 & \cdots & \frac{h_{N-1}}{6} & \frac{1}{3}(h_{N-1}+h_N) \end{bmatrix}$$

and the $N \times (N+2)$ matrix B is given by

$$B = \begin{bmatrix} \frac{1}{h_0} & -(\frac{1}{h_0}+\frac{1}{h_1}) & \frac{1}{h_1} & 0 & \cdots & 0 & 0 & 0 \\ 0 & \frac{1}{h_1} & -(\frac{1}{h_1}+\frac{1}{h_2}) & \frac{1}{h_2} & \cdots & 0 & 0 & 0 \\ \cdots & \cdots & \cdots & \cdots & \cdots & \cdots & \cdots & \cdots \\ 0 & 0 & 0 & 0 & \cdots & \frac{1}{h_{N-1}} & -(\frac{1}{h_{N-1}}+\frac{1}{h_N}) & \frac{1}{h_N} \end{bmatrix}.$$

is of order $N \times (N+2)$.

We assume that $s_0 = s_{N+1} = 0$.

It is easily seen that the specification of the mesh $\{x_k\}$ and the values $\{u_k\}$ define the function u_h uniquely. In this case the following relations hold:

$$u_h(x_k-0) = u_h(x_k+0) = u_h, \quad u_h'(x_k-0) = u_h'(x_k+0),$$

$$u_h''(x_k-0) = u_k''(x_k+0) = s_k;$$

that is, the function $u_h(x)$ is an interpolating function which is continuous together with its derivatives up to the second order.

We shall denote by S_3^h the class of interpolating piecewise-cubic functions u_h on a fixed mesh of nodes $\{x_k\}$. By the 'step' of the mesh $\{x_k\}$ we shall mean the quantity $\bar{h} = \max_k h_k$.

The following basic properties of the functions $u_hk \in S_3^h$ are well-known.

<u>The Convergence Property</u>. For any twice continuously differentiable function $\bar{u}(x)$, $a \le x \le b$ we have

$$\lim_{\bar{h}\to 0} ||\bar{u}-\bar{u}_h||_C = \lim_{\bar{h}\to 0} ||\bar{u}'-\bar{u}_h'||_C$$

$$= \lim_{\bar{h}\to 0} ||\bar{u}''-\bar{u}_h''||_{L_2} = 0,$$

where $\bar{u}_h \in S_3^h$, $\bar{u}_h(x_k) = \bar{a}_k$.

The Minimality Property. Let u_k, $k = 0,1,\ldots,N+1$, be given on nodes of the mesh $\{x_k\}$. Also, let u be an arbitrary twice continuously differentiable function satisfying the conditions $u(x_k) = u_k$.

Then, if $u_h(x) \in S_3^h$ and $u_h(x_k) = u_k$, we have

$$\int_a^b \bar{u}''^2 dx \le \int_a^b u''^2 dx. \tag{26}$$

The minimality property shows that a solution of the variational problem

$$\min \int_a^b u''^2(x)dx, \quad u(x_k)-u_k, \quad k = 0,1,\ldots,N+1,$$

where twice continuously differentiable functions are taken as comparison functions, is the interpolating function $u_h \in S_3^h$. Since this function is defined uniquely by values u_k at the nodes of the mesh x_k, we have in Eq. (26) strict inequality for each interpolating function of the class considered which is different from u_h.

5.2. Approximate Solution of the Problem (2). Let $Lu = d^2u/dx^2$ and let the function f_δ on $[a,b]$ be given on the mesh $\{x_k\}$ by its values $f_k = f_\delta(x_k)$.

We now define the functional Φ_α^h by

$$\Phi_\alpha^h[u] = \sum_{k=0}^{N+1} \mu_k \rho_k (u_k - f_k)^2 + \alpha \int_a^b \left(\frac{d^2u}{dx^2}\right)^2 dx. \tag{27}$$

Furthermore, we consider the problem of minimizing the function Φ_α^h, if twice continuously differentiable functions u are taken as comparison functions and also $u(x_k) = u_k$.

Let u_α be a solution of the problem (27) and let $u_\alpha(x_k) = u_k^\alpha$. Let us construct a function $u_h^\alpha \in S_3^h$ such that $u_h^\alpha(x_k) = u_k^\alpha$. In conjunction with (26), we have

$$\Phi_\alpha^h[u_h^\alpha] \le \Phi_\alpha^h[u_\alpha] = \min_u \Phi_\alpha^h[u].$$

Therefore, the function u_h^α is a solution of the problem (27). However, these two functions coincide; that is, $u_\alpha(x) \equiv u_h^\alpha(x)$. Therefore, it is possible to seek a solution of the problem (27) in the class S_3^h.

We now find formulas in order to determine a solution of the problem (27). As earlier, let $\hat{u} = (u_0,u_1,\ldots,u_{N+1})$, $\hat{f} = (f_0,f_1,\ldots,f_{N+1})$.

Introducing an auxiliary diagonal matrix P, where $P = \text{diag}(\mu_0 P_0, \ldots, \mu_{N+1} P_{N+1})$, we have

$$\Phi_\alpha^h[u] \equiv (P(\hat{u}-\hat{f}), \hat{u}-\hat{f}) + \alpha(Cs, s),$$

if we take into account the easily verifiable equality

$$\int_a^b u_h''^2(x)\,dx = (Cs, s).$$

Next, taking into account (25), we obtain

$$\Phi_\alpha^h[u] \equiv (P(\hat{u}-\hat{f}), \hat{u}-\hat{f}) + \alpha(B^T C^{-1} B\hat{u}, \hat{u}) \equiv \Phi_\alpha^h[\hat{u}].$$

Therefore, $\hat{u}_\alpha$ satisfies the system of algebraic equations

$$(\alpha B^T C^{-1} B + P)\hat{u} = P\hat{f}. \tag{28}$$

In solving Eq. (28) and defining the vector $\hat{s}_\alpha$ as a solution of Eq. (25), we can define completely the function u_h^α using the vectors $\hat{u}_\alpha$ and $\hat{s}_\alpha$. However, it is not convenient to solve Eq. (28) because it is necessary to know the matrix C^{-1}. Hence in [108] the following transformation has been suggested:* multiplying (28) on the left first by P^{-1} and then by B and, in addition, taking into account (25), yields the equation for $\hat{s}_\alpha$:

$$(\alpha B P^{-1} B^T + C)\hat{s}_\alpha = Bf. \tag{29}$$

We define a solution of (28) having first found $\hat{s}_\alpha$, using

$$\hat{u}_\alpha = \hat{f} - \alpha P^{-1} B^T \hat{s}_\alpha. \tag{30}$$

We can obtain approximate numerical values for (3)-(6) using

$$\begin{aligned} \rho(\alpha) &= (P(\hat{u}_\alpha - \hat{f}), \hat{u}_\alpha - \hat{f}), \quad \gamma(\alpha) = (C\hat{s}_\alpha, \hat{s}_\alpha), \\ \theta(\alpha) &= (C(\hat{s}_\alpha - \hat{s}_{\tau\alpha}), \hat{s}_\alpha - \hat{s}_{\tau\alpha}). \end{aligned} \tag{31}$$

We note that using the criterion for choosing the parameter from the values $\gamma(\alpha)$, we need not compute $\hat{u}_\alpha$ by the formula (30) for all parameter values considered. This can make the computation less time-consuming.

A disadvantage of the method is the need for repeated solution of

*In [108], Eq. (28) was obtained, however, via a more cumbersome technique.

the system (29) and computations using (30), which naturally require considerable computer time. However, this extra computer time is compensated by the possibility of tabulating values of the function u_h^α and its two first derivatives at any point $x \in [a,b]$.

6. Approximation of Experimental Data by Piecewise Cubic Functions of Class $C^2[a,b]$ (continued).

6.1. The Problem of Generalized Interpolation. Let the function f_δ be continuous on $[a,b]$ and let it satisfy the inequality

$$|\hat{u}-\hat{f}| \leq \delta, \tag{32}$$

which is short for denoting the system of inequalities

$$|u(x_k) - f_\delta(x_k)| \leq \delta, \quad k = 0,\dots,N+1,$$

where x_k are nodes of some mesh on an interval $[a,b]$.

Considerations similar to those in the preceding section in regard to the functional (27) lead us to the following theorem.

Theorem 67. A solution of the problem of generalized interpolation (if it exists) is a function of class S_3^h.

Therefore, only functions of class S_3^h need be regarded as comparison functions in the problem of generalized interpolation. Since

$$\int_a^b u_h''^2(x)\,dx = (C\hat{s},\hat{s}) = (B\hat{u},C^{-1}B\hat{u}) = (B^TC^{-1}B\hat{u},\hat{u}),$$

the problem of generalized interpolation is reduced to the problem of minimizing

$$(B^TC^{-1}B\hat{u},\hat{u}) \quad \text{for} \quad |\hat{u}-\hat{f}| \leq \delta; \tag{33}$$

that is, the problem of minimizing a quadratic function on a closed bounded set. Hence a solution of the problem (33), and, therefore, the initial problem of generalized interpolation, exists.

Next, let $\tilde{u}_h(x)$ and $\tilde{\tilde{u}}_h(x)$ be two solutions of the problem in question. From the equality

$$\int_a^b \left(\frac{\tilde{u}_h''-\tilde{\tilde{u}}_h''}{2}\right)^2 dx = \frac{1}{2}\int_a^b \tilde{u}_h''^2 dx + \frac{1}{2}\int_a^b \tilde{\tilde{u}}_h''^2 dx - \int_a^b \left(\frac{\tilde{u}_h''+\tilde{\tilde{u}}_h''}{2}\right)^2 dx$$

as well as properties of the variational problem, it follows that

$$\int_a^b \left(\frac{\tilde{u}_h''-\tilde{\tilde{u}}_h''}{2}\right)^2 dx = \int_a^b \tilde{u}_h''^2 dx - \int_a^b \left(\frac{\tilde{u}_h''+\tilde{\tilde{u}}_h''}{2}\right)^2 dx \leq 0,$$

yielding

$$\tilde{u}_h(x) - \tilde{\tilde{u}}_h(x) = C_h x + d_h.$$

By virtue of Eq. (32) we have

$$|C_h x_k + d_h| \le 2\delta, \qquad x_k \in [a,b].$$

We obtain the following theorem.

Theorem 68. A solution of the problem of generalized interpolation exists and is defined to within linear functions $C_h x + d_h$ satisfying the inequality

$$|C_h x + d_h| \le 2h, \qquad x \in [a,b].$$

6.2. Approximation Properties of Solutions of the Problem of Generalized Interpolation. Let $u_{h,\delta} \in S_3^h$ be a solution of the problem of generalized interpolation.

We show that

$$\int_a^b u''^2_{h,\delta}(x)\,dx \le \int_a^b \bar{u}''^2(x)\,dx. \tag{34}$$

To this end, we construct a function $\bar{u}_h(x) \in S_3^h$ such that $\bar{u}_h(x_k) = \bar{u}(x_k)$. Then, by the minimality property,

$$\int_a^b \bar{u}''^2_h(x)\,dx \le \int_a^b \bar{u}''^2(x)\,dx.$$

On the other hand, $|\bar{u}_h(x_k) - f_k| = |\bar{u}(x_k) - f_\delta(x_k)| \le \delta$. By the definition of $u_{h,\delta}$ we have

$$\int_a^b u''^2_{h,\delta}(x)\,dx \le \int_a^b \bar{u}''^2_h(x)\,dx,$$

thus proving (34).

Let

$$s_k = u''_{h,\delta}(x_k), \qquad k = 0,\dots,N+1.$$

We have

$$\begin{aligned}\int_a^b u''^2_{h,\delta}(x)\,dx &= (C\hat{s},\hat{s}) = \sum_{k=0}^{N} \frac{s_k^2 + s_k s_{k+1} + s_{k+1}^2}{3} h_k \\ &\ge \underline{h} \sum_{k=0}^{N} \frac{s_k^2 + s_k s_{k+1} + s_{k+1}^2}{3} = \underline{h}(C_0\hat{s},\hat{s}),\end{aligned} \tag{35}$$

where $\underline{h} = \min_k h_k$ and the matrix C_0 is given by

$$C_0 = \begin{vmatrix} 4 & 1 & 0 & \dots & 0 & 0 \\ 1 & 4 & 1 & \dots & 0 & 0 \\ . & . & . & \dots & . & . \\ . & . & . & \dots & . & . \\ 0 & 0 & 0 & \dots & 1 & 4 \end{vmatrix} .$$

It is possible to show that the eigenvalues of the matrix C_0 are not less than one. Therefore,

$$(C_0\hat{s},\hat{s}) \geq (\hat{s},\hat{s}). \tag{36}$$

It follows from the inequalities (34)-(36) that

$$|s_k| \leq \underline{h}^{-1/2}M, \quad k = 0,\dots,N+1,$$

where $M = ||\bar{u}''||_{L_2}$.

Further, we estimate the quantity

$$\max_{x\in[a,b]} |\bar{u}(x) - u_{h,\delta}(x)|.$$

We have

$$|\bar{u}(x)-u_{h,\delta}(x)| \leq |\bar{u}(x)-f_\delta(x)| + |f_\delta(x)-f_k| + |f_k-u_{h,\delta}(x_k)|$$
$$+ |u_{h,\delta}(x)-u_{h,\delta}(x_k)| \leq \delta + \Delta_{\delta h} + \delta + |u_{h,\delta}(x)-u_{h,\delta}(x_k)|,$$
$$x \in [x_k,x_{k+1}],$$

where

$$\Delta_{\delta h} = \max_k \max_{x\in[x_k,x_{k+1}]} |f_\delta(x)-f_k| \to 0 \quad \text{as} \quad \bar{h},\delta \to 0.$$

In conjunction with (34),

$$|u_{h,\delta}(x)-u_{h,\delta}(x_k)| \leq |\Delta u_{h,\delta}(x_k)| + 2\left(\frac{|s_k|}{2} + \frac{|\Delta s_k|}{6}\right)h_k^2$$
$$\leq 2\delta + 2(\tfrac{1}{2}\underline{h}^{-1/2}M + \tfrac{1}{6}\cdot 2\underline{h}^{-1/2}M)\bar{h}^2$$
$$= 2\delta + \frac{5}{3}\underline{h}^{-1/2}\bar{h}^2M, \quad x \in [x_k,x_{k+1}].$$

Therefore

$$\max_{x\in[a,b]} |\bar{u}(x)-u_{h,\delta}(x)| \leq 4\delta + \Delta_{\delta,h} + \frac{5}{3}\underline{h}^{-1/2}\bar{h}^2M,$$

yielding the following theorem.

Theorem 69. If $\lim_{\bar{h}\to 0} \underline{h}^{-1/2}\bar{h}^2 = 0$, then

$$\lim_{\delta,\bar{h}\to 0} ||\bar{u}-u_{h,\delta}||_C = 0. \tag{38}$$

It follows from the estimate (34) that the family $\{u''_{\delta,h}\}$ is bounded in $L_2[a,b]$ and therefore is weakly compact. Let $\{u''_{\delta',h'}\}$ be an arbitrary weakly convergent subfamily of $\{u''_{\delta,h}\}$ and let v be the weak limit of this subfamily. Since the function $\bar{u}$ is twice continuously differentiable, $v = \bar{u}''$. Using the properties of a weakly convergent sequence and the inequality (34), we obtain

$$||\bar{u}''||^2_{L_2} \le \varliminf_{\delta',h'\to 0} ||\bar{u}''_{\delta',h'}||^2 \le \varlimsup_{\delta',h'\to 0} ||\bar{u}''_{\delta',h'}||^2_{L_2} \le ||\bar{u}''||^2_{L_2};$$

that is,

$$\lim_{\delta',h'\to 0} \int_a^b \bar{u}''^2_{\delta',h'}(x)dx = \int_a^b \bar{u}''^2(x)dx.$$

It follows from the limiting relation (38) that $u_{\delta',h'}$ converges uniformly to $\bar{u}$, which together with the weak convergence property yields

$$\lim_{\delta',h'\to 0} \int_a^b [\bar{u}''_{\delta',h'}(x) - \bar{u}''(x)]^2 dx = 0$$

for any weakly convergent subfamily of $\{u''_{\delta,h}(x)\}$. But then

$$\lim_{\delta,h\to 0} \int_a^b [\bar{u}''_{\delta,h}(x) - \bar{u}''(x)]^2 dx = 0.$$

Therefore, the following theorem holds.

Theorem 70. Let $\lim_{\bar{h}\to 0} \underline{h}^{-1/2}\bar{h}^2 = 0$. Then

$$\lim_{\delta,\bar{h}\to 0} ||u_{\delta,h} - \bar{u}|| = 0,$$

where

$$||u(x)|| = \max\left\{\max_x |u(x)|, \left(\int_a^b u''^2(x)dx\right)^{1/2}\right\}.$$

As a corollary of Theorem 5, we also have that the sequence $\{u'_{\delta,h}\}$ converges uniformly to $\bar{u}'$ as $\delta,\bar{h} \to 0$.

Remark. The method of generalized interpolation can be extended to solving integral equations. In fact, let $\hat{u} \in W_2^{(1)}[a,b]$, with norm

$||u||_{W^{(1)}} = (||u||^2_{L_2} + ||u'||^2_{L_2})^{1/2}$ be a solution (the unique solution, although this assumption is not necessary) of the equation

$$K[u](x) = A(x)u(x) + \int_a^b k(x,\xi)u(\xi)d\xi = f(x), \quad a \le x,\xi \le b, \tag{39}$$

where A, K, and f are continuous functions of their arguments. We assume that there is known a continuous function $\tilde{f}$ related to f by

$$|f(x) - \tilde{f}(x)| \le \delta(x),$$

where δ is continuous and $\int_a^b \delta^2(x)dx \to 0$ as the accuracy of specification of the function f increases.

We now define the set

$$\Omega_\delta = \left\{u \in W_2^{(1)}: |A(x)u(x) + \int_a^b K(x,\xi)u(\xi)d\xi - \tilde{f}(x)| \le \delta(x)\right\}.$$

Obviously, $\hat{u} \in \Omega_\delta$. We define an approximate solution of (39) to be an element $\tilde{u} \in W_2^{(1)}$ which solves the extremal problem:

$$||\tilde{u}-u''||_{W_2^{(1)}} = \inf_{u \in \Omega_\delta} ||u-u^*||_{W_2^{(1)}}, \tag{40}$$

where $u^* \in W_2^{(1)}$ is a "testing" function, being some approximation to the sought function.

Since Ω_δ is closed and convex in $W_2^{(1)}$, the problem (40) has a unique solution. Arguing in the usual way, we prove that

$$\lim_{||\delta||_{L_2} \to 0} ||\tilde{u}-\hat{u}||_{W_2^{(1)}} = 0,$$

that is, the approximation method (40) is convergent.

If $A \equiv 0$ and $k(x,\xi) = \delta(x-\xi)$ is the Dirac function, the operator K is the identity. We have thus arrived at a problem similar to that considered above.

We shall cast (40) in the discrete form. On the interval [a,b] let two meshes of modes ($m \ge n$) be defined: $a \le x_1 < x_2 < \ldots < x_i < x_{i+1} < \ldots < x_m = b$ and $a \le \xi_1 < \xi_2 < \ldots < \xi_j < \xi_{j+1} < \ldots < \xi_n \le b$, which may coincide. We write the approximation

$$\int_a^b k(x,\xi)u(\xi)d\xi \cong \sum_{j=1}^m \zeta_j k(x,\xi_j)u_j, \quad u_j = u(\xi_j),$$

where ζ_j are coefficients of some quadrature formula.

We now define the set

$$\Omega_h = \left\{u_j;\ j = 1,\ldots,n:\ \left|\sum_{j=1}^{k} \zeta_j k_{ij} u_j - \tilde{f}_i\right| \le \Delta_j,\quad 1 \le i \le m\right\},$$

where $k_{ij} = k(x_i,\xi_j)$, $\tilde{f}_i = \tilde{f}(x_i)$ and $\Delta_i \ge \delta_i$ are chosen so that $\hat{u}_j = \hat{u}(\xi_j) \in \Omega_h$. We have arrived at the following discrete problem: find the quantities $\tilde{u}_j \in \Omega_h$ for which the conditions

$$\phi[\tilde{u}_1,\tilde{u}_2,\ldots,\tilde{u}_n] = \inf_{\tilde{u}\in\Omega_h} \phi[u_1,u_2,\ldots,u_n],$$

$$\phi[u_1,u_2,\ldots,u_n] = \sum_{j=1}^{n-1} \left\{\left(\frac{u_{j+1}-u_j}{\xi_{j+1}-\xi_j}\right)^2 + \frac{u_{j+1}^2+u_j^2}{2}\right\}(\xi_{j+1}-\xi_j),$$

are satisfied, that is to say, we have the problem of dynamic programming. The presence of a priori restrictions on the required solution can be seen in the definition of the set of admissible mesh functions Ω_h.

This method is applicable to the case of the multi-dimensional equation (39), and the considerations are similar.

Section 21. The Theory of Splines and the Problem of Stable Computation of Values of an Unbounded Operator

1. One can formulate the problem of stable computation of values of an unbounded closed linear operator L acting on a separable Hilbert space H into a similar space G as follows.

We introduce the space H_L with scalar product

$$(u,v)_L = (u,v)_{H_L} = (u,v)_H + (Lu,Lv)_G, \quad u,v \in D_L,$$

where D_L is the region of definition of the operator L. Obviously, H_L is a Hilbert space. Let $\hat{u} \in D_L$ and $\hat{g} = L\hat{u}$. Using the sequence $\{u_\delta\} \subset H$:

$$0 < \delta \le \delta_0; \quad \lim_{\delta\to 0} ||u_\delta - \hat{u}||_H = 0,$$

we are to find an 'algorithm' S_δ having the properties:

1) $S_\delta u_\delta \in D_L$ 2) $\lim_{\delta\to 0} ||\hat{u} - S_\delta u_\delta||_L = 0.$

We call the algorithm S_δ satisfying conditions 1) and 2) a *smoothing algorithm*. Then, an element $LS_\delta u_\delta$ can be taken as the approximation to the value of the operator.

2) We now consider a technique for constructing the smoothing operator S_δ.

We define the functional ϕ by

$$\phi[u;v] \equiv ||u-v||_H^2 + \alpha||Lu||_G^2, \quad u \in D_L,$$

where $v \in H$ is a specified element and $\alpha > 0$ is a parameter.

Let $u_v^\alpha \in D_L$ be an element minimizing $\phi[u;v]$;

$$\inf_{u \in D_L} \phi[u;v] = \phi[u_v;v]. \tag{1}$$

We have shown earlier that for any $v \in H$ the problem (1) has a unique solution, that is, we may define $S_\alpha v = u_v^\alpha$. Using an appropriate technique for choosing the parameter $\alpha = \alpha(\delta)$ (satisfying, for example, the so-called criterion ρ) we obtain the smoothing operator $S_{\alpha(\delta)}$. We note that we have assumed that the initial information about the element $\hat{u} \in D_L$ is specified in a "continuous" form, that is, it has been assumed that the element u_δ is such that $||u_\delta-\hat{u}||_H < \delta$.

However, the initial information about the element $\hat{u}$ is often specified in a "discrete" form: we are given values of some functionals (the number of which is finite or infinite) of the element u_δ approximating $\hat{u}$ on H. On the basis of this initial information it is required to reconstruct the value $L\hat{u}$ sufficiently closely. We shall consider such a problem below and see how it is related to the problem of stable computation of values of an unbounded operator.

3. Let

$$H_L \subset B \subset H,$$

where B is a Banach space into which the space H_L is imbedded and which itself is imbedded in H; that is, there exist constants $m > 0$, $M > 0$ such that

$$m||u||_H \leq ||u||_B \leq M||u||_L \quad \text{for all} \quad u \in H_L.$$

We shall specify n linearly independent functionals ℓ_i on B. Assuming that $v \in B$, we replace then the problem (1) by the problem of finding the element $s_v^\alpha \in D_L$ minimizing

$$\phi_\alpha^n[u;v] = \sum_{i=1}^n \ell_i^2(u-v) + \alpha||Lu||_G^2, \quad u \in D_L,$$

that is,

$$\inf_{u \in D_L} \phi_\alpha^n[u;v] = \phi_\alpha^n[s_v^\alpha;v]. \tag{2}$$

The element $s_v^\alpha \in D_L$ satisfying (2) is said to be a *generalized spline*.

Next, we investigate the problem of existence of generalized splines and some of their properties. To do this, we make the assumptions on the operator L and the functionals ℓ_i more specific. To wit, let $N = \{u \in D_L: Lu = 0\}$ be the null space of the operator L, let $N^\perp$ be the orthogonal complement to N (in H), and let Q_L be the range of the operator L. We assume the following conditions:

(a) the kernel N is finite-dimensional with dimension $q \leq n$;

(b) the range Q_L is G;

(c) if $u \in N$ and $\ell_i(u) = 0$, $i = 1,\ldots,n$, then $u = 0$.

Conditions (a) and (b) imply the existence of a bounded linear operator L^+ which is the inverse of L on the set $D_L \cap N^\perp$. This fact follows from the well-known open mapping theorem.

We shall prove some auxiliary assertions which we need for our future discussion.

<u>Lemma 33</u>. The functionals ℓ_i are bounded on H_L. In fact,

$$|\ell_i(u)| \leq M_i||u||_B,$$

where M_i is constant, $i = 1,2,\ldots,n$, by assumption.

<u>Proof</u>: Since H_L is imbedded in B, then

$$|\ell_i(u)| \leq M_i M||u||_L \quad \text{for all} \quad u \in H_L;$$

that is, the functional ℓ_i is bounded on H_L.

<u>Corollary</u>. From the foregoing and the Riesz theorem we derive the representation

$$\ell_i(u) = (u,k_i)_L, \quad u \in H_L,$$

where $k_i \in H_L$.

<u>Remark</u>. The elements k_i are linearly independent since the functionals ℓ_i are linearly independent.

<u>Lemma 34</u>. If $u \in N$, there exists a constant $m_1 > 0$ not depending on the choice of the element u and such that

$$\sum_{i=1}^{n} \ell_i^2(u) \geq m_1||u||_H^2.$$

<u>Proof</u>: Suppose the opposite is true. Then there is a sequence $\{u_j\} \subset N$, $j = 1,2,\ldots$, such that $||u_j||_H = 1$, with

$$\lim_{j\to\infty} \sum_{i=1}^{n} \ell_i^2(u_j) = 0, \qquad \lim_{j\to\infty} \ell_i(u_j) = 0, \qquad i = 1,\ldots,n.$$

Since $||u_j||_L = 1$, the sequence $\{u_j\}$ is weakly compact in H_L.

Let $\{u_j\}$ converge weakly to the element $u_0 \in H_L$.

It is clear that $u_0 \in N$. Since N is finite-dimensional,

$$||u_0||_H = \lim_{j\to\infty} ||u_j||_H = 1.$$

On the other hand,

$$\ell_i(u_0) = (u_0,k_i)_L = \lim_{j\to\infty} \ell_i(u_j) = \lim_{j\to\infty} (u_j,k_i)_L = 0, \qquad i = 1,\ldots,n.$$

Using condition (c), we then have $u_0 = 0$, contradicting the fact that $||u_0||_H = 1$.

The contradiction proves our assertion. □

Next, for any $u,v \in D_L$ let

$$(u,v)_n = \sum_{i=1}^{n} \ell_i(u)\ell_i(v) + (Lu,Lv)_G.$$

It is easily seen that the expression $(u,v)_n$ defines an inner product on the set D_L. Let H_n denote the inner product space this defines.

Lemma 35. The space H_n is complete.

Proof: Let u_j, $1 \le j < \infty$, be an arbitrary Cauchy sequence on H_n:

$$||u_{j+p} - u_j||_n \to 0 \quad \text{as} \quad n \to \infty,$$

p an arbitrary integer.

Then, obviously,

$$||Lu_{j+p} - Lu_j||_G \to 0,$$

$$\ell_i(u_{j+p}) - \ell_i(u_j) \to 0, \qquad i = 1,\ldots,n, \text{ as } j \to \infty, \tag{3}$$

p being arbitrary.

Let

$$u_j = u_j' + u_j'', \qquad u_j' \in N, \quad u_j'' \in N^{\perp}.$$

It is then possible to rewrite Eq. (3) as

$$||Lu_{j+p}'' - Lu_j''||_G \to 0,$$

$$\ell_i(u_{j+p}' - u_j') + \ell_i(u_{j+p}'' - u_j'') \to 0 \quad \text{as} \quad j \to \infty, \tag{4}$$

p being arbitrary.

By (a) and (b) we have:

$$||u''_{j+p} - u''_j||_H \leq ||L^+|| \; ||L(u''_{j+p} - Lu''_j)||_G \to 0 \quad \text{as} \quad j \to \infty,$$

independently of p. Hence, the sequence $\{Lu''_j\}$ is Cauchy on G and, therefore, it is convergent. Let

$$u'' = \lim_{j\to\infty} u''_j \quad \text{on} \quad H.$$

Since the sequence $\{Lu''_j\}$ is convergent on G, $u'' \in D_L$ because the operator L is closed.

Next, from Eq. (4) we have

$$\ell_i(u'_{j+p}) - \ell_i(u'_j) \to 0 \quad \text{as} \quad j \to \infty,$$

p being arbitrary.

Since $u'_j \in N$, $n \geq q$, due to the linear independence of the system $\{k_i\}$ each element u'_j is representable as

$$u'_j = \sum_{r=1}^{n} \mu_r^j k_r.$$

Therefore,

$$\sum_{r=1}^{n} (\mu_r^{j+p} - \mu_r^j)(k_r,k_i)_L \to 0 \quad \text{as} \quad j \to \infty,$$

p being arbitrary.

Due to the linear independence of the elements k_i, $i = 1,2,\ldots,n$,

$$\mu_r^{j+p} - \mu_r^j \to 0 \quad \text{as} \quad j \to \infty, \quad r = 1,\ldots,n,$$

p being arbitrary; or,

$$\lim_{j\to\infty} \mu_r^j = \mu_r, \qquad r = 1,\ldots,n.$$

Assuming that

$$u' = \sum_{r=1}^{n} \mu_r k_r,$$

we have

$$\lim ||u'_j - u'||_H = \lim_{j\to\infty} ||\sum_{r=1}^{n} (\mu_r^j - \mu_r)k_r||_H = 0;$$

that is,

$$\lim_{j\to\infty} u_j' = u' \quad (\text{in } H).$$

We note that $u' \in N$ since N is a subspace of H. Assuming that $\bar{u} = u' + u''$, we derive from the foregoing that

$$\lim_{j\to\infty} u_j = \bar{u} \quad (\text{in } H).$$

Then, due to the linearity of the functionals ℓ_i,

$$\lim_{j\to\infty} ||u_j - \bar{u}||_h = 0,$$

which was to be proved. □

From the preceding considerations it follows that convergence in the space H_L implies the convergence in the space H_n and vice versa. Hence the following lemma holds.

Lemma 36. The norms in the spaces H_n and H_L are equivalent, that is, there exist constants $c > 0$ and $C > 0$ (not depending on u) such that for all $u \in H_L$,

$$c||u||_n \leq ||u||_L \leq C||u||_n.$$

4. We proceed to investigating the question of the existence and uniqueness of generalized splines and also of some of their limiting properties.

Theorem 71. Under the above assumptions, the generalized spline exists and is uniquely defined.

Proof: The proof of the theorem is based on the identity

$$\sum_{i=1}^{n} \ell_i^2\left(\frac{u-h}{2}\right) + \alpha||L\,\frac{u-h}{2}||_F^2 - \frac{1}{2}\,\phi_\alpha^n[u;v] + \frac{1}{2}\,\phi_\alpha^n[h;v] - \phi_\alpha^n\left[\frac{u+h}{2};v\right]$$

which holds for any $u,h \in D_L$ and follows from the principal points of the proof of Theorem 2. In this case, Lemma 36 on the equivalence of the norms in H_n and H_L is crucially used.

We next examine the limiting properties of generalized splines as $\alpha \to 0$ and as $\alpha \to \infty$ (n is fixed); first, though, we give the following definition.

Definition. Let $v \in B$. We say that $u_v \in D_L$ *interpolates the element* v if $\ell_i(u_v) = \ell_i(v)$, $i = 1,2,\ldots,n$. Let U_v be the set of possible interpolating elements for v (we assume that this set is non-empty for any $v \in B$). We call the element $s_v \in D_L$ an *interpolating spline* if

1) $s_v \in U_v$;

2) $\inf\limits_{u \in U_v} ||Lu||_G = ||Ls_v||_G$,

that is, among all the elements interpolating v, it has the minimal "L-norm".

Lemma 37. The interpolating spline is defined uniquely.

In fact, let s_v and $\hat{s}_v$ be two interpolating splines for the element $v \in B$; that is,

$$\ell_i(s_v) = \ell_i(\hat{s}_v) \quad \text{and} \quad ||Ls_v||_G = ||L\hat{s}_v||_G \leq ||Lu||_G$$

for all $u \in U_v$.

Then

$$||L(\frac{s_v - \hat{s}_v}{2})||^2 = \frac{1}{2}||Ls_v||^2 + \frac{1}{2}||L\hat{s}_v||^2 - ||L(\frac{s_v + \hat{s}_v}{2})||^2$$

that is, $s_v - \hat{s}_v \in N$.

By (c) we have $\hat{s}_v = s_v$. □

Theorem 72. The interpolating spline exists and can be constructively defined as the limit

$$\lim_{\alpha \to 0} s_v^\alpha = s_v \quad (\text{in } H_n \text{ or } H_L).$$

Proof: For any $u_v \in U_v$ we have

$$\phi^n[s_v^\alpha; v] \leq \phi_\alpha^n[u_v; v] = \alpha||Lu_v||_G^2.$$

Therefore,

$$\sum_{i=1}^{n} \ell_i^2(s_v^\alpha - v) = \sum_{i=1}^{n} \ell_i^2(s_n^\alpha - u_v) \leq \alpha||Lu_v||_G^2, \tag{5}$$

$$||Ls_v^2||_G \leq ||Lu_v||_G \tag{6}$$

for any $u_v \in U_v$.

The relations (5) and (6) imply the boundedness (therefore, weak compactness) of the family $\{s_v^\alpha\}$ with respect to the parameter α, $0 < \alpha \leq \bar{\alpha} < +\infty$.

Let $s_v^\alpha \xrightarrow{\text{weakly}} s_v$ on H_n as $\alpha \to 0$.

We shall show that s_v is an interpolating spline for the element v. In fact, taking in (5) the limit as $\alpha \to 0$, we have

$$\ell_i(s_v - v) = 0, \quad i = 1,\dots,n, \tag{7}$$

that is, $s_v \in U_v$.

Next, we have

$$||Ls_v|| \le \varliminf_{\alpha\to 0} ||Ls_v^\alpha|| \le \varlimsup_{\alpha\to 0} ||Ls_v^\alpha|| \le ||Lu_v||_G, \tag{8}$$

for any $u_v \in U_v$. (We have used well-known properties of norms of weakly convergent sequences and also the relation (4).) From Eqs. (6) and (7) it follows that s_v is the interpolating spline for v. We note that by Lemma 37 s_v is uniquely defined.

In order to complete the proof, we assume in Eq. (8) that $u_v = s_v$. Then we have

$$||Ls_v|| = \lim_{\alpha\to 0} ||Ls_v^\alpha||,$$

which together with the weak convergence $Ls_v^\alpha \xrightarrow{\text{weakly}} Ls_v$ yields

$$\lim_{\alpha\to 0} Ls_v^\alpha = Ls_v. \tag{9}$$

The latter relation holds for any weakly convergent sequence of the family $\{s_v^\alpha\}$. Therefore, it holds for the entire family as well.

This argument completes proving the theory since it follows from Eqs. (7) and (9) that

$$\lim_{\alpha\to 0} s_v^\alpha = s_v \quad \text{in } H_n \quad (\text{and in } H_L).$$

<u>Corollary</u>. Letting

$$\rho(\alpha) = \sum_{i=1}^{n} \ell_i^2(s_v^\alpha - v),$$

we have

$$\lim_{\alpha\to 0} \rho(\alpha) = 0. \tag{10}$$

We next investigate the behavior of s_v^α as $\alpha \to +\infty$; first, though, we give the following definition.

<u>Definition</u>. We call the element s_g^∞, $g \in B$, a *mean-square spline* if:

1) $s_v^\infty \in N$, the null space of the operator L;
2) $\inf_{u\in N} \sum_{i=1}^{n} \ell_i^2(u-v) = \sum_{i=1}^{n} \ell_i^2(s_v^\infty - v)$.

Lemma 38. The mean-square spline exists for any $v \in B$ and is uniquely defined.

In fact, the functional ϕ given by

$$\phi(u) = \sum_{i=1}^{n} \ell_i^2(u), \quad u \in N,$$

is positive definite (see Lemma 34). This implies that the respective variational problem has a unique solution.

Theorem 73. We have the relation

$$\lim_{\alpha\to\infty} s_v^\alpha = s_v^\infty \text{ in } H_n \quad (\text{and in } H_L).$$

Proof: We have

$$\phi_\alpha^n[s_v^\alpha;v] \leq \phi_\alpha^n[s_v^\infty;v] = \sum_{i=1}^{n} \ell_i^2(s_v^\infty - v).$$

Therefore, for any $\alpha > 0$,

$$\sum_{i=1}^{n} \ell_i^2(s_v^\alpha - v) \leq \sum_{i=1}^{n} \ell_i^2(s_v^\infty - v) \leq \sum_{i=1}^{n} \ell_i^2(u-v), \quad \text{for all } u \in N, \tag{11}$$

and

$$||Ls_v^\alpha|| \leq \frac{1}{\alpha} \sum \ell_i^2(s_v^\infty - v). \tag{12}$$

It follows from Eqs. (11), (12) that the family $\{s_v^\alpha\}$ is weakly compact on H_n for α satisfying the condition $0 < \underline{\alpha} \leq \alpha$. Let this family $\{s_v^\alpha\}$ converge weakly to some element $\hat{s}_v$. We shall show that $\hat{s}_v = s_v^\infty$. Indeed, in conjunction with (11) and (12), we have

$$\lim_{\alpha\to\infty} \sum_{i=1}^{n} \ell_i^2(s_v^\alpha - v) = \sum_{i=1}^{n} \ell_i^2(\hat{s}_v - v) \leq \sum_{i=1}^{n} \ell_i^2(s_v^\infty - v) \leq \sum_{i=1}^{n} \ell_i^2(u-g), \qquad u \in N,$$

$$||L\hat{s}_v|| \leq \lim_{\alpha\to+\infty} ||Ls_v^\alpha|| = 0.$$

Therefore, $\hat{s}_v \in N$. Furthermore, assuming in the foregoing relation that $u = \hat{s}_v$, we obtain

1) $\hat{s}_v \in N$;

2) $\sum_{i=1}^{n} \ell_i^2(\hat{s}_v - v) = \sum_{i=1}^{n} \ell_i^2(s_v^\infty - v)$.

Taking into account the definition of the element s_v^∞ as well as the preceding lemma, we obtain that $\hat{s}_v = s_v^\infty$ and in addition,

$$\lim_{\alpha\to\infty} ||Ls_v^\alpha|| = 0,$$

thus proving the theorem. □

<u>Corollary</u>. It follows from the foregoing that

$$\lim_{\alpha\to\infty} \rho(\alpha) = \sum_{i=1}^{n} \ell_i^2(s_v^\infty - v). \tag{13}$$

5. We now investigate some optimal properties of interpolating splines.

<u>Theorem 74</u>. If $v \in D_L$, then

$$\inf_{u\in B} ||Ls_u - Lv||^2 = ||Ls_v - Lv||^2; \tag{14}$$

that is, with the element s_v we obtain an optimal (in the sense of the norm of the space G) approximation to the value of the operator L on the element v. We have in this case the Pythagorean equality

$$||Lv||^2 = ||Ls_v||^2 + ||Lv - Ls_v||^2. \tag{15}$$

<u>Proof</u>: Since s_v is a solution of the problem (14), for any $v_0 \in V_0 = \{v_0 \in D_L: \ell_i(v_0) = 0, \ i = 1,\ldots,n\}$ the Euler identity holds:

$$(Ls_v, Lv_0) = 0, \quad \text{for all } v_0 \in V_0. \tag{16}$$

Since $s_v - v \in V_0$,

$$(Ls_v, Ls_v - Lv) = 0. \tag{17}$$

It follows from Eqs. (16), (17) that the element s_v is a solution of the problem (14) and the equality (15) holds. □

We consider some linear functional ℓ on H_L. By the Riesz representation theorem and Lemma 36, there exists a unique element $v \in H_L$ such that

$$\ell(z) = (z,v)_n = \sum_{i=1}^{n} \ell_i(z)\ell_i(v) + (Lz,Lv).$$

Let $u \in B$. Then, as is easily seen,

$$\ell(z) - \ell(s_u) = (z - s_u, v)_n.$$

Now we consider the problem: find the element $\hat{u} \in B$ for which

$$\sup_{v:||v||_n \leq 1} |\ell(z) - \ell(s_u)|, \quad u \in B \tag{18}$$

is minimized.

The following theorem holds.

Theorem 75. The interpolating spline s_z is a solution of the problem (18).

Proof: It is obvious that

$$\sup\{|\ell(z)-\ell(s_u)|^2: ||v||_n \leq 1\} = \sum_{i=1}^{n} \ell_i^2(z-s_u) + ||Ls_u-Lz||^2.$$

It remains only to apply Theorem 4. □

Corollary. The optimal approximation (in the sense (18)) to the value of the linear (on H_L) functional $\ell(z)$ is $\ell(s_z)$.

We note that the value $\ell(s_z)$ can be found using the formula

$$\ell(s_z) = \sum_{i=1}^{n} \ell_i(z)\ell_i(v) + (Ls_z, Lv),$$

knowing only $\ell_i(z)$, $i = 1,2,\ldots,n$ and Ls_z; that is, it is not necessary to know the element s_z.

Remark. The quantity

$$||Ls_z-Lz|| = \inf_{u\in B} ||Ls_u-Lz||$$

defines the approximation error of the value Lz by means of the element Ls_z and depends crucially on the choice of the functionals $\ell_i(u)$, $i = 1,2,\ldots$. It would be very important to solve the following optimization problem: for fixed n (being a measure of the computational work) find functionals ℓ_i, $i = 1,2,\ldots,n$, for which

$$\sup_{z\in Z} ||Ls_z - Lz||,$$

is minimized, where Z is a nonempty set in H_L; that is, the highest accuracy is achieved.

6. We assume next that $v = \tilde{u}$, where $||\hat{u}-\tilde{u}||_H < \delta$, $\hat{u} \in Z$. Also, we investigate the behavior of generalized splines as $n \to \infty$ and $\delta \to 0$. In this connection, we assume that the following conditions are satisfied.

The approximation condition: for any $u \in B$,

$$\left| \sum_{i=1}^{n} \ell_i^2(u) - ||u||_H^2 \right| \leq r_n ||u||_B^2, \tag{19}$$

where $r_n \to 0$ as $n \to \infty$ and the r_n do not depend on the choice of $u \in B$.

The consistency condition:

$$r_n ||\tilde{u}||_B^2 < \varepsilon_{n,\delta} \to 0 \quad \text{as} \quad n \to \infty,\ \delta \to 0.$$

We assume that $\ell_i(u) = \ell_i^n(u)$.

We note that the condition (19) is quite convenient for verification and is satisfied in all familiar cases for the appropriate choice of the space B.

Lemma 39. Under the assumptions made above, the family $s_{\tilde{u}}^{\infty}$ is bounded in n and δ, $0 < \delta \leq \delta_0$.

Remark. Since the subspace N is finite-dimensional, it is seen that all the norms are equivalent.

Proof: Using this remark, it suffices to prove the boundedness of $s_{\tilde{u}}^{\infty}$ in the space H.

We have

$$\begin{aligned} ||s_{\tilde{u}}^{\infty}||_H &\leq ||s_{\tilde{u}}^{\infty}-\tilde{u}||_H + ||\tilde{u}-\hat{u}||_H + ||\hat{u}||_H \\ &\leq \left(\sum_{i=1}^{n} \ell_i^2(s_{\tilde{u}}^{\infty}-\tilde{u}) + r_n ||s_{\tilde{u}}^{\infty}-\tilde{u}||_B^2 \right)^{1/2} + \delta + ||\hat{u}||_H \\ &\leq \left[\sum_{i=1}^{n} \ell_i^2(\tilde{u}) + r_n(||s_{\tilde{u}}^{\infty}||_B + ||\tilde{u}||_B) \right]^{1/2} + \delta_0 + ||\hat{u}||_H \\ &\leq [||\tilde{u}||_H^2 + \varepsilon_{n,\delta} + r_n(||s_{\tilde{u}}^{\infty}||_B + ||\tilde{u}||_B)^2]^{1/2} + \delta_0 + ||\hat{u}||_H. \end{aligned}$$

where we have used the fact that for any $u \in N$,

$$\sum_{i=1}^{n} \ell_i^2(s_{\tilde{u}}^{\infty}-\tilde{u}) \leq \sum_{i=1}^{n} \ell_i^2(u-\tilde{u}),$$

and put $u = 0$. This leads to the required inequality

$$||s_{\tilde{u}}^{\infty}||_H \leq \text{const},$$

where the const does not depend on n and δ. □

Lemma 40. If u' is the projection of the element $\hat{u}$ onto the subspace N, then

$$\lim_{\delta\to 0, n\to\infty} \sum_{i=1}^{n} \ell_i^2(s_{\tilde{u}}^{\infty}-\tilde{u}) = ||u'-\tilde{u}||_H^2. \tag{20}$$

Proof: We have

$$\sum_{i=1}^{n} \ell_i^2(s_{\tilde{u}}^{\infty}-\tilde{u}) \leq \sum \ell_i^2(u'-\tilde{u}) \leq ||u'-\tilde{u}||_H^2 + r_n||u'-\tilde{u}||_H^2.$$

Therefore,

$$\overline{\lim_{\delta\to 0, n\to\infty}} \sum_{i=1}^{n} \ell_i^2(s_{\tilde{u}}^{\infty}-\tilde{u}) \leq ||u'-\tilde{u}||_H. \tag{21}$$

Next,

$$\begin{aligned} ||u'-\hat{u}||_H &\leq ||s_{\tilde{u}}^{\infty}-\hat{u}||_H \leq ||s_{\tilde{u}}^{\infty}-\tilde{u}||_H + \delta \\ &\leq \left(\sum_{i=1}^{n} \ell_i^2(s_{\tilde{u}}^{\infty}-\tilde{u}) + r_n||s_{\tilde{u}}^{\infty}-\tilde{u}||_B^2\right)^{1/2} + \delta. \end{aligned}$$

Using the uniform boundedness of $||s_{\tilde{u}}^{\infty}||_B$ in n and δ, we obtain

$$||u'-\hat{u}||_H \leq \underline{\lim_{\delta\to 0, n\to\infty}} \sum_{i=1}^{n} \ell_i^2(s_{\tilde{u}}^{\infty}-\tilde{u}). \tag{22}$$

Comparing (21) and (22), we obtain (20).

We assume now that

$$||u'-\hat{u}||_H > 0 \tag{23}$$

(the case $\hat{u} = u'$ will be considered separately).

Using (10) and (13), we obtain

$$0 \leq \rho^2(\alpha) = \sum_{i=1}^{n} \ell_i^2(s_{\tilde{u}}^{\infty}-\tilde{u}) \leq \sum_{i=1}^{n} \ell_i^2(s_{\tilde{u}}^{\infty}-\tilde{u}), \quad 0 \leq \alpha \leq +\infty.$$

It is possible to prove (see Section 9) that the function $\tilde{\rho}$ is strictly increasing and continuous. Therefore, for each ρ,

$$0 \leq \rho^2 < \sum_{i=1}^{\infty} \ell_i^2(s_{\tilde{u}}^{\infty}-\tilde{u})$$

the equation

$$\tilde{\rho}(\alpha) = \rho$$

has a unique solution α_ρ; $\tilde{\rho}(\alpha_\rho) = \rho$.

Let $\rho = \rho_{n,\delta}$ be such that

$$\delta^2 + r_n||\hat{u}-\tilde{u}||_B \le \rho^2 < \sum_{i=1}^{n} \ell_i^2(s_{\tilde{u}}^{\infty}-\tilde{u}), \qquad \lim_{n\to\infty,\delta\to 0} \rho = 0.$$

The possibility of choosing such ρ for sufficiently small δ and large n follows from Lemma 40 and the condition (23).

We denote by α_ρ the unique root of the equation

$\tilde{\rho}(\alpha) = \rho$.

The following theorem holds.

Theorem 76. If $s_{n,\delta} = s_{\tilde{u}}^{\alpha_\rho}$, then

$$\lim_{\delta\to 0,n\to\infty} ||s_{n,\delta} - \hat{u}||_L = 0 ;$$

that is, the algorithm $S_{\delta,n}\tilde{u} = s_{n,\delta}$ is smoothing.

Remark. If n is sufficiently large, so that one can ignore the approximation error compared to the quantity δ, we can then assume that $\rho_{n,\delta} = \delta$.

Proof: For any u D_L we have

$$\sum_{i=1}^{n} \ell_i^2(s_{\tilde{u}}^{\alpha}-\tilde{u}) + \alpha||Ls_{\tilde{u}}^{\alpha}||^2 \le \sum_{i=1}^{n} \ell_i^2(u-\tilde{u}) + \alpha||Lu||^2. \tag{24}$$

Assuming in (24) that $\alpha = \alpha_\rho$ and $u = \hat{u}$, we obtain

$$\begin{aligned}\rho^2 + \alpha_\rho||Ls_{n,\delta}||^2 &\le \sum_{i=1} \ell_i^2(\hat{u}-\tilde{u}) + \alpha_\rho||L\hat{u}||^2 \\ &\le ||\hat{u}-\tilde{u}||^2 + r_n||\hat{u}-\tilde{u}||_B^2 + \alpha_\rho||L\hat{u}||^2 \le \rho^2 + \alpha_\rho||L\hat{u}||^2\end{aligned}$$

and therefore

$$||Ls_{n,\delta}|| \le ||L\hat{u}||. \tag{25}$$

Next, using (19), we obtain

$$\begin{aligned}||s_{n,\delta}-\hat{u}||_H &\le ||s_{n,\delta}-\tilde{u}||_H + ||\tilde{u}-\hat{u}||_H \\ &\le \delta + \left(\sum_{i=1}^{n} \ell_i^2(s_{n,\delta}-\tilde{u}) + r_n||s_{n,\delta}-\tilde{u}||_B^2\right)^{1/2} \\ &\le \delta + \left[\rho^2 + r_n(||s_{n,\delta}||_B + ||\tilde{u}||_B)^2\right]^{1/2}.\end{aligned}$$

We show that $||s_{n,\delta}||_B$ are uniformly bounded in n and δ. In fact, using the imbedding of the space H_L into B and the relation (25), we have

$$||s_{n,\delta}||_B^2 \leq M^2(||s_{n,\delta}||_H^2 + ||Ls_{n,\delta}||_G^2) \leq M^2\Big(\sum_{i=1}^n \ell_i^2(s_{n,\delta}) + r_n||s_{n,\delta}||_B^2 + ||L\hat{u}||_G^2\Big).$$

Then we have

$$\sum_{i=1}^n \ell_i^2(s_{n,\delta}) \leq \left\{\left[\sum_{i=1}^n \ell_i^2(s_{n,\delta}-\tilde{u})\right]^{1/2} + \left[\sum_{i=1}^n \ell_i^2(\tilde{u})\right]^{1/2}\right\}^2 \leq [\rho + (||\tilde{u}||_H^2 + \varepsilon_{n,\delta})^{1/2}]^2$$

and therefore

$$||s_{n,\delta}||_B^2 \leq M^2\Big\{[\rho + (||\tilde{u}||_H^2 + \varepsilon_{n,\delta})^{1/2}]^2 + r_n||s_{n,\delta}||_B^2 + ||L\hat{u}||^2\Big\}.$$

Taking into account that r_n, ρ and $\varepsilon_{\delta,n}$ tend to zero as $n \to \infty$ and $\delta \to 0$, we obtain that the norms of the elements $s_{n,\delta}$ are bounded in B.

But then

$$\lim_{\delta\to 0, n\to\infty} ||s_{n,\delta}-\hat{u}||_H = 0. \tag{26}$$

Using standard considerations, we derive from Eqs. (25), (26) the required assertion. □

Remark. Arguing in a similar way, we can show that for any $v \in B$

$$\lim_{n\to\infty} ||s_v^\alpha - u_v^\alpha||_L = 0,$$

where u_v^α is a solution of the problem (1); that is, the sequence $\{s_v^\alpha\}$ with respect to the parameter n can be regarded as a convergent minimizing sequence for the functional $\phi_\alpha[u;v]$, $u \in D_L$.

We now consider the case where

$$||u'-\hat{u}||_H = 0;$$

that is, $\hat{u} \in N$, which we have earlier excluded from our consideration.

Theorem 77. The following relation holds:

$$\lim_{\delta\to 0,\alpha,n\to\infty} ||s_{\tilde{u}}^\alpha - \hat{u}||_L = 0.$$

Proof: Assuming that $u = \hat{u}$ in Eq. (24), we obtain

$$\sum_{i=1}^{n} \ell_i^2(s_{\tilde{u}}^{\alpha}-\tilde{u}) + \alpha||Ls_{\tilde{u}}^{\alpha}||^2 \leq \sum_{i=1} \ell_i^2(\hat{u}-\tilde{u})$$

and therefore

$$\sum_{i=1}^{n} \ell_i^2(s_{\tilde{u}}^{\alpha}-\tilde{u}) \leq \sum_{i=1}^{n} \ell_i^2(\hat{u}-\tilde{u}), \qquad ||Ls_{\tilde{u}}^{\alpha}|| \leq \frac{1}{\alpha} \sum_{i=1}^{n} \ell_i^2(\hat{u}-\tilde{u}).$$

It is easy to show that

$$\lim_{n\to\infty,\delta\to 0} \sum_{i=1}^{n} \ell_i^2(\hat{u}-\tilde{u}) = 0.$$

Hence

$$\lim_{\delta\to 0,\alpha,n\to\infty} ||Ls_{\tilde{u}}^{\alpha}|| = 0. \tag{27}$$

Next we have

$$||s_{\tilde{u}}^{\alpha}-\hat{u}||_H \leq ||s_{\tilde{u}}^{\alpha}-\tilde{u}||_H + \delta \leq \left(\sum_{i=1}^{n} \ell_i^2(s_{\tilde{u}}^{\alpha}-\tilde{u}) + r_n||s_{\tilde{u}}^{\alpha}-\tilde{u}||_B^2\right)^{1/2} + \delta$$
$$\leq \left[\sum_{i=1}^{n} \ell_i^2(\hat{u}-\tilde{u}) + r_n(||s_{\tilde{u}}^{\alpha}||_B + ||\tilde{u}||_B)^2\right]^{1/2} +$$

If $||s_{\tilde{u}}^{\alpha}||_B$ is bounded uniformly in n, δ and α, then

$$\lim_{\delta\to 0,n,\alpha\to\infty} ||s_{\tilde{u}}^{\alpha}-\hat{u}||_H = 0. \tag{28}$$

The assertion of the theorem follows from Eqs. (27), (28). We can prove the assumption made in the same way as the similar assertion in Theorem 76. Therefore we omit the proof. □

7. Let ℓ be any linear functional on H_L whose value we need to calculate on the element $\hat{u} \in D_L$. We assume that instead of $\hat{u}$ we know its δ-approximation in H; that is, the element $\tilde{u}$. Since the expression $\ell(\tilde{u})$ may not make sense, it is natural to take $\ell(s_{\tilde{u}}^{\alpha})$ as an approximation to $\ell(\hat{u})$ for the appropriate choice of the parameter α (see Theorems 76 and 77). This is a stable technique for approximating any linear functional on H_L. From the results obtained in Subsection 5, one might expect that this technique of approximating the functionals is close to optimal.

Since

$$\ell(s_{\tilde{u}}^{\alpha}) = \sum_{i=1}^{n} \ell_i(s_{\tilde{u}}^{\alpha})\ell_i(v) + (Ls_{\tilde{u}}^{\alpha}, Lv),$$

where $v \in D_L$ is known, in order to calculate $\ell(s_{\tilde{u}}^{\alpha})$ it suffices to know only the values of the functionals $\ell_i(s_{\tilde{u}}^{\alpha})$, $i = 1,2,\ldots,n$, and of the operator $Ls_{\tilde{u}}^{\alpha}$. The technique for determining these values will be considered below (without direct determination of the element $s_{\tilde{u}}^{\alpha}$). A similar remark obviously holds in the case where one needs to calculate only the value of the operator L on the spline $s_{\tilde{u}}^{\alpha}$. Calculating $\ell_i(s_{\tilde{u}}^{\alpha})$ and $Ls_{\tilde{u}}^{\alpha}$ without direct calculation of the spline $s_{\tilde{u}}^{\alpha}$ essentially reduces the calculations while seeking appropriate values of the parameter α (for example, those satisfying the residue criterion $\tilde{\rho}(\alpha) = \rho$).

The condition for choosing the level of the residue ρ, namely,

$$\delta^2 + r_n||\hat{u}-\tilde{u}||_B^2 \leq \rho^2,$$

is rather difficult in practice because it depends on the deviation $||\hat{u}-\tilde{u}||_B$, which is not always easily estimable. It is natural to indicate when we may assume that $\rho = \delta$, thus simplifying the problem of choosing ρ. We assume that the system of functionals $\{\ell_i : i = 1,2,\ldots\}$, is complete in H, that is, the system $\{k_i : i = 1,2,\ldots\}$ is complete in H and, furthermore,

$$||u||_H^2 = \sum_{i=1}^{\infty} \ell_i^2(u), \qquad \text{for all } u \in H.$$

Let $\alpha = \alpha_\delta$ be defined from the condition

$$\tilde{\rho}(\alpha) = \delta.$$

We assume $s_{n,\delta} = s_{\tilde{u}}^{\alpha_\delta}$. Then, obviously,

$$\left(\sum_{i=1}^{n} \ell_i^2(s_{n,\delta}-\hat{u})\right)^{1/2} \leq \left(\sum_{i=1}^{n} \ell_i^2(s_{n,\delta}-\tilde{u})\right)^{1/2} + \left(\sum_{i=1}^{n} \ell_i^2(\tilde{u}-\hat{u})\right)^{1/2} \leq \delta + \delta = 2\delta. \tag{29}$$

Assuming $u = \hat{u}$, $\alpha = \alpha_\delta$ in Eq. (24), we have

$$\delta^2 + \alpha_\delta||Ls_{n,\delta}||^2 \leq \sum_{i=1}^{n} \ell_i^2(\hat{u}-\tilde{u}) + \alpha_\delta||L\hat{u}||^2 \leq ||\hat{u}-\tilde{u}||^2 + \alpha_\delta||L\hat{u}||^2$$

and therefore

$$||Ls_{n,\delta}||_G^2 \leq ||L\hat{u}||_G^2. \tag{30}$$

It follows from (29) that

$$s_{n,\delta} \xrightarrow{\text{weakly}} \hat{u} \text{ (in } H) \text{ as } \delta \to 0, n \to \infty. \tag{31}$$

Using (30) and (31), we obtain in the usual way that

$$\lim_{n\to\infty,\delta\to 0} ||s_{n,\delta}-\hat{u}||_L = 0.$$

The assertion is proved.

The conditions imposed on the system of functionals $\{\ell_i\}$ are satisfied a priori if the system $\{k_i\}$ is orthonormal on H.

8. We now consider the generalized method of interpolation (Section 20) to be used in the general case. Let a system of functionals $\{\ell_i^{(n)}\}$ be complete in the sense that the closure of the span of elements $\{k_i^{(n)},$ $i = 1,2,\ldots,n\ldots\}$ coincides with the space H. Let $\hat{u} \in B$ and let

$$|\ell_i^{(n)}(\hat{u}) - \ell_i^{(n)}(\tilde{u})| \leq \delta, \quad i = 1,\ldots,n. \tag{32}$$

We define $\hat{u}_\delta$ to be a solution of the problem

$$\inf_{u\in N_\delta} ||Lu|| = ||L\tilde{u}_\delta||, \tag{33}$$

where $N_\delta = \{u \in D_L: |\ell_i^{(n)}(u) - \ell_i^{(n)}(\tilde{u})| \leq \delta\} \neq \emptyset$ by (32). We can prove the existence of a solution (at least one solution) of the variational problem (33) in the following way.

Let $\tilde{m} = \inf_{u\in N_\delta} ||Lu||$. Also, let $\{u_s\} \subset D_L$ be a sequence such that

$$\tilde{m} \leq ||Lu_s|| \leq \tilde{m} + \frac{1}{s}, \quad s = 1,2,\ldots .$$

It is obvious that $|\ell_i(u_s)| \leq \delta + |\ell_i(\tilde{u})| \leq \delta + \max_i|\ell_i(u)|$, that is, $\Sigma_{i=1}^n \ell_i^2(u_s) < +\infty$ uniformly in s. Then, however, $||u_s||_n < +\infty$ uniformly in s. This yields weak compactness of $\{u_s\}$ in H_n as well as H_L. Assuming that the sequence $\{u_s\}$ converges weakly in H_L (if need be we choose a subsequence) to some element $\hat{u}_\delta \in H_L$, we have, because L is closed, that $\hat{u}_\delta \in D_L$, $\hat{u}_\delta \in N_\delta$; that is, is an "admissible" element and, in addition,

$$\tilde{m} \leq ||L\hat{u}_\delta|| \leq \lim_{s\to\infty} ||Lu_s|| = \tilde{m}.$$

This implies that the set of solutions of the problem (33) is nonempty.

The assertion is proved.

We can guarantee the uniqueness of the elements $\hat{u}_\delta$ if the null space of the operator L consists only of the zero element.

Proof of the convergence method follows the scheme given in Subsection 7 of this section.

We note that while constructing the interpolating spline $s_{\hat{u}_\delta}$ for the element $\hat{u}_\delta$, we convince ourselves that $s_{\hat{u}_\delta} \in N_\delta$ and $||Ls_{\hat{u}_\delta}|| \leq ||Ls_{\hat{u}_\delta}||$; that is, one can seek a solution of the problem (33) only on the set of interpolating splines satisfying the admissibility condition $|\ell_i(u) - \ell_i(\hat{u})| \leq \delta$, $i = 1,\dots,n$.

A similar remark holds also for the case of generalized splines (2).

We note that all the results given above still hold if we consider instead of the set D_L a convex set $D \subseteq D_L$ on which the operator L is closed. This remark enables us to take into account the presence of known boundary conditions imposed on the element $\hat{u}$. The main results will also hold if we replace the form $||Lu||^2$ by $||Lu-g||^2$, where g is a fixed element of G.

It is not hard to see that the method (33) defines some smoothing algorithm $\hat{S}_\delta$. We can show that this algorithm is optimal in the sense close to the one considered in Section 19. We need to determine the estimate

$$\tilde{\omega}_B(\delta,R) = \sup_u ||Bu||_V, \quad \text{where} \quad u \in D_L, \quad \max_{1\leq i\leq r} |\ell_i(u)|, \quad ||Lu||_G \leq R.$$

If the algorithm $T = T(\ell_1(\tilde{u}),\dots,\ell_n(\tilde{u}))$ is any mapping of a n-dimensional Euclidean space into D_L, it is easy to prove the relation (2) from Section 19; that is,

$$\tilde{\omega}_B(\delta,R;T) \geq \tilde{\omega}_B(\delta,R),$$

where

$$\tilde{\omega}_B(\delta,R;T) = \sup_{\substack{\xi=\tilde{u}-\hat{u} \\ |\ell_i(\xi)|\leq\delta \\ \hat{u}\in\hat{U}_R}} ||B\hat{u} - T(\ell_1(\tilde{u}),\dots,\ell_n(\tilde{u}))||_V.$$

The proof of this fact is similar to that of Theorem 65.

We note that a solution of the problem (33) has the following properties:

$$|\ell_i(\hat{u}_\delta) - \ell_i(\hat{u})| \leq 2\delta, \quad ||L(\hat{u}_\delta - \hat{u})|| \leq 2R.$$

Therefore, we have the error estimate

$$||B\hat{u} - B\hat{u}_\delta||_V \leq 2\tilde{\omega}_B(\delta, R).$$

In all the consideration above the number of functionals n is assumed to be fixed.

Section 22. Approximate Solution of Operator Equations Using Splines

1. Let A be a linear operator with domain $D_A \subseteq H$ into F, where H and F are Hilbert spaces. We assume that the operator A has a bounded inverse defined on all of F so the equation

$$Au = f \tag{1}$$

has a unique solution for any $f \in F$.

We define another linear and closed operator $L: H \to F$ with domain $D_L \subseteq D_A$, $\overline{D}_L = H$ and assume the following conditions:

1) $\hat{u} := A^{-1}f \in D_L$;

2) $||Au||_F^2 \leq \gamma^2(||u||_H^2 + ||Lu||_F^2)$, $u \in D_L$,

where $\gamma > 0$ is constant.

In D_L we give n linearly independent (in the norm $||u||_L = (||u||^2 + ||Lu||^2)^{1/2}$ linear functionals $\ell_i = \ell_i^{(n)}$, $i = 1,\ldots,n$. Also, we assume, as we did in Section 21, that:

3) the null space $N = \{u \in D_L: Lu = 0\}$ of the operator L has finite dimension q;

4) $n \geq q$ and, if $u \in N$ and $\ell_i(u) = 0$, $i = 1,\ldots,n$, then $u = 0$;

5) the approximation condition is satisfied: for each $u \in D_L$,

$$\left| ||\ell(u)||^2 - ||u||_H^2 \right| \leq r_n^2 ||u||_L^2, \tag{2}$$

where $r_n \to 0$ as $n \to \infty$, and the r_n do not depend on $u \in D_L$, $\ell(u) = (\ell_1(u),\ell_2(u),\ldots,\ell_n(u))$, $||\ell(u)||^2 = \sum_{i=1}^{n} \zeta_{ij}\ell_i(u)\ell_j(u)$, $\zeta_{ij} = \zeta_{ij}^{(n)}$, $i,j = 1,\ldots,n$ form a positive definite matrix ζ (in the particular case $\zeta_{ij} = 0$, $i \neq j$, so that the matrix ζ is diagonal, this reduces to the scheme given in the previous section);

6) if $\hat{g} = (g_1,g_2,\ldots,g_n)$ is an arbitrary vector of dimension n, there exists at least one element $g \in D_L$ such that $\ell(g) = \hat{g}$, that is,

an interpolating element (an interpolant).

Let $U_{\hat{g}} = \{g \in D_L: \ell(g) = \hat{g}\}$ and let $s_{\hat{g}}$ be a spline interpolating the values $\hat{g}$. If there exists $g \in D_L$ such that $\ell(g) = \hat{g}$, the given vector, we assume that $s_{\hat{g}} = s_g$. Let

$$S_n = \{s_{\hat{g}}: -\infty < g_i < +\infty, \quad i = 1,\ldots,n\}.$$

The basic problem to be considered is the following: Define the spline $\hat{s}_n \in S_n$ satisfying the condition

$$\inf_{s_n \in S_n} ||As_n - f||_F^2 = ||A\hat{s}_n - f||_F \tag{3}$$

and investigate its approximative properties in relation to a solution of Eq. (1).

Let $e_i = (0,\ldots,1,\ldots,0)$, $i = 1,\ldots,n$ (1 is used in the ith entry). Let $s_i = s_{e_i}$. The spline s_{e_i} is said to be a *basis spline*.

Theorem 78. We can express each spline $s_{\hat{g}}$ uniquely as a linear combination of basis splines:

$$s_{\hat{g}} = \sum_{k=1}^{n} g_k s_k. \tag{4}$$

Proof: The fact that $s_{\hat{g}}$ is representable as (4) follows from the fact that $\ell_i(\Sigma_{k=1}^{n} g_k s_k) = g_i = \ell_i(s_{\hat{g}})$, $i = 1,\ldots,n$, and from the Euler equation

$$(Ls, Lv)_F = 0, \quad \forall v \in D_L \text{ such that } \ell(v) = 0,$$

which holds for any spline s. We prove that the representation (4) is unique. To this end, it suffices to prove that the basis splines s_i are linearly independent in H_L or H_n respectively (see Section 21).

Let $\Sigma_{k=1}^{n} g_k s_k = 0$ and let $g_{i_0} \neq 0$. We have

$$0 = ||\sum_{k=1}^{n} g_k s_k||_n^2 = \sum_{i=1}^{n} \ell_i^2 \left(\sum_{k=1}^{n} g_k s_k\right) + ||L(\sum_{k=1}^{n} g_k s_k)||^2$$

$$\geq \sum_{i=1}^{n} g_i^2 \geq g_{i_0}^2 > 0.$$

This contradiciton proves linear independence of s_i, $i = 1,\ldots,n$ in H_n as well as in H_L. □

Corollary. S_n is a linear space of dimension n whose basis is given by the basis splines s_i, $i = 1,\ldots,n$.

In order to determine the coefficients $g_1, g_2, \ldots, g_n$ of the solution $\hat{s}_n$ of the problem (3), we have the following system of linear algebraic equations:

$$\sum_{j=1}^{n} g_j (As_i, As_j)_F = (As_i, f)_F, \quad i = 1,2,\ldots,n. \tag{5}$$

Theorem 79. The problem (5) has a unique solution, with

$$||\hat{u}-\hat{s}_n||_H + ||A\hat{u}-A\hat{s}_n||_F \to 0, \quad n \to \infty, \tag{6}$$

where $\hat{s}_n = \Sigma_{j=1}^n g_j s_j$ and, in addition, g_j is defined as a solution of (5); that is, the spline method is convergent.

Proof: For the problem (5) to have a unique solution, it suffices to show that the system of basis splines As_i is linearly independent; that is, the elements As_i, $i = 1,\ldots,n$ are linearly independent in F. Assuming this is not true, we obtain, because A is invertible, that the basis splines are linearly dependent in H_L, which is impossible by what has been proved above.

Next we have, using the extremal property of the element $\hat{s}_n$,

$$||\hat{u}-\hat{s}_n||_H^2 \le ||A^{-1}||^2 ||A\hat{s}_n-f||_F^2 \le ||A^{-1}||^2 ||As_{\hat{u}}-f||_F^2,$$

where $s_{\hat{u}}$ is the spline defined by values of the functionals $\ell_1(\hat{u})$, $\ell_2(\hat{u}),\ldots,\ell_n(\hat{u})$ on the exact solution of (1). Using the condition (2), we derive then

$$||\hat{u}-\hat{s}_n||_H \le ||A^{-1}|| \; ||A\hat{s}_n-f|| \le \gamma ||A^{-1}|| \; ||s_{\hat{u}}-\hat{u}||_L \to 0 \quad \text{as} \quad n \to \infty \tag{7}$$

due to convergence of interpolating splines (Section 22). □

3. We examine now the rate of convergence of the spline method for solving the operator equation (1). It is obvious that this rate is essentially determined by the rate of convergence of the spline $s_{\hat{u}}$ which interpolates the solution $\hat{u}$ of Eq. (1). Hence, we shall examine the latter.

Obviously, we have:

$$||Ls_{\hat{u}}||_F \le ||L\hat{u}||F, \quad \text{for all} \quad \hat{u} \in D_L. \tag{8}$$

On the other hand, using the condition (2), we obtain, assuming in (2) that $u = \hat{u}-s_{\hat{u}}$:

$$||\hat{u}-s_{\hat{u}}||_H^2 \leq r_n^2(||\hat{u}-s_{\hat{u}}||_H^2 + ||L(\hat{u}-s_{\hat{u}})||_F^2);$$

that is, for sufficiently large n,

$$||\hat{u}-s_{\hat{u}}||_H^2 \leq \frac{r_n^2}{1 - r_n^2} ||L(\hat{u}-s_{\hat{u}})||_F^2 = q_n^2 \to 0 \quad \text{as} \quad n \to \infty. \tag{9}$$

We have thus obtained a representation of the rate of convergence of the spline $s_{\hat{u}}$ to the element $\hat{u}$ in the space H.

Assume that $\hat{u} \in D_{L*L}$; that is, the element $\hat{u}$ has a higher degree of smoothness. Then, using the Euler equation we have

$$\begin{aligned} ||L(\hat{u}-s_{\hat{u}})||_F^2 &= (\hat{u}-s_{\hat{u}}, L^*L\hat{u})_H + (L\hat{u}-Ls_{\hat{u}}, Ls_{\hat{u}})_F \\ &= (L^*L\hat{u}, \hat{u}-s_{\hat{u}})_H \leq ||L^*L\hat{u}||_H ||\hat{u}-s_{\hat{u}}||_H. \end{aligned}$$

Taking (9) into account, we have

$$q_n^2 \leq \frac{r_n^4}{(1 - r_n^2)^2} ||L^*L\hat{u}||_H^2, \quad \text{for all} \quad \hat{u} \in D_{L*L}, \tag{10}$$

$$||L\hat{u}-Ls_{\hat{u}}||_F^2 \leq ||L^*L\hat{u}|| q_n, \quad \text{for all} \quad \hat{u} \in D_{L*L}. \tag{11}$$

Also, we note that the Pythagorian equality implies

$$||L\hat{u}-Ls_{\hat{u}}||_F \leq ||L\hat{u}||_F, \quad \text{for all} \quad \hat{u} \in D_L. \tag{12}$$

Now let E_λ, $0 \leq \lambda < +\infty$, be the spectral decomposition of the self-adjoint operator L^*L. Define for $u \in D_L$ the quadratic form

$$||Lu||_F^2 = \int_{\lambda_0}^{\infty} \lambda(dE_\lambda u,u) < +\infty, \qquad \lambda_0 \geq 0.$$

We denote by L_ϕ an arbitrary operator of H into F for which the quadratic form

$$||L_\phi u||_F^2 = \int_{\lambda_0}^{\infty} \phi(\lambda)(dE_\lambda u,u),$$

is finite for $u \in D_\phi = D_L$, where $\phi(\lambda)$, $\lambda \geq \lambda_0$ is a continuous positive function satisfying the following conditions:

(a) $\phi(\lambda)$, $\lambda \geq \lambda_0$ is strictly increasing;
(b) $\phi^{-1}(\lambda)$ is downward convex for $\lambda \geq \lambda_0$;
(c) the function $\lambda^2\phi(1/\lambda^2)$ is increasing.

The following theorem holds.

Theorem 80. Under the above conditions on the function ϕ we have

$$||L(\hat{u}-s_{\hat{u}})||_F^2 \leq q_n^2\phi(||L(\hat{u}-s_{\hat{u}})||_F^2/q_n^2) \leq q_n^2\phi\left(\frac{||L\hat{u}||_F^2}{q_n^2}\right), \tag{13}$$

where q_n^2 is defined by (9) if it is only known that the element $\hat{u} \in D_L$, and by (10) when $\hat{u} \in D_{L*L}$. In particular, if $\phi(\lambda) = \lambda^\sigma$, $0 < \sigma < 1$, conditions (a) - (c) are satisfied and, furthermore, we have

$$||L_\phi(\hat{u}-s_{\hat{u}})||_F \leq q_n^{1-\sigma}||L(\hat{u}-s_{\hat{u}})||_F^\sigma \leq q_n^{1-\sigma}||L\hat{u}||_F^\sigma. \tag{14}$$

Proof: For any $u \in D_L$, applying Jensen's inequality, we obtain

$$\phi^{-1}\left(\frac{\int_{\lambda_0}^\infty \phi(\lambda)(dE_\lambda u,u)}{\int_{\lambda_0}^\infty (dE_\lambda u,u)}\right) \leq \frac{\int_{\lambda_0}^\infty \lambda(dE_\lambda u,u)}{\int_{\lambda_0}^\infty (dE_\lambda u,u)}$$

and therefore, using (a) - (c):

$$||L_\phi u||_F^2 = \int_{\lambda_0}^\infty \phi(\lambda)(dE_\lambda u,u) \leq \int_{\lambda_0}^\infty (dE_\lambda u,u)\,\phi\left(\frac{||Lu||_F^2}{\int_{\lambda_0}^\infty (dE_\lambda u,u)}\right)$$

$$\leq ||u||_H^2\phi\left(\frac{||Lu||_F^2}{||u||_H^2}\right). \tag{15}$$

Assuming here that $u = \hat{u}-s_{\hat{u}}$ and also, using (10) - (12), we derive (13), (14). □

Remark. We can apply the inequality (15) in order to estimate the computational error of the operator L_ϕ if we assume that $u = \hat{u}_\delta - u_\delta$, where $\hat{u}_\delta$ is the smoothed value of the element $u_\delta : ||\hat{u}-u_\delta|| \leq \delta$. Then, for the smoothing algorithms considered above,

$$||\hat{u}_\delta-\hat{u}||_H \leq 2\delta, \qquad ||L\hat{u}_\delta-L\hat{u}||_F \leq 2||L\hat{u}||_F \leq 2R$$

and therefore

$$||L_\phi(\hat{u}_\delta-\hat{u})||_F^2 \leq 4\delta^2\phi\left(\frac{R^2}{\delta^2}\right). \tag{16}$$

For $\phi(\lambda) = \lambda^\sigma$ the estimate (16) assumes the form

$$||L_\phi(\hat{u}-\hat{u}_\delta)||_F^2 \leq 4\delta^2\left(\frac{R^2}{\delta^2}\right)^\sigma = 4\delta^{2(1-\sigma)}R^{2\sigma}.$$

We note that the case $\phi(\lambda) = \lambda^\sigma$ corresponds to the differentiation operator.

The following assertion is a corollary of our analysis above: if the solution $\hat{u}$ of the problem (1) is such that $\hat{u} \in D_{L*L}$, we have

$$||A\hat{s}_n - f||_H \leq O(r_n^2) \quad \text{as} \quad n \to \infty.$$

If the condition

$$||Au||_F^2 \leq \gamma_\phi^2(||u||_H^2 + ||L_\phi u||_F^2), \quad u \in D_\phi, \quad \gamma_\phi = \text{const}$$

is satisfied, we can, using the estimates (13), (14) proved above, obtain estimates of the rate of convergence of the deviation $||\hat{u}-\hat{s}_n||_H$ to zero. We omit them here because they are immediate.

Let the quadratic form $||Au||_F^2$ be equivalent to the quadratic form $||u||_L^2$; that is,

$$\gamma_0^2||u||_L^2 \leq ||Au||_F^2 \leq \gamma^2||u||_L^2, \quad \text{for all} \quad u \in D_L$$

where $||u||_L^2 = ||u||_H^2 + ||Lu||_F^2$ and γ_0 and γ are positive constants. Using the approximation condition (2), it is easy to show that for sufficiently large n the spaces H_L and H_n having the norm $||u||_n$ defined by

$$||u||_n^2 = ||\ell(u)||^2 + ||Lu||_F^2, \quad \text{for all} \quad u \in D_L,$$

are equivalent, that is; there exist positive m and M (not depending on $u \in D_L$) such that

$$m||u||_n \leq ||u||_L \leq M||u||_n, \quad \text{for all} \quad u \in D_L.$$

Then, however, the form $||Au||_F^2$ is equivalent to the form $||u||_n^2$ by the inequalities

$$m^2\gamma_0^2 \leq ||Au||_F^2 \leq M^2\gamma^2||u||_n^2.$$

Let $Z = \hat{u}-\hat{s}_n$. Then, using the extremal property of the element $\hat{s}_n$, we obtain

$$||A\hat{u}-A\hat{s}_n||_F^2 \leq ||A\hat{u}-As_{\hat{u}}||_F^2 \leq M^2\gamma^2||\hat{u}-s_{\hat{u}}||_n^2 = M^2\gamma^2||L\hat{u}-Ls\hat{u}||_F^2.$$

On the other hand, using Theorem 74, we have:

$$||A\hat{u}-A\hat{s}_n||_F^2 \geq m^2\gamma_0^2||\hat{u}-\hat{s}_n||_n^2 \geq m^2\gamma_0^2||\hat{u}-s_{\hat{u}}||_n^2 = m^2\gamma_0^2||L\hat{u}-Ls_{\hat{u}}||_F^2$$

where $s_{\hat{u}}$ is the spline defined by the element $\hat{u}$.

The above relations yield an important two-sided relation:

$$m\gamma_0 ||L\hat{u}-Ls_{\hat{u}}||_F \leq ||A\hat{u}-A\hat{s}_n||_F \leq M\gamma ||L\hat{u}-Ls_{\hat{u}}||_F,$$

which shows that the residue of the solution obtained by means of the spline method is completely defined by the accuracy of the approximation of the value of the operator L on the solution $\hat{u}$ of (1) by the value of the operator L on the interpolating spline $s_{\hat{u}}$.

4. We dwell now on some conditions for the given approach to be realized. For example, let

$$||Au||_F^2 = ||A_0u||_F^2 + ||A_1u||_F^2,$$

where the quadratic form $||A_0u||_F^2$ defines the "main" part in the following sense:

$$||A_1u||_F^2 \leq K^2(||u||_H^2 + ||A_0u||_F^2), \quad K > 0, \quad \text{constant}.$$

Then, obviously,

$$||Au||_F^2 \leq (1 + K^2)(||u||_H^2 + ||A_0u||_F^2)$$

and the inequality (2) is satisfied if $L = A_0$ is defined.

Next, we discuss the verification of conditions (3) - (6). As we shall see, in some cases it is not difficult to do.

For example, let

$$Lu = \frac{d^ku}{dx^k}, \quad k \geq 2, \quad a \leq x \leq b,$$

where the differentiation is understood in the Sobolev sense. Then $q = k$. We define on $[a,b]$ the mesh of nodes $x_i = a + (i-1)h_n$, $i = 1,\ldots,n+1$, $h_n = b-a/n$, $n > 1$. Also, let $\ell_i(u) = u(x_{i-1/2})$, where $x_i - 1/2 = a + (i-1/2)h_n$. The functionals $\ell_i(u)$ are defined for any sufficiently smooth function. The fact that (6) is satisfied follows from the existence of an interpolation polynomial passing through the given points $(x_i-1/2, u_i)$. Since the set N consists of polynomials of degree not higher than $k-i$, then (4) is satisfied, obviously, for $n \geq k$. The fact that (5) is satisfied can be verified directly; applying the Taylor formula up to the second order with the remainder in integral form, we obtain the estimate

$$\left|\sum_{i=1}^{n} h_n u^2(x_{i-1/2}) - \int_a^b u^2(x)dx\right| \le C_K h_n^2(||u||_{L_2}^2 + ||Lu||_{L_2}^2),$$

where C_k is a constant not depending on n.

Therefore, all the conditions are satisfied.

Next, we consider the choice of approximating functionals $\ell_i^{(n)}(u)$, $i = 1,\ldots,n$. It is seen from the above considerations that the accuracy of an approximate solution is essentially defined by the quantity r_n characterizing the approximation accuracy in the condition (2). It seems natural to choose a system of functionals $\ell_i(u)$, $i = 1,\ldots,n$ so that the order of approximation to be possibly higher. This can be achieved by satisfying the condition

$$\sup_{u\in D_1} \frac{|\,||\ell(u)||^2 - ||u||_H^2|}{||Lu||_F^2} \text{ minimized in } \ell_i, \zeta_{ij}. \tag{17}$$

If $H = L_2(\Omega)$, where Ω is a sufficiently regular domain in a finite-dimensional space and $\ell_i(u) = u(P_i)$, $P_i \in \Omega$, $i = 1,2,\ldots,n$, (that is, are values of sufficiently smooth functions at nodal points) the problem (17) is reduced, as is easily seen, to choosing an optimal quadrature formula. If the nodes $\{P_i\}$ are fixed, the problem is reduced to defining only coefficients of the best quadrature formula.

Section 23. Recovering the Solution of the Basic Problem From Approximate Values of the Functionals

1. Statement of the Problem. First we consider the basic problem (linear case). We assume that its characteristic is $(0,\nu_L)$; that is, the equation

$$Au = f \tag{1}$$

is solvable in the usual sense. Furthermore, we assume that $D = D_{AL} = D_A \cap D_L$. In addition to the main assumptions we require that the range Q_A of the operator A lie in some Banach space B imbedded in F; that is, $Q_A \subseteq B \subseteq F$ and for any element $f \in B$ we have the inequality

$$||f||_F \le k||f||_B. \tag{2}$$

We assume that the operator $A: H_{AL} \to B$ is bounded:

$$||Au||_B^2 \le k_1|u| \quad \text{for all } u \in H_{AL}, \tag{3}$$

where, as before, $|u|^2 = ||Au||^2 + ||Lu||^2$, $u \in D$, and H_{AL} is the Hilbert space consisting of the elements of D with scalar product

$$[u,v] = (Au,Av)_F + (Lu,Lv)_G, \quad \text{for all } u,v \in D; \text{ and}$$
$$|u| = [u,u]^{1/2}. \tag{4}$$

It is not hard to see that the joint closure of the operators A and L implies completeness of the space H_{AL}.

The right side of (1) is assumed to be in B; that is, $f \in B$. Let the linear functionals $\ell_i = \ell_i^{(n)}$, $i = 1,\ldots,n$, be defined on the space B for each n. The problem considered in this section consists of constructing a stable algorithm for approximate solution of the operator equation (1) under the condition that we are given values of the functionals $\ell_i(f) = f_i$ at an element $\tilde{f} \in B$ such that $||f-\tilde{f}||_F < \delta$. In this case the knowledge of the element $\tilde{f}$ is not assumed.

We note that the problem was considered earlier mainly under the assumption that $\tilde{f}$ is given and $B = F$. In Section 21 we considered the particular case with $H = F$ and $A = E$, the identity operator. The results obtained in Section 21 will be generalized here to a wider class of problems. Moreover, we obtain finite-dimensional problems for constructing an approximate solution of the basic problem which are, in our opinion, are rather convenient to implement.

Example. Let $H = F = G = L_2[a,b]$, $B = C[a,b]$. Also, let

$$(Au)(x) = \int_a^b K(x,\xi)u(\xi)d\xi = f(x), \quad a \le x \le b, \tag{5}$$

where K is a continuous function of its variables and f is a continuous function. The operator $L = d^Rx/dx^R$ is the operator of R-fold differentiation, defined on the set D of functions $u \in C^{(R-1)}[a,b]$ whose $(R-1)$th derivative is absolutely continuous and, in addition, $d^Ru/dx^R \in L_2$, $R \ge 1$.

Then, as a system of functionals $\{\ell_i\}$ one can take values of the continuous function $\tilde{f}$ at the nodes of some mesh $x_i \in [a,b]$, $i = 1,2,\ldots,n$. We assume that

$$||\tilde{f}-f||_{L_2} = \left[\int_a^b (f-\tilde{f})^2 dx\right]^{1/2} < \delta.$$

2. Additional conditions and auxiliary assertions. In the sequel we shall need the following conditions to be satisfied (compare with Section 21).

The approximation condition: for any $f \in B$,

$$\Big| ||\ell(f)|| - ||f||_F^2 \Big| \le r_n^2 ||f||_B^2, \tag{6}$$

where $r_n \to 0$ as $n \to \infty$ and the r_n do not depend on $f \in B$, and in addition, $||\ell(f)||$ is defined in Section 22.

The consistency condition:

$$\lim_{n\to\infty,\delta\to 0} r_n ||\delta f||_B = 0,$$

where $\delta f = f - \tilde{f}$.

Let

$$(u,v)_n = \sum_{i=1}^{n} \zeta_{ij} \ell_i(Au)\ell_j(Au) + (Lu,Lv), \quad u,v \in D,$$

$$||u||_n = (u,u)_n^{1/2}$$

The following assertions are valid (we assume that n is sufficiently large).

Lemma 41. There exists a constant $K_2 > 0$ such that

$$||Au||_B^2 \le K_2 ||u||_n^2 \quad \text{for all } u \in D. \tag{7}$$

In fact, using (4) and (6), we obtain

$$||Au||_B^2 \le K_1(||Au||_F^2 + ||Lu||_G^2) \le K_1\Big[(||\ell(Au)||^2 + r_n^2||Au||_B^2) + ||Lu||_G^2\Big].$$

Solving this inequality with respect to $||Au||_B^2$, we obtain (7).

Similarly, using (7), we obtain $|u|^2 \le M_n ||u||_n^2$ and $||u||^2 \ge m_n ||u||_n^2$, where m_n, $M_n \to 1$ as $n \to \infty$ and do not depend on $u \in D$. This implies the following.

Lemma 42. The Hilbert space H_n generated by the scalar product $(u,v)_n$, $u,v \in D$, and the space H_{AL} have equivalent norms.

3. Determination of approximate solutions by means of the residue method. Let $U_A = \{u \in D: A_u = f\} \ne \emptyset$. Then, as was shown in Section 1, the solution $\hat{u} \in U_A$ of (1) is uniquely defined by

$$||L\hat{u}-g||_G = \inf_{u\in U_A} ||Lu-g||_G$$

Now we define the set U_θ of "formal" approximate solutions of the problem (1):

$$U_\theta = \{u \in D: \rho_n(u) \leq \theta\},$$

where

$$\rho_n(u) \equiv ||\ell(Au-\tilde{f})||, \quad u \in D,$$

and the function $\theta = \theta(n,\delta)$ is such that $\rho_n(\hat{u}) < \theta$ and $\lim_{n\to\infty,\delta\to 0} \theta(n,\delta) =$
The possibility of choosing this function θ follows from the relation

$$\rho_n^2(\hat{u}) \leq ||f-\tilde{f}||_F^2 + r_n^2||\delta f||_B^2,$$

which is a consequence of the approximation condition (6). Then

$$\rho_n^2(\hat{u}) < \delta^2 + r_n^2||\delta f||_B^2 \to 0, \quad n \to \infty, \quad \delta \to 0.$$

In particular, for sufficiently large n we may assume that $\theta(n,\delta) \equiv \delta$. Therefore, $U_\theta \neq \emptyset$.

By a meaningful approximate solution of the problem (1) we shall mean any element $u_\theta \in U_\theta$ for which the L-norm is minimal among all formal solutions:

$$||Lu_\theta-g|| = \inf_{u\in U_\theta} ||Lu-g||. \tag{8}$$

We note that among all formal solutions of the problem (1), there may exist elements which do not converge in H_{AL} (nor even in H) to the element u as $n \to \infty$, $\delta \to 0$.

Let

$$\hat{U}_\theta = \{u_\theta \in U_\theta: ||Lu_\theta-g|| = \inf_{u\in U_\theta} ||Lu-g||\}.$$

<u>Theorem 81</u>. The set $\hat{U}_\theta$ is not empty.

<u>Proof</u>: Let

$$m_\theta = \inf_{u\in U_\theta} ||Lu-g||.$$

There exist $u_p \in U_\theta$ such that $m_\theta \leq ||Lu-g|| \leq m_\theta + 1/p$, $p = 1,2,\ldots$. Then, obviously, $||Lu_p|| \leq ||g|| + m_\theta + 1$ for any $p \geq 1$. Furthermore, since $u_p \in U_\theta$, then

$$||\ell(Au_p)|| \leq ||\ell(Au_p-\tilde{f})|| + ||\ell(\tilde{f})|| \leq \text{constant},$$

where the constant does not depend on p.

Therefore, the norms $||u_p||_n$ are uniformly bounded in p. By Lemma 42, $\{|u_p|\}$ are also uniformly bounded in p. Hence the sequence

$\{u_p\}$ is weakly compact in H_{AL}. Let u_θ be the weak limit of this sequence (or, if necessary, take a subsequence). Since the operators A and L are continuous in H_{AL}, they are weakly continuous as well. Hence

$$\rho_n(u_\theta) \leq \underline{\lim}_{p\to\infty} \rho_n(u_p) \leq \theta,$$

$$||Lu_\theta - g|| \leq \underline{\lim}_{p\to\infty} ||Lu_p - g|| \leq m_\theta.$$

This implies that $u_\theta \in \hat{U}_\theta$; that is, it is the meaningful approximate solution. □

Theorem 82. If $u_1, u_2 \in \hat{U}_\theta$, then the difference $u_2 - u_1 \in N_L$, the null space of L, and furthermore,

$$|u_1-u_2|^2 = ||(Au_1-Au_2)||^2 \leq K^2||A(u_1-u_2)||_B^2 \leq 4K^2K_1M_n\theta^2.$$

Proof: Since u_1 and u_2 are solutions of the problem (8),

$$||L(\frac{u_1-u_2}{2})||^2 = \frac{1}{2}||Lu_1-g||^2 + \frac{1}{2}||Lu_2-g||^2 - ||L(\frac{u_1+u_2}{2}) - g||^2$$

$$= m_\theta - ||L(\frac{u_1+u_2}{2}) - g||^2 \leq 0,$$

that is, $u_1 - u_2 \in N_L$.

Next,

$$||\ell(Au_1-Au_2)|| \leq ||\ell(Au_1-\tilde{f})|| + ||\ell(Au_2-\tilde{f})|| \leq 2\theta.$$

Then, by virtue of (1), (4) and Lemma 42, we have

$$||A(u_1-u_2)||_F^2 \leq K^2||A(u_1-u_2)||_B^2 \leq K^2K_1|u_1-u_2|^2$$

$$\leq K^2K_1M_n||u||_n^2 \leq 4K^2K_1M_n\theta^2. \quad □$$

Remark 1. The set of solutions $\hat{U}_\theta$ may consist of more than one element. However, if $N_L = \{0\}$, the set of meaningful solutions consists, obviously, of a single element: $\hat{U}_\theta = \{u_\theta\}$.

4. Convergence of approximate solutions in H_{AL}.

Let

$$\hat{U}_\theta^\varepsilon = \{u \in U_\theta: ||Lu-g|| \leq m_\theta + \varepsilon\}$$

where ε, $0 < \varepsilon \leq \varepsilon_0 < +\infty$, is a number. Let

$$\Delta(\delta,n,\varepsilon) = \sup_{u\in\hat{U}_\theta^\varepsilon} ||u-\hat{u}||.$$

Theorem 83. We have

$$\lim_{n\to\infty,\delta,\varepsilon\to 0} \Delta(\delta,n,\varepsilon) = 0;$$

that is, the method (8) of approximate solution of (1) is convergent, being stable with respect to perturbations of the right side of (1) and the error of solution of the variational problem (8). The quantity $\Delta(\delta,n,\varepsilon)$ characterizes the error of approximate solution of the problem (1).

Proof: By definition, for any $u \in \hat{U}^{\varepsilon}_{\theta}$

$$||Lu-g|| \le ||L\hat{u}-g|| + \varepsilon, \quad \rho_n(u) \le \theta. \tag{9}$$

Using the approximation condition, we obtain

$$||Au-\tilde{f}||_F^2 \le \rho_n^2(u) + r_n^2||Au-\tilde{f}||_B^2 \le \theta^2 + r_n^2||Au-\tilde{f}||_B^2 \quad \text{for all} \quad u \in \hat{U}^{\varepsilon}_{\theta}.$$

Then

$$||Au-f||_F \le ||Au-\tilde{f}||_F + ||f-\tilde{f}||_F \le \delta + (\theta^2 + r_n^2||Au-\tilde{f}||_B^2)^{1/2}$$
$$\le \delta + [\theta^2 + r_n^2(||\delta f||_B + ||Au-f||_B)^2]^{1/2} \quad \text{for all} \quad u \in \hat{U}^{\varepsilon}_{\theta}.$$

If $||Au-f||_B$ are uniformly bounded in $u \in \hat{U}^{\varepsilon}_{\theta}$, the previous inequality yields

$$\lim_{n\to\infty,\varepsilon\to 0} \sup_{u\, \hat{U}^{\varepsilon}_{\theta}} ||Au-f||_F = 0. \tag{10}$$

In fact, using Lemmas 41 and 42, we obtain

$$||Au-f||_B^2 \le \frac{K_2}{m}|u-\hat{u}|^2 = \frac{K_2}{m}(||Au-f||_F^2 + ||Lu-L\hat{u}||_G^2). \tag{11}$$

But

$$||Au-f||_F^2 \le ||\ell(Au-f)||^2 + r_n^2||Au-f||_B^2$$
$$\le (||\ell(Au-\tilde{f})|| + ||\ell(A\hat{u}-\tilde{f})||)^2 + r_n^2||Au-f||_B^2 \tag{12}$$
$$\le 4\theta^2 + r_n||Au-f||_B^2.$$

Taking into account the first inequality in (9), we derive from (11) and (12) that for sufficiently large n

$$\sup_{u\in\hat{U}^{\varepsilon}_{\theta}} ||Au-f||_B \le \text{constant},$$

where the constant does not depend on the choice of $u \in \hat{U}_\theta^\varepsilon$ and n, δ and ε. These arguments suffice to justify (10). Eq. (9) yields also

$$\overline{\lim_{n\to\infty;\delta,\varepsilon\to 0}} ||Lu-g|| \leq \nu_L. \tag{13}$$

Considering the inequalities (10) and (13), we see that the method (8) is regular and, therefore, convergett (Theorem 31). □

Remark 2. It follows from (11) that

$$\lim_{n\to\infty,\delta\to 0} \sup_{u\in\hat{U}_\theta^\varepsilon} ||Au-f||_B = 0,$$

although the approximations to f were given in the space F!

3. The consistency condition requires that

$$\lim_{n\to\infty,\delta\to 0} r_n ||\delta f||_B = 0$$

which tends to zero as $n \to \infty$ and for fixed $\tilde{f}$. Hence we can recommend that one take a sufficiently large number of functionals, $n = 1,2,\ldots$. This can, however, lead to difficulties in computation. To avoid undesirable consequences, it seems natural to slightly smooth the element $\tilde{f}$ so that the element δf (small only in the norm F) is bounded by a reasonable quantity in the norm of the space B. To this end, we may use various smoothing procedures. In particular, the algorithms of smoothing considered earlier enable one to have $||\tilde{f}_{smooth} - f||_B \to 0$ for $||\tilde{f}-f||_F \to 0$. In this case one should replace the element $\tilde{f}$ by $\tilde{f}_{smooth}$. Of course, the procedure for solving the problem becomes more cumbersome; however, the computations are simpler.

5. Determination of approximate solutions by means of the regularization method. Following the regularization method, we define the parameter functional ϕ_α by

$$\phi_\alpha[u;\tilde{f}] \equiv \rho_n^2(u) + \alpha||Lu-g||^2, \quad u \in D,$$

where $\alpha > 0$ is the regularization parameter.

It is not hard to show, following the scheme of proving Theorem 81, that for any $\alpha > 0$ there exists a solution u_θ^α of the variational problem

$$\inf_{u\in D} \phi_\alpha[u;\tilde{f}] = \phi_\alpha[u_\theta^\alpha;\tilde{f}]. \tag{14}$$

We note that the element u_θ^α is uniquely defined.

Thus the following theorem holds.

Theorem 84. For any $\alpha > 0$ a solution of the problem (14) exists and is uniquely defined.

It is possible to show that the function $\rho_n^2(\alpha) := \rho_n^2(u_\theta^\alpha)$ is a continuous, strictly increasing function. If the equation $\rho_n(\alpha) = \theta$ has a solution, the solution of this equation is obviously defined uniquely.

Theorem 85. Let the equation $\rho_n(\alpha) = \theta$ have a solution α_θ, $0 < \alpha_\theta < +\infty$. Then, the element $u_\theta \in U_\theta$ is a solution of the problem (8).

In fact, for any $u \in \hat{U}_\theta$ we have

$$\phi_{\alpha_\theta}[u_\theta;\tilde{f}] \leq \phi_\alpha[u;\tilde{f}] = \rho_n^2(u) + \alpha_\theta||Lu-g||^2 \leq \theta^2 + \alpha_\theta||Lu-g||^2.$$

Since $\rho_n^2(\theta) = \theta^2$ and $\alpha_\theta > 0$, it follows from the foregoing inequality that $||Lu_\theta-g|| \leq ||Lu-g||$ $\forall u \in \hat{U}_\theta$; that is, $u_\theta \in \hat{U}_\theta$.

The assertion is proved.

Corollary. If the conditions of Theorem 85 are satisfied, the limiting relation

$$\lim_{n\to\infty,\delta\to 0} |u_\theta-\hat{u}| = 0$$

holds by Theorem 83.

We now formulate the condition under which the equation

$$\rho_n(\alpha) = \theta \tag{15}$$

has a root α_θ, $0 < \alpha_\theta < +\infty$. We asume for simplicity that $g = 0$.

Let u^∞ be a solution of the following problem: $u^\infty \in N_L$,

$$m = \inf_{u\in N_L} ||Au-f||_F = ||Au^\infty-f||_F > 0. \tag{16}$$

The solution u^∞ of the problem (16) exists and is uniquely defined, which follows from the general considerations given above.

Similarly as above, we define elements $u_\theta^\infty \in N_L$ such that

$$\rho_n(u_\theta^\infty) = \inf_{u\in N_L} \rho_n(u) = \tilde{m}. \tag{17}$$

A solution of the problem (17) exists and is uniquely defined. Furthermore, we have the relation

$$\lim_{n\to\infty,\delta\to 0} \tilde{m} = m.$$

Theorem 86. Let the function $\theta = \theta(n,\delta)$, in addition to the properties listed above, satisfy the condition

$$\theta(n,\delta) < \tilde{m}. \tag{18}$$

Then Eq. (15) has the unique root α_θ, $0 < \alpha_\theta < \infty$.

Proof: We prove this theorem following the proof of the theorem on convergence of method $\tilde{\rho}$ (Theorem 25).

Remark 4. If $N_L = \{0\}$ and the number of functionals ℓ_i is infinite, $u_\theta^\infty = 0$ will be a solution of the problem (16), and (18) assumes the form $\delta < ||\tilde{f}||_F$; that is, it is assumed that the error of specification of the right side of (5) is to be sufficiently small compared with the given approximation of $\tilde{f}$. Therefore, the requirement (18) is asymptotically satisfied (always) for $m > 0$.

6. Numerical algorithms for determination of approximate solutions. We denote by S the linear operator on H_{AL} into a Euclidean space E_n of dimension n defined by

$$Su = (\ell_1(Au), \ell_2(Au), \ldots, \ell_n(Au)), \quad u \in D.$$

Let $Su = r \in E_n$ for some $u \in D$. Following the scheme of proving Theorem 81, we can show that there exists a unique element $\hat{u}_r \in D$ such that

$$||L\hat{u}_r||^2 = \inf_{u\in D:\ Su=r} ||Lu||^2. \tag{19}$$

Next, let $u_\theta \in \hat{U}_\theta$ be any solution of the variational problem (8). Also, let $r_\theta = (\ell_1(Au_\theta), \ell_2(Au_\theta), \ldots, \ell_n(Au_\theta))$. In accord with Lemma 43, we construct the element u_{r_θ} which is a solution of the problem (19). Then, obviously, $\rho_n(u_{r_\theta}) = \rho_n(u_\theta)$ and

$$||Lu_{r_\theta}|| \le ||Lu_\theta|| \le \inf_{u\in U_\theta} ||Lu||.$$

Therefore, the element u_{r_θ} is also a solution of the problem (8): that is, $u_{r_\theta} \in \hat{U}_\theta$. These arguments confirm the validity of the following theorem.

Theorem 87. We may seek a solution of the problem (8) as a solution of the problem (19). A similar assertion holds for the problem (14).

For our further discussion we shall require that some additional conditions are satisfied. Thus we assume that the domain of the operator L is dense in G, and also that the range of the adjoint operator L^* coincides with the whole space H; that is, the equation

$$L^*g = u$$

has a solution (not necessarily a unique solution) for any $u \in H$. Furthermore, we assume that the functionals $\ell_i(Au)$ are representable as

$$r_i = \ell_i(Au) = (k_i,u)_H \quad \forall u \in D, \quad i = 1,\dots,n,$$

where $k_i \in H$.

Now we define the elements $g_i \in G$ as solutions of the equations

$$L^*g_i = k_i, \quad i = 1,2,\dots,n.$$

Then, the functionals $\ell_i(Au)$ are, obviously, representable as

$$\ell_i(Au) = (g_i,Lu)_G, \quad u \in D, \quad i = 1,\dots,n.$$

Using a theorem on orthogonal projection, we then derive that the element Lu_r, where u_r is the solution of the problem (19), is representable as a linear combination of elements g_j, $j = 1,2,..,n$. Let

$$g_j = \sum_{k=1}^{n} \gamma_{jk}h_k, \quad j = 1,\dots,n$$

where $h_k \in G$ and γ_{ik}, $i = 1,\dots,n$, are elements of a square matrix Γ. We note that the elements h_k can be always chosen to be linearly independent. Then we seek the element Lu_r to be

$$Lu_r = \sum_{j=1}^{n} a_jh_j. \tag{20}$$

Since

$$r_i = (g_i,Lu_r)_G = \sum_{j=1}^{n} a_j(g_i,h_j)_G = \sum_{j=1}^{n}\sum_{k=1}^{n} \gamma_{ik}(h_k,h_j),$$

then, defining the matrix $\overline{H}$ to have elements $h_{kj} = (h_k,h_j)_G$, we obtain that the coefficients a_j of the linear combination (20) necessarily satisfy the system of equations

$$\Gamma\overline{H}a = r,$$

where $a = (a_1,a_2,\dots,a_n)$. In this case, $||Lu_r||_G^2 = a^T\overline{H}a$. Further, using Theorem 87, we obtain that the elements $g_\theta = Lu_\theta$, where $u_\theta \in \hat{U}_\theta$ (the set of solutions of the problem (8)), are also representable as in (20):

$$g_\theta = Lu_\theta = \sum_{j=1}^{n} \tilde{a}_j h_j;$$

the vector $\tilde{a} = (\tilde{a}_1, \tilde{a}_2, \ldots, \tilde{a}_n)$ is the solution of the problem

$$\lim_{a \in E_n:\ ||\Gamma\overline{H}a - \tilde{r}||^2 \le \theta^2} a^T\overline{H}a = \tilde{a}^T\overline{H}a \qquad (21)$$

where $||r||^2 = \Sigma_{i,j=1}^{n} \zeta_{ij} r_i r_j$ for any vector $r \in E_n$, and in addition, $\tilde{r} = (\ell_1(\tilde{f}), \ldots, \ell_n(\tilde{f}))$.

The solution of the problem (21) defines the numerical algorithm for determination of $g_\theta = Lu_\theta$, $u_\theta \in \hat{U}_\theta$, with $\ell_i(Au_\theta) = (\Gamma\overline{H}\tilde{a})$.

The element u_θ is, obviously, uniquely defined by the values of the functionals $\ell_i(Au_\theta)$ as well as the elements $g_\theta = Lu_\theta$.

Similar considerations are valid for the problem (14). Thus, $g^\alpha = Lu_\theta^\alpha$ can be written as

$$g^\alpha = \sum_{j=1}^{n} a_j^\alpha h_j,$$

where the vector $a^\alpha = (a_1^\alpha, a_2^\alpha, \ldots, a_n^\alpha)$ is the solution of the problem

$$\inf_{a \in E_n} J(a) = J(a^\alpha), \qquad J(a) = ||\Gamma\overline{H}a - \tilde{r}||^2 + \alpha a^T\overline{H}a. \qquad (22)$$

The parameter α_θ is defined from the equation $\rho_n(\alpha) = ||r^\alpha - \tilde{r}|| = \theta$, where $r^\alpha = \Gamma\overline{H}a^\alpha$, $r_i^\alpha = \ell_i(Au_\theta^\alpha)$.

The solution of (22) is reduced to solution of the system of equations

$$(\alpha\overline{H} + \overline{H}\Gamma'\zeta\Gamma\overline{H})a = \overline{H}\Gamma'\,\tilde{r}$$

where ζ is a positive definite matrix with elements ζ_{ij}, $i,j = 1,\ldots,n$.

If $\Gamma = E$, the vector a^α is defined from the system

$$(\alpha\zeta^{-1} + \overline{H})a^\alpha = \tilde{r},$$

with $r^\alpha = \tilde{r} - \alpha\zeta^{-1}a^\alpha$. If the system of vectors $\{h_j\}$ is linearly independent, then we have, assuming that $b^\alpha = \overline{H}a^\alpha$, that

$$(\alpha\overline{H}^{-1} + \Gamma'\zeta\Gamma)b^\alpha = \Gamma'\zeta\tilde{r}, \qquad r^\alpha = \Gamma b^\alpha.$$

In particular, for $\overline{H} = E$ (that is, when the elements h_j are mutually orthogonal), the vector a^α is defined from the equations

$$(\alpha E + \Gamma'\zeta\Gamma)a^\alpha = \Gamma'\zeta\tilde{r},$$

with $r^\alpha = \Gamma a^\alpha$.

Remark 5. The method for determination of solutions of the problems (8) and (14) considered above is close to the method suggested in [95] for determination of generalized splines. However, our method does not require the knowledge of a basis of the subspace spanned by the elements $\{g_j\}$. If this basis is known, it is not hard to modify the scheme of solving the problem presented above. We now give such a modification.

Let $\{\phi_k\}$, $k = 1,2,\ldots,p$, $p \leq n$, be a basis of the subspace spanned on the elements $\{g_j\}$. Then, obviously,

$$\phi_k = \sum_{j=1}^{n} b_{jk}g_j, \quad k = 1,2,\ldots,p. \tag{23}$$

We denote by B the matrix with elements b_{jk}, $j = 1,2,\ldots,n$, $k = 1,2,\ldots,p$. It is easy to see that in this case the element Lu_r, where u_r is the solution of the problem (19), is uniquely representable as the linear combination

$$Lu_r = \sum_{k=1}^{p} s_k\phi_k. \tag{24}$$

Taking the inner product of (24) with ϕ_ℓ yields

$$\sum_{k=1}^{p} s_k(\phi_k,\phi_\ell)_G = (\phi_\ell,Lu_r)_G, \quad \ell = 1,2,\ldots,p, \tag{25}$$

where, by (23),

$$(\phi_\ell,Lu_r)_G = \sum_{j=1}^{n} b_{j\ell}(g_j,Lu_r)_G = \sum_{j=1}^{n} b_{j\ell}r_j. \tag{26}$$

Next, denoting by $C = \{(\phi_k,\phi_\ell)_G\}$ a positive definite $p \times p$-matrix and by s the vector $s = (s_1,s_2,\ldots,s_p)$, we derive from (25), (26) the system of equations for s

$$Cs = B'r.$$

Assuming, as before, that $r^\alpha = (\ell_1(Au_\theta^\alpha),\ldots,\ell_n(Au_\theta^\alpha))$, we obtain for the solution u_θ^α of the problem (14):

$$g^\alpha = Lu_\theta^\alpha = \sum_{k=1}^{p} s_k^\alpha \phi_k,$$

where the vector $s^\alpha = (s_1^\alpha, s_2^\alpha, \ldots, s_p^\alpha)$ is related to r^α by

$$Cs^\alpha = B'r^\alpha. \tag{27}$$

It is easy to verify that

$$||Lu_\theta^\alpha||_G^2 = (Cs^\alpha, s^\alpha)_p = (BC^{-1}B'r^\alpha, r^\alpha)_n,$$

where the subscript p or n shows that we have taken the Euclidean inner product in the space E_p or E_n respectively.

Using the foregoing relations as well as the assertion on the relationship between solutions of the problems (14) and (19), for determination of the vector r^α we have the problem

$$\inf_{r \in E_n} \{||r-\tilde{r}||_\zeta^2 + \alpha(BC^{-1}B'r,r)_n\} = ||r^\alpha-\tilde{r}||_\zeta^2 + \alpha(BC^{-1}B'r^\alpha, r^\alpha)_n$$

which implies that the vector r^α satisfies the following system of linear equations:

$$(\alpha BC^{-1}B' + \zeta)r^\alpha = \zeta\tilde{r} \tag{28}$$

From (28) and (27) we obtain the system of equations for the vector s^α

$$(\alpha B'\zeta^{-1}B + C)s^\alpha = B'\tilde{r}. \tag{29}$$

Since (28) contain the inverse matrix of C (C is sometimes a banded matrix), it is preferable to compute the vector s^α using the system (29) and then to determine the vector r^α using the relation

$$r^\alpha = \tilde{r} - \alpha\zeta^{-1}Bs^\alpha.$$

We can write the equation for determination of the parameter α_θ from the criterion (15) as follows:

$$\rho_n^2(\alpha) = ||r^\alpha-\tilde{r}||^2 = \theta^2, \tag{30}$$

or

$$\rho_n^2(\alpha) = \alpha^2||Bs^\alpha||_n^2 = \theta^2 \tag{31}$$

(compare with Subsection 20.5.2).

The last equation for determination of the parameter α_θ is especially preferable if we are interested only in knowing the approximate value of the operator L on the solution $\hat{u}$ of the equation $Au = f$.

We note that the function ρ_n is a monotone increasing function. In order to apply fast techniques of the Newton type for solving (31) (or (30)), it is useful to take a new variable $\lambda = 1/\alpha$. We can show that the function $\rho_n(\lambda)$ is convex in λ, $0 \leq \lambda < +\infty$, which ensures the application of these techniques. We shall consider this problem in more detail in Section 26.

We now note that the methods mentioned are applicable, obviously, in each case when the functionals $\ell_i(Au)$ are representable as $\ell_i(Au) = (g_i, Lu)_G$, where g_i, $i = 1,2,\ldots,n$, are elements of G; in particular, if

$$\ell_i(Au) = \sum_{j=1}^{n} a_{ij} t_j(u), \quad m \geq n,$$

where each t_j is a linear functional on H_{AL} (and the t_j are not necessarily linearly independent).

7. Relation to Theory of Splines. It is not hard to see that for $A = E$, $H = F$ we arrive at the problem of constructing splines and investigating their behavior in the case when the values of functionals of the approximation to the required element are exact or approximate.

It is crucial to note that all the results of the general theory presented in this section will still hold if instead of the functionals $\ell_i^{(n)}$, $i = 1,2,\ldots,n$, we consider linear operators acting from B into some Hilbert (and even normed) spaces F_i, $i = 1,\ldots,n$. Then we can rewrite the approximation condition (6) as

$$\left| \sum_{i,j=1}^{n} \zeta_{ij} \|\ell_i(f)\|_{F_i} \|\ell_j(f)\|_{F_j} - \|f\|_F^2 \right| \leq r_n^2 \|f\|_B^2, \tag{32}$$

where the coefficients $\zeta_{ij} = \zeta_{ij}^{(n)}$ have the same meaning as before. In particular, for $\zeta_{ij} = 0$, $i \neq j$, the condition (32) becomes

$$\left| \sum_{i=1}^{n} \zeta_i \|\ell_i(f)\|_{F_i}^2 - \|f\|_F^2 \right| \leq r_n^2 \|f\|_B^2. \tag{33}$$

The values $\ell_i(f)$, $i = 1,2,\ldots,n$, can be regarded as the "traces" of the element f in the spaces F_i.

The case $A = E$, $H = F$ provides a theory of the so-called "operator splines" introduced in [105]. We note that in [105] the authors deal only with the problems of existence and uniqueness of interpolating and smoothing (generalized) operator splines without considering their approximation properties. The methods discussed in the present section can be completely extended to the theory of operator splines.

Chapter 5

Regular Methods for Special Cases of the Basic Problem. Algorithms for Choosing the Regularization Parameter

Section 24. Pseudosolutions

1. We consider a special case of the basic problem where $D = H$; that is, it coincides with the whole space H. It is natural to regard the operator A as bounded. Moreover, we assume that $H = G$ and $L = E$, and in addition, $g = 0$. Then the solution $\hat{u}$ of the basic problem is obviously an element with minimal norm ensuring the minimal residual $||Au-f||_F$, $u \in H$. In the case $||A\hat{u}-f||_F = 0$, the element u_f is said to be a *normal* [89] solution of the equation

$$Au = f. \tag{1}$$

In the general case the element $\hat{u}$, the solution of the basic problem under the conditions considered is said to be a *pseudo-solution* (or *least-squares solution*) and is denoted by u_f.

It is obvious that if $f \in Q_A$, the range of the operator, the pseudo-solution u_f exists and is defined uniquely (Theorem 1). Therefore, on some non-empty set $Q \supset Q_A$ the operator A^+ is defined by $A^+f = \hat{u}$. We note that if the operator A has inverse A^{-1}, then $D_{A^{-1}} \subseteq D_{A^+}$ and $A^+f = A^{-1}f$, $\forall f \in D_{A^{-1}}$; that is, the operator A^+ is an extension of the operator A^{-1}. However, the existence of theoperator A^+ is not a necessary condition for the operator A^{-1} to exist. In fact, as is seen from systems of linear algebraic equations, the operator A^+ exists and is defined uniquely on the entire F even if the operator A is non-invertible; that is, the system of equations (1) is inconsistent or has a non-unique solution.

The operator A^+ is said to be the *pseudoinverse*.

2. Now we are interested in the existence and stability of pseudosolutions of (1). It is easy to see that in the case where H and F are finite-dimensional spaces, the problem of finding pseudosolutions has a solution for any $f \in F$. The following theorem is a generalization of this fact to a wider class of problems.

Theorem 88. In order that for any $f \in F$ there exists a pseudosolution u_f (that is, $D_{A^+} = F$), it is necessary and sufficient that the operator A be normally solvable.*

In fact, let $N_A = \{u: Au = 0\}$. It is obvious that N_A is a closed subspace. Let $H = N_A + H_A$, where H_A is the orthogonal complement of N_A to H. Then the problem of finding pseudosolutions is equivalent to the problem of minimizing

$$||Au - f||_F, \quad u \in H_A \tag{2}$$

Sufficiency. Let A be a normally solvable operator. We denote by P the projection operator of F onto Q_A. Then, obviously, the problem (2) is equivalent to the problem of minimizing

$$||Au - Pf||_F, \quad u \in H_A,$$

which, obviously, has a solution $u_f = \hat{A}^{-1}Pf$ for any $f \in F$, where $\hat{A}$ is the operator mapping H_A onto Q_A and coinciding with the operator A on H_A. Thus,

$$u_f = \hat{A}^{-1}Pf = A^+f \quad \text{for all} \quad f \in F; \tag{3}$$

that is, $D_{A^+} = F$, thus proving the sufficient condition of the theorem.

Necessity. Let $D_{A^+} = F$. Then the operator $P = AA^+$ is given on all of F. We show that P is a projection operator. As is well known, to do this we need only to prove that the following conditions are satisfied:

(a) $P^* = P$, that is, the operator P is self-adjoint;
(b) $P^2 = P$.

For any f and g of F we have

$$(f-Pf,Pg)_F = (f-AA^+f,AA^+g)_F = (A^*(f-AA^+f),A^+g)_H = 0,$$

since, obviously, $A^*(f-Au_f) = 0$. Therefore, $(f,Pg)_F = (Pf,Pg)_F$. This

*The operator A is said to be *normally solvable* if its range Q_A is closed in F.

implies that $(f,Pg)_F = (Pf,g)_F$; that is, condition (a) is satisfied. Also, $(f,Pf)_F = (Pf,Pf) = (f,P^2f)$ $\forall f$, which proves (b).

Therefore, P is a projection operator. Let G be the subspace onto which P projects F. We show that $Q_A = G$. In fact, let $g \in G$. Then $g = Pg = AA^+g \in Q_A$; that is, $G \subseteq Q_A$. We note next that the operator equality $AA^+A = A$ holds, which in turn implies that $PAu = AA^+Au = Au \in G$ for any $u \in H$. This yields that $Q_A \subseteq G$. Therefore, $Q_A = G$ is a subspace.

We have proved the assertion and thereby the theorem. □

We can formulate the result thus obtained as follows: the problem of finding pseudosolutions is well-posed in the Hadamard sense if and only if the operator A is normally solvable. In this case, it follows from Eq. (3) that

$$||A^+|| = ||\hat{A}^{-1}P|| \leq ||\hat{A}^{-1}|| < +\infty. \tag{4}$$

Example. Let $H = F$ and let Eq. (1) be written as

$$Au \equiv u - \lambda A_0 u = f,$$

where λ is a numerical parameter and A_0 is a completely continuous linear operator acting on H into H. Then a necessary and sufficient condition for this equation to have a solution for the given f becomes

$$(f,\omega_i)_H = 0, \quad i = 1,2,\ldots,s,$$

where $\omega_i \in H$, $\omega_i - \lambda A_0^* \omega_i = 0$ are linearly independent eigenelements of the operator A_0^* corresponding to the eigenvalue λ. Since the operator A_0^* is completely continuous as well, there can be only a finite number s of these elements. It is seen that the set of all f with $(f,\omega_i) = 0$, $i = 1,2,\ldots,s$, forms a subspace in H and, therefore, the operator $A = A - \lambda A_0$ is normally solvable for any λ. If the equation $\omega - \lambda A_0^* \omega = 0$ has only the trivial solution $\omega = 0$, by the Fredholm theorems the equation $u - \lambda A_0 u = f$ has a unique solution for any $f \in H$.

The problem considered belongs to the so-called class of *spectral problems*.

3. The calculation of pseudosolutions using the definition given above is difficult. The regularization method is quite effective for obtaining pseudosolutions. Let u_α be the solution of the variational problem of minimizing

$$\Phi_\alpha[u] = \Phi_\alpha[u;A,f] = ||Au-f||_F^2 + \alpha||u||_H^2, \quad u \in H, \tag{5}$$

where $\alpha > 0$ is a parameter. Then

$$u_\alpha = (\alpha E + A^*A)^{-1}A^*f =: R_\alpha f,$$

where A^* is the adjoint operator of A.

Lemma 43. If the operator A is normally solvable, then

(a) $||R_\alpha|| \le ||A^+|| \quad \forall\alpha > 0;$

(b) $||R_\alpha - A^+|| \le \alpha||\hat{A}^{-1}||^3.$

Proof: For any $u \in H$ we have

$$\Phi_\alpha[u] \le \Phi_\alpha[u].$$

Assuming $u = u_f$ and using the main property of pseudosolutions

$$||Au_f - f|| \le ||Au-f|| \quad \forall u \in H,$$

we obtain

$$||Au\ -f||^2 + \alpha||u_\alpha||^2 \le ||Au_f-f||^2 + \alpha||u_f||^2 \le ||Au_\alpha-f||^2 + \alpha||u_f||^2;$$

that is,

$$||u_\alpha|| = ||R_\alpha f|| \le ||u_f|| = ||A^+f|| \quad \text{for all } f \in F,$$

which proves (a).

Next we note that Eq. (5) is equivalent to the problem of minimizing

$$||\hat{A}u-Pf||^2 + \alpha||u||^2, \quad u \in H_A,$$

and therefore

$$u_\alpha = (\alpha E + \hat{A}^*\hat{A})^{-1}\hat{A}^*Pf \equiv R_\alpha f.$$

Using (3), we have

$$||R_\alpha - A^+|| = ||\{(\alpha E + \hat{A}^*A)^{-1}\hat{A}^* - \hat{A}^{-1}\}P||$$

$$\le \alpha||(\alpha E + \hat{A}^*\hat{A})^{-1}\hat{A}^{-1}|| \le \alpha||\hat{A}^{-1}||^3$$

thus proving (b). □

Lemma 44. If the operator A is normally solvable, for any $u \in H_A$ the equations

$$A^*f = u, \quad A^*Av = u \tag{6}$$

are solvable (perhaps not uniquely).

In fact, if the continuous operator A is normally solvable, the range Q_{A*} of the adjoint operator A^* equals H_{A*}. This implies that each equation in (6) has a solution. □

Theorem 89. Let A be a normally solvable operator, let $\tilde{f}$ be a given element of F such that $||f-\tilde{f}||_F < \delta||\tilde{f}||_F$ and let $\tilde{u}_\alpha = R_\alpha\tilde{f}$ be a solution of the problem (5) when $f = \tilde{f}$.

Then for any $\alpha > 0$, for the deviation $||\tilde{u}_\alpha - u_f||_H$ we have the following:

$$\begin{aligned} &(a) \quad ||\tilde{u}_\alpha-u_f||_H \leq \alpha||\hat{A}^{-3}||\,||\tilde{f}|| + ||A^+||\,||\tilde{f}||_F\delta \\ &(b) \quad ||\tilde{u}_\alpha-u_f||_H \leq \sqrt{\alpha}||f_0||_F + ||A^+||\,||\tilde{f}||_F\delta \\ &(c) \quad ||\tilde{u}_\alpha-u_f||_H \leq \alpha||v_0|| + ||A^+||\,||\tilde{f}||_F\delta \end{aligned} \tag{7}$$

where $f_0 \in F$ and $v_0 \in H$ are normal solutions, respectively, of the first and second equations in (6) when the right side $u = u_f$.

Proof: We have, using the estimate (b) of Lemma 43:

$$||\tilde{u}_\alpha-u_f|| \leq ||\tilde{u}_\alpha-u_{\tilde{f}}|| + ||u_{\tilde{f}}-u_f|| = ||(R_\alpha-A^+)\tilde{f}|| + ||A^+(\tilde{f}-f)|| \leq \alpha||\hat{A}^{-1}||^3||\tilde{f}||_F + ||A^+||\,||\tilde{f}||\delta.$$

This proves (7a).

Further, we can write

$$||\tilde{u}_\alpha-u_f|| \leq ||\tilde{u}_\alpha-u_\alpha|| + ||u_\alpha-u_f|| = ||R_\alpha(\tilde{f}-f)|| + ||u_\alpha-u_f|| \leq \delta||A^+||\,||\tilde{f}||_F + ||u_\alpha-u_f||$$

(due to the inequality (a) in Lemma 43). The estimates

$$||u_\alpha-u_f||_H \leq \sqrt{\alpha}||f_0||_F, \quad ||u_\alpha-u_f||_H \leq \alpha||v_0||_H$$

can be derived in the same way as similar estimates were obtained in Section 3. □

From (7a) and (7b) it follows that for an appropriate choice of the value of the regularization parameter the deviation of the approximation obtained from the required pseudosolution will be of the same order as the error of specification of f.

The following theorem is interesting.

Theorem 90. Let the conditions of Theorem 89 be satisfied. Furthermore, let u_f be the normal solution of (1). Then if the parameter α is defined by the equation

$$||A\tilde{u}_\alpha - \tilde{f}|| = \delta||\tilde{f}||, \tag{8}$$

for the deviation $||\tilde{u}_\alpha - u_f||$ we have

$$||\tilde{u}_\alpha - u_f|| \leq 2||\hat{A}^{-1}||\ ||\tilde{f}||\delta.$$

The existence of a solution of (8) follows from the general properties of the residual on regularized solutions which were proved earlier. Next we have

$$||\tilde{u}_\alpha - u_f|| = ||\hat{A}^{-1}(\hat{A}u_\alpha - \hat{A}u_f)|| \leq ||\hat{A}^{-1}||\ ||A\tilde{u}_\alpha - Au_f||$$
$$\leq ||\hat{A}^{-1}||(||A\tilde{u}_\alpha - P\tilde{f}|| + ||P\tilde{f} - Pf||) \leq 2||\hat{A}^{-1}||\ ||\tilde{f}||\delta$$

since $||P|| \leq 1$, and

$$||A\tilde{u}_\alpha - P\tilde{f}|| = (||A\tilde{u}_\alpha - \tilde{f}||^2 - ||P\tilde{f}||^2)^{1/2} \leq ||A\tilde{u}_\alpha - \tilde{f}|| = \delta||\tilde{f}||.$$

We have proved the theorem. □

4. We proceed to investigating the influence of errors in specifying the operator A on the approximations obtained by the regularization method. To wit, instead of A, let a continuous operator A_h be defined on H with values in F and such that $||A-A_h|| \leq h < +\infty$. Let $\tilde{u}_\alpha^h$ be a solution of the variational problem of minimizing

$$\Phi_\alpha[u;A_h,\tilde{f}], \quad u \in H. \tag{9}$$

Then we have

$$(\alpha E + A_h^* A_h)(\tilde{u}_\alpha^h - \tilde{u}_\alpha) = (A_h^* - A^*)(\tilde{f} - A\tilde{u}_\alpha) + A_h^*(A - A_h)\tilde{u}_\alpha$$

and therefore

$$||\tilde{u}_\alpha^h - \tilde{u}_\alpha|| \leq hB(\alpha,h)||\tilde{f} - A\tilde{u}_\alpha|| + hC(\alpha,h)||\tilde{u}_\alpha||$$

where

$$B(\alpha,h) = ||(\alpha E + A_h^* A_h)^{-1}||, \qquad C(\alpha,h) = ||(\alpha E + A_h^* A_h)^{-1} A_h^*||.$$

Using the Cauchy-Schwarz inequality, we obtain

$$||\tilde{u}_\alpha^h - \tilde{u}_\alpha|| \leq h[B^2 + C^2/\alpha]^{1/2}\Phi_\alpha^{1/2}[\tilde{u}_\alpha;A,\tilde{f}] \leq h[B^2 + C^2/\alpha]^{1/2}\Phi_\alpha^{1/2}[u_f;A,\tilde{f}].$$

This together with the estimates $B \leq 1/\alpha$, $C \leq 1/2\sqrt{\alpha}$ yields

$$||\tilde{u}_\alpha^h - \tilde{u}_\alpha|| \leq \frac{h\sqrt{\delta}}{2\alpha} \Phi_\alpha^{1/2}[u_f;A,f]. \tag{10}$$

The estimates (7) and (10) imply the following theorem.

Theorem 91. If the operator A is normally solvable, then

$$\lim_{\alpha,\delta,h,h/\alpha\to 0} ||\tilde{u}_\alpha^h - u_f|| = 0;$$

that is, in order to obtain stable approximations using the regularization method (9), it suffices to adapt the regularization parameter α only to the error h of specifying the operator.

Remark. If the element u_f is a normal solution of (1), then, obviously,

$$\Phi_\alpha[u_f;A,\tilde{f}] \leq \delta^2||\tilde{f}||^2 + \alpha||u_f||^2.$$

The estimate (10) has improved:

$$||\tilde{u}_\alpha^h - \tilde{u}_\alpha|| \leq \frac{h\sqrt{\delta}}{2\alpha}[\delta^2||\tilde{f}|| + \alpha||u_f||^2]^{1/2}.$$

Now we establish the possible order of accuracy of approximations to the normal solution of the problem (1) which have been obtained by the regularization method. To this end, we estimate the deviation $||\tilde{u}_\alpha^h - u_\alpha||$:

$$(\alpha E + A_h^* A_h)(\tilde{u}_\alpha^h - u_\alpha) = A_h^*(f - \tilde{f}) + (A_h^* - A_h)(f - Au_\alpha) + A_h^*(A - A_h)u_\alpha$$

and therefore

$$||\tilde{u}_\alpha - u_\alpha|| \leq \delta C||\tilde{f}|| + hB||f - Au_\alpha|| + hC||u_\alpha|| \leq \frac{\delta}{2\sqrt{\alpha}}||\tilde{f}|| + \frac{h\sqrt{\delta}}{2\alpha} \Phi_\alpha^{1/2}[u_\alpha;A,f].$$

Noting that

$$\Phi_\alpha[u_\alpha;A,f] \leq \Phi_\alpha[u_f;A,f] = \alpha||u_f||^2$$

in the case where u_f is a normal solution of (1), we have

$$||\tilde{u}_\alpha^h - u_\alpha|| \leq \frac{\delta}{2\sqrt{\alpha}}||\tilde{f}|| + \frac{h\sqrt{\delta}}{2\sqrt{\alpha}}||u_f||$$

and therefore

$$||\tilde{u}_\alpha^h - u_f|| \leq ||u_\alpha - u_f|| + \frac{\delta}{2\sqrt{\alpha}}||\tilde{f}|| + \frac{h}{2\sqrt{\alpha}}||u_f||. \tag{11}$$

The estimates (11) and (7b) imply the following theorem.

Theorem 92. If the operator A is normally solvable and u_f is a normal solution of (1), then

$$\min_{\alpha>0}||\tilde{u}_\alpha - u_f|| \le 3\left(\frac{\delta||\tilde{f}|| + h\sqrt{\delta}||u_f||}{2}\right)^{2/3}||v_0||^{1/3}. \qquad (12)$$

We note that (10) and (11) hold regardless of any assumption on normal solvability of the operator A. In the last case the estimate (12) still holds if the second equation in (6) has a solution.

Let $\{A_h, f_\delta\}$ be approximate quantities in the problem (1), where the linear operator $A_h: H \to F$, $||A-A_h|| \le h$, $||f_\delta - f|| < \delta$ and the problem (1) is replaced by the problem

$$A_h u = f_\delta. \qquad (13)$$

We regard the operator A as normally solvable.

We now denote by P the projection of F onto Q_A, the range of A.

We assume that $PA_h = A_h$; that is, $Q_{A_h} = Q_A$, where Q_{A_h} is the range of the operator A_h.

Lemma 45. For a sufficiently small h we have $Q_{A_h} = Q_A$.

Proof: We need to use the fact that Q_A is a closed subspace and, in addition, to apply the open mapping theorem.

Corollary. For sufficiently small h the operator A_h is normally solvable.

We denote by u_σ the pseudosolution (13), $\sigma = (h,\delta)$.

It is easy to prove the following lemma.

Lemma 46. We have the estimate

$$||u_\sigma - u_f|| \le ||A_h^+||\delta + ||A_h^+ - A^+||\ ||f||,$$

where A_h^+, A^+ are the inverse operators of A_h and A, respectively.

Therefore

$$\lim_{|\sigma|\to 0} ||u_\sigma - \hat{u}|| = 0,$$

where $|\sigma| = (h^2+\delta^2)^{1/2}$.

We now consider the problem

$$\Phi_\alpha^\sigma[u] \equiv \Phi_\alpha[u, A_h; f_\delta] = ||A_h u - f_\delta||^2 + \alpha||u||^2, u \in H \text{ -min} \qquad (14)$$

where $\alpha > 0$ is a regularization parameter; also, let u_σ^α be a solution of this problem. Then

$$\Phi_\alpha^\sigma[u_\alpha^\sigma] \leq \Phi_\alpha^\sigma[u] \quad \forall u \in H.$$

Assuming $u = u_\sigma$, we have

$$||A_h u_\alpha^\sigma - f_\delta||^2 + \alpha||u_\alpha^\sigma||^2 \leq ||A_h u_\sigma - f_\delta||^2 + \alpha||u||^2.$$

Furthermore, let

$$f_\delta = f_\delta' + f_\delta'', \quad f_\delta' = Pf_\delta, \quad f_\delta'' + f_\delta'.$$

Using the Pythagorian equality, we have

$$||A_h u_\alpha^\sigma - f_\delta'||_F^2 + \alpha||u_\alpha^\sigma||^2 + ||f''||^2 \leq ||A_h u_\sigma - f_\delta'||^2 + \alpha||u_\delta||^2$$
$$+ ||f_\delta''||^2 = \alpha||u_\sigma||^2 + ||f_\delta''||^2$$

since $||A_h u_\sigma - f_\delta'|| = 0$.

Therefore, for any $\alpha > 0$

$$||u_\alpha^\sigma|| \leq ||u_\sigma||.$$

Furthermore,

$$||A_h u_\alpha^\sigma - f_\delta'||^2 \leq \alpha||u_\sigma||^2 - \alpha||u_\alpha^\sigma||^2$$
$$\leq \alpha(||u_\sigma|| + ||u_\alpha^\sigma||)||u_\sigma - u_\sigma^\alpha|| \leq 2\alpha||u_\sigma|| \, ||u_\sigma - u_\alpha^\sigma||.$$

Assuming $z = u_\alpha^\sigma - u_\sigma$, we obtain

$$||A_h z||^2 \leq 2\alpha||u_\sigma|| \, ||z||$$

and

$$||z|| \leq ||A_h^+|| \, ||A_h z|| \leq ||A_h^+||(2\alpha||u_\sigma|| \, ||z||)^{1/2}.$$

Therefore,

$$||z|| = ||u_\alpha^\sigma - u_\sigma|| \leq 2\alpha||A_h^+||^2||u_\sigma||$$

which together with Lemma 46 yields the following.

<u>Theorem 93</u>. Let $\alpha, |\sigma| \to 0$. Then, $u_\delta^\alpha \to \hat{u}$. In this case

$$||u_\alpha^\sigma - \hat{u}|| \leq ||A_h^+||\delta + ||A_h^+ - A^+|| \, ||\overline{f}|| + 2\alpha||A_h^+||^2||u_\sigma||.$$

Therefore, for sufficiently small α we have

$$||u_\alpha^\sigma-\hat{u}|| \leq ||A_h^+||\delta + ||A_h^+-A^+|| \, ||\bar{f}||.$$

We note that since $\alpha, |\sigma| \to 0$ independently, the problem of choosing a regularization parameter does not arise.

Our analysis is valid for a system of linear algebraic equations, integral equations with degenerate kernels, as well as for some classes of differential equations having a non-unique solution (for example, the Neumann problem).

Under the approximate realization of the foregoing technique for solving the problem, let additional errors be involved in the operator A_h and the right side f_δ, so that $\tilde{A}_h = A_h + \delta A_h$, $\tilde{f}_\delta = f_\delta + \delta f_\delta$ and $||\delta A_h|| \leq \tau$, $||\delta f_\delta|| \leq \tau$. We assume that $\tau << h,\delta$.

Let $\tilde{u}_\alpha^\sigma$ be the exact solution of the problem of minimizing

$$\Phi_\alpha^\sigma[u] \equiv \Phi_\alpha[u;\tilde{A}_h,\tilde{f}_\delta], \quad u \in H.$$

Choosing $\alpha = \tau^{2/3}$ and using (12), we obtain $||\tilde{u}_\alpha^\sigma-u_\sigma|| \leq C_\sigma\tau^{2/3}$, where $\lim_{\sigma\to 0} C_\sigma = C$, that is, C_σ are uniformly bounded in σ.

Therefore, if we assume that $\alpha = \tau^{2/3}$ while solving the problem (14) approximately, then

$$||\hat{u}-\tilde{u}_\alpha^\sigma|| \leq ||u_\sigma-\hat{u}|| + C_\sigma\tau^{2/3}.$$

Since $\tau << h,\delta$, the error of the realistically computed solution is seen to be

$$||\tilde{u}_\sigma^\alpha-\hat{u}|| \approx ||u_\sigma-\hat{u}||.$$

Let (1) be a system of linear algebraic equations with an $n \times n$ coefficient matrix. Suppose we use a computer with floating point arithmetic for our computations. If t is the number of binary digit bits of the mantissa, using the algorithm described in [8], we obtain $\tau = O(2^{-t})$ and, therefore, we must have $O(2^{-t}) << h$, which is natural to require. The choice of the parameter value $\alpha = 2^{-(2/3)t}$ ensures, as a rule, numerical stability of the method.

6. The methods described above can be applied to numerical solution of linear integral equations of the first kind having a degenerate kernel, as well as to some classes of differential equations having a non-unique solution (for example, the Neumann problem).

The computational experiment of numerical solution of the degenerate systems of linear algebraic equations, using standard programming, showed that regular solutions "stabilize" for $\alpha \approx 2^{-(2/3)t} \approx 10^{-7}$ whereas for $\alpha = 0$ we have overflow, yielding thereby an unacceptable result. A stabilization effect was observed also while solving degenerate systems by iteration type-methods (for example, the gradient method).

7. We now consider the problem of finding pseudosolutions of (1) under an additional assumption: $H = F$, $A = A^*$, $A \geq 0$. However, we do not assume that the operator A is normally solvable. We require only the pseudosolution u_f of the problem (1) exist.

Let v_α be a (unique) solution of the equation

$$(\alpha E + A)_{v_\alpha} = f, \quad \alpha > 0.$$

It is easy to show that when $Au_f \neq f$ the solutions v_α of this equation do not converge to the pseudosolution u_f as $\alpha \to 0$.

Therefore, let

$$u_\alpha = v_\alpha + \alpha dv_\alpha/d\alpha$$

where $dv_\alpha/d\alpha$ denotes the strong derivative of the element v_α as the function of the parameter α. It is not hard to see that

$$u_\alpha(\alpha E + A)^{-2} = Af;$$

that is, is a solution of the equation

$$(\alpha E + A)^2 u = Af. \tag{15}$$

Since in this case the pseudosolution u_f satisfies in the usual sense the equation $A^2 u_f = Af$, then $u_\alpha = (\alpha E + A)^{-2} A^2 u_f$ and the difference

$$u_f - u_\alpha = \alpha(\alpha E + 2A)(\alpha E + A)^{-2} u_f. \tag{16}$$

Theorem 94. We have the limit relation

$$\lim_{\alpha\to 0} ||u_f - u_\alpha|| = 0.$$

Proof: Let E_λ, $0 \leq \lambda \leq M < +\infty$, be the spectral decomposition of unity, generated by the operator A:

$$Au = \int_0^M \lambda \, dE_\lambda u.$$

Then

$$(\alpha E + 2A)(\alpha E + A)^{-2}u = \int_0^M \frac{\alpha+2\lambda}{(\alpha+\lambda)^2}\, dE_\lambda u$$

and therefore

$$||u_f-u_\alpha||^2 = \alpha^2 \int_0^M \frac{(\alpha+2\lambda)^2}{(\alpha+\lambda)^4}\,(dE_\lambda u_f,u_f).$$

It is seen that

$$\frac{(\alpha+2\lambda)^2}{(\alpha+\lambda)^4} \le \frac{4}{(\alpha+\lambda)^2} \le \frac{4}{\lambda^2}\,.$$

Splitting the integral into two parts within the range from 0 to ε, $0 < \varepsilon < M$ and from ε to M, we obtain:

$$||u_f-u_\alpha||^2 \; 4\int_0^\varepsilon (dE_\lambda u_f,u_f) + 4\alpha^2 \int_\varepsilon^M (dE_\lambda u_f,u_f)$$

$$\le 4\int_0^\varepsilon (dE_\lambda u_f,u_f) + 4\,\frac{\alpha^2}{\varepsilon^2}\,||u_f||^2.$$

Assuming $\alpha = \alpha(\varepsilon) = \varepsilon^{3/2}$, we see that

$||u_f-u_\alpha|| \to 0$ as $\varepsilon \to 0$. □

We assume that $u_f = Av$, $v \in H$ (this is satisfied a priori if the operator A is normally solvable).

Eq. (16) thus becomes

$$u_f - u_\alpha = \alpha(\alpha E + 2A)(\alpha E + A)^{-2}Av.$$

Using the spectral decomposition of the operator A, we obtain:

$$||u_f-u_\alpha||^2 = \alpha^2 \int_0^M \frac{(\alpha+2\lambda)^{2\;2}}{(\alpha+\lambda)^4}\,(dE_\lambda v,v).$$

Since

$$\frac{(\alpha+2\lambda)^2\lambda^2}{(\alpha+\lambda)^4} \le \frac{(\alpha+2\lambda)^2}{(\alpha+\lambda)^2} \le 4,$$

the convergence rate is seen to be

$||u_f-u_\alpha|| \le 2\alpha||v||$, for all u_f such that $u_f = Av$, $v \in H$

If the operator A and the element f are perturbed, that is, the linear (not necessarily self-adjoint) operator $\tilde{A}$ with $||\tilde{A}-A|| \leq h$ and the element $\tilde{f}$ with $||\tilde{f}-\bar{f}|| \leq \delta$ are given, it is possible to determine the approximation to the pseudosolution u_f by solving the equation

$$(\alpha E + A_h)\tilde{v}_\alpha = \tilde{f} \tag{17}$$

and assuming that

$$\tilde{u}_\alpha = \tilde{v}_\alpha + \alpha(d\tilde{v}_\alpha/d\alpha). \tag{18}$$

Here the operator $A_h(\tilde{A} + \tilde{A}^*)/2$ is self-adjoint and satisfies the estimate $||A-A_h|| \leq h$. Eq. (17) has always a unique solution for $\alpha > h$.

Using the triangle inequality, we have:

$$||\tilde{u} - u_f|| \leq ||\tilde{u}_\alpha - u_\alpha|| + ||u_\alpha - u_f||.$$

It is easy to show that there exists $\alpha = \alpha(\delta,h) > 0$ such that

$$\lim_{\delta,h} ||\tilde{u}_\alpha - u_\alpha|| = 0.$$

Then

$$||\tilde{u}_\alpha - u_f|| \to 0, \quad |\sigma| \to 0.$$

This method has definite advantages because it does not require Gaussian elimination to be applied, the latter disturbing as frequently happens, the particular structure of the operator A (for example, the band-structured matrix A), that makes the application of the regularization method in a "pure" form difficult.

We note that it is convenient to use an approximate formula instead of Eq. (18). Let α and $\tau\alpha$ be two neighboring values of the parameter, $\tau \approx 1$. We then have:

$$\tilde{u}_\alpha \simeq \tilde{v}_\alpha + \alpha \frac{\tilde{v}_\alpha - \tilde{v}_{\varepsilon\alpha}}{\alpha - \tau_\alpha} = \tilde{v}_\alpha - \frac{\tilde{v}_\alpha - \tilde{v}_{\tau\alpha}}{1-\tau}.$$

This formula essentially facilitates the determination of the required approximations $\tilde{u}_\alpha$ to the pseudosolution u_f.

This method is certainly applicable when A is symmetric nonnegative definite. The matrix A may be singular.

These properties hold for the matrix of the system of difference equations for numerical solution of the Neumann problem.

As was noted earlier, for $Au_f \neq f$ the elements $v_\alpha \not\to u_f$, $\alpha \to 0$. If $Au_f = f$, we have the limit relation

$$\lim_{\alpha \to 0} ||u_f - v_\alpha|| = 0. \tag{19}$$

This can easily be shown applying the spectral decomposition of the operator A. Furthermore, let $A = E - A_0$, where $A_0 = A^*$ and $(A_0u,u) \leq ||u||^2$. Then, obviously, the condition $A \geq 0$ is satisfied. We demand that the operator A_0 be continuous on a Banach space B inbedded in H; that is,

$$||A_0u||_B \leq ||A_0|| \; ||u||_H \quad \forall u \in H.$$

If $u_f \in B$, we have instead of Eq. (19) the stronger relation

$$\lim_{\alpha\to 0} ||u_f-v_\alpha||_B = 0. \tag{20}$$

Let us prove this. It is easily seen that

$$v_\alpha - u_f = [A_0(v_\alpha-u_f) + \alpha u_f]/(1+\alpha), \quad \alpha > 0.$$

Then

$$||v_\alpha-u_f|| \leq \frac{1}{1+\alpha}(||A_0|| \; ||v_\alpha-u_f||_H + \alpha||u_f||_B) \to 0 \quad \text{as} \quad \alpha \to 0,$$

which was to be proved.

We note that the conditions imposed are satisfied a priori if, for example, A_0 is an integral operator with a sufficiently smooth kernel.

If the element f is approximate, so that $||\tilde{f}-f||_B \leq \delta_0$, then, obviously, because B is imbedded in H,

$$||\tilde{f}-f||_H \leq \gamma||\tilde{f}-f||_B \leq \gamma\delta_0 = \delta.$$

Hence, if $\tilde{v}_\alpha$: $(\alpha E+A)\tilde{v}_\alpha = \tilde{f}$, then

$$||\tilde{v}_\alpha-v_\alpha||_H \leq ||(\alpha E+A)^{-1}|| \; ||f-\tilde{f}||_H \leq \frac{\delta}{\alpha}.$$

On the other hand,

$$\tilde{v}_\alpha-v_\alpha = \frac{1}{1+\alpha}[A_0(\tilde{v}_\alpha-v_\alpha) + (\tilde{f}-f)];$$

therefore

$$||\tilde{v}_\alpha-v_\alpha||_B \leq \frac{1}{1+\alpha}[||A_0|| \; \frac{\delta}{\alpha} + \delta_0].$$

From the triangle inequality it follows that

$$||v_\alpha-u_f||_B \to 0 \quad \text{as} \quad \alpha + \delta_0/\alpha \to 0, \quad \delta_0 \to 0.$$

If $A = E-A_0$ and the operator A does not satisfy the condition $A = A^*$, similar results hold true if the operator A_0^* acts continuously on B. In this case one should take as approximations to u_f the solutions u_α of the problem (5).

Section 25. Optimal Regularization

1. Statement of the Problem. Let A be a completely linear, continuous operator on H into F, where H and F are Hilbert spaces.

We consider the problem of minimizing

$$||Au-f||_F^2, \quad u \in H \tag{1}$$

where f is an element of F. We assume that the problem (1) has a solution.

By a *pseudosolution* u_f of the problem (1) we mean a solution of (1) with minimal norm (see Section 24).

Let there be given in H a class $\{T\}$ of nonnegative definite self-adjoint operators T such that

$$||Au||_F^2 + (Tu,u)_H \geq k^2||u||_H^2, \quad u \in D_T \tag{2}$$

where $k = k(A,T) > 0$ is a constant not depending on the choice of $u \in D_T$.

We now define the functional

$$\Phi_T[u;g] = ||Au-g||_F^2 + (Tu,u)_H, \quad u \in D_T$$

where $g \in F$ is arbitrary. It is easy to see that the element $u_T \in D_T$ minimizing $\Phi_T[u,g]$ satisfies the equation

$$(T+A^*A)u = A^*g, \tag{3}$$

where A^* is the adjoint operator of A. By (2), (3) has a unique solution

$$u_T = R_Tg, \quad R_T = (T+A^*A)^{-1}A^*.$$

Let the element f be approximate: $\tilde{f} := f+\xi$, where ξ is a "small", in the sense defined below, random process with values in F for which the condition $M\xi = 0$ is satisfied, where M is the operator of mathematical expectation. Let $\tilde{u}_T = R_T\tilde{f}$; let, also,

$$\varepsilon_T^2(u_f;\xi) = M||R_T\tilde{f}-u_f||_H^2.$$

Let $\{u_f\}$ and $\{\xi\}$ be classes of "admissible" solutions of (1) and the perturbations, respectively:

$$\varepsilon_T^2(\{u_f\}, \{\xi\}) = \sup_{u_f\in\{u_f\},\xi\in\{\xi\}} \varepsilon_T^2(u_f;\xi).$$

Further, we consider the following problem.

Problem 1. Construct the operator $T_{opt} \in \{T\}$ defined by the condition

$$\varepsilon_{T_{opt}}(\{u_f\},\{\xi\}) = \inf_{T\in\{T\}} \varepsilon_T(\{u_f\},\{\xi\}). \tag{4}$$

If this problem has a solution, the element $\tilde{u}_{opt} = R_{T_{opt}}\tilde{f}$ is said to be a $(\{T\},\{u_f\},\{\xi\})$-*optimal (approximate)* solution of the problem (1). The quantity $\varepsilon_{opt}(\{u_f\},\{\xi\}) = \varepsilon_{T_{opt}}(\{u_f\},\{\xi\})$ is said to be the error of the $(\{T\},\{u_f\},\{\xi\})$-optimal solution.

In the case where $\{T\} = \{T: T = \alpha E\}$, $\alpha > 0$, is a parameter, the problem (4) reduces to that of determining the optimal value α_{opt} of the regularization parameter.

Problem 2. Estimate errors of optimal (in the sense (4)) solutions of the problem (1).

Problem 3. Show possible effective realizations of optimal approximations (construction of quasi-optimal approximations).

2. Auxiliary facts and assertions. Let $\{e_i\}$, $i = 1,2,\ldots$, be the ortho ormal system of eigenelements of the operator A^*A, $A^*Ae_i = \lambda_i e_i$, $i = 1,2,\ldots$, $\lambda_{i+1} \le \lambda_i$, $\lambda_i > 0$. By the Hilbert-Schmidt theory, each element $u \in H$ can be represented as

$$u = u'+u'', \quad A^*Au' = 0, \quad u' \perp u'', \quad u'' = \sum_{i=1}^{\infty} u_i e_i, \quad u_i = (u,e_i)_H. \tag{5}$$

We note that the condition $A^*Au = 0$ is equivalent to $Au = 0$.

Assuming that $\omega_i = Ae_i/\sqrt{\lambda_i}$, we note that

$$AA^*\omega_i = \sqrt{\lambda_i}\,\omega_i, \quad (\omega_i,\omega_j)_F = \delta_{ij},$$

where δ_{ij} is the Kronecker delta.

Lemma 47. Each element $g \in F$ is representable as

$$g = g' + g'', \quad g' \perp g'', \quad A^*g' = 0, \quad g'' = \sum_{i=1}^{\infty} g_i w_i, \quad g_i = (g,w_i)_F. \tag{6}$$

Eqs. (5) and (6) imply the following.

Lemma 48. If u_f is the pseudosolution of Eq. (1),

$$u_f = \sum_{i=1}^{\infty} \hat{u}_i e_i, \quad \hat{u}_i = \frac{f_i}{\sqrt{\lambda_i}}, \quad f_i = (f,w_i)_F. \tag{7}$$

We shall assume from now on that the class

$$\{T\} = \{T: Te_i = \alpha_i e_i, \quad i = 1,2,\ldots\infty\}.$$

Lemma 49. We have the representation

$$\tilde{u}_T = R_T\tilde{f} = \sum_{i=1}^{\infty} \left(\frac{\lambda_i}{\alpha_i+\lambda_i}\right)\frac{\tilde{f}_i}{\sqrt{\lambda_i}}\, e_i \tag{8}$$

where

$$\tilde{f}_i = (\tilde{f},w_i)_F, \quad \tilde{f}_i = f_i + \xi_i, \quad \xi_i = (\xi,w_i)_F, \quad M\xi_i = 0.$$

We note that $\alpha_i + \lambda_i \geq K^2 > 0$ by Eq. (2).

Next, we assume that

$$\xi \in \{\xi\}_{\hat{\sigma}} = \{\xi: M\xi_i^2 = \sigma_i^2, \quad \sup_i \sigma_i^2 \leq \hat{\sigma}^2\}$$

where $\hat{\sigma}^2$ is a positive constant. Therefore, ξ can be a "generalized" random process.

Lemma 50. For any $T \in \{T\}$, $\xi \in \{\xi\}_{\hat{\sigma}}$ the quantity

$$\varepsilon_T^2(u_f,\xi) = \sum_{i=1}^{\infty} (\lambda_i^2\sigma_i^2 + \alpha_i^2 f_i^2)/\lambda_i(\alpha_i + \lambda_i)^2 \tag{9}$$

3. Construction of optimal approximations. $(\{T\},u_f,\xi)$-optimal approximations. Minimizing (9) with respect to α_i, we obtain

$$\varepsilon_{opt}^2(u_f,\xi) = \sum_{i=1}^{\infty} \frac{\sigma_i^2 f_i^2}{\lambda_i(\sigma_i^2+f_i^2)} \tag{10}$$

for $\alpha_i = \alpha_i^{opt} = \lambda_i\sigma_i^2/f_i^2$, $i = 1,2,\ldots$. Therefore, the $(\{T\},u_f,\xi)$-optimal approximation is seen to be

$$\tilde{u}_{opt} = \sum_{i=1}^{\infty} \left(\frac{f_i^2}{\sigma_i^2 + f_i^2}\right)\frac{\tilde{f}_i}{\sqrt{\lambda_i}}\, e_i \tag{11}$$

Theorem 95. Let

$$\hat{\varepsilon}_{opt}^2(u_f,\{\xi\}_{\hat{\sigma}}) = \sup_{\xi\in\{\xi\}_{\hat{\sigma}}} \varepsilon_{opt}^2(u_f,\xi)$$

Then

$$\lim_{\hat{\sigma}\to 0} \hat{\varepsilon}_{opt}(u_f,\{\xi\}_{\hat{\sigma}}) = 0.$$

Proof: We have

$$\hat{\varepsilon}^2_{opt}(u_f,\{\xi\}_{\hat{\sigma}}) = \sum_{i=1}^{\infty} \frac{\hat{\sigma}^2 f_i^2}{\lambda_i(\hat{\sigma}^2+f_i^2)} = \sum_{i=1}^{\infty} \frac{\hat{\sigma}^2\hat{u}_i^2}{\hat{\sigma}^2+\lambda_i\hat{u}_i^2} \le \frac{N}{\lambda_N}\hat{\sigma}^2 + \sum_{i=N+1}^{\infty} \hat{u}_i^2.$$

Since $\sum_{i=N+1}^{\infty} \hat{u}_i^2 \to 0$, $N \to \infty$, we can choose $N = N(\hat{\sigma})$ such that

$$\lim_{\hat{\sigma}\to 0} N(\hat{\sigma}) = +\infty, \qquad \lim_{\hat{\sigma}\to 0} \frac{N}{\lambda_N}\hat{\sigma}^2 = 0.$$

The theorem is proved.

$(\{T\},u_f,\{\xi\}_{\hat{\sigma}})$-optimal approximations. It is easy to see that

$$\varepsilon_T^2(u_f,\{\xi\}_{\hat{\sigma}}) = \sum_{i=1}^{\infty} \frac{\lambda_i^2\hat{\sigma}^2+\alpha_i^2 f_i^2}{\lambda_i(\alpha_i+\lambda_i)^2}$$

Minimizing $\xi_T(u_f,\{\xi\}_{\hat{\sigma}})$ with respect to α_i, we obtain

$$\varepsilon^2_{opt}(u_f,\{\xi\}_{\hat{\sigma}}) = \sum_{i=1}^{\infty} \frac{\hat{\sigma}^2 f_i^2}{\lambda_i(\hat{\sigma}^2+f_i^2)}$$

for

$$\alpha_i = \alpha_i^{opt} = \frac{\lambda_i\hat{\sigma}^2}{f_i^2}, \qquad i = 1,2,\dots .$$

Then $(\{T\},u_f,\{\xi\}_{\hat{\sigma}})$-optimal approximations are defined by

$$\hat{u}_{opt} = \sum_{i=1}^{\infty} \left(\frac{f_i^2}{\hat{\sigma}^2+f_i^2}\right)\frac{\tilde{f}_i}{\sqrt{\lambda_i}}\, e_i . \tag{12}$$

It is easily seen that $\varepsilon_{opt}(u_f,\{\xi\}_{\hat{\sigma}}) = \hat{\varepsilon}_{opt}(u_f,\{\xi\}_{\hat{\sigma}})$, and therefore, as was proved earlier,

$$\lim_{\hat{\sigma}\to 0} \varepsilon_{opt}(u_f,\{\xi\}_{\hat{\sigma}}) = 0.$$

$(\{T\},\{u_f\}_C,\{\xi\}_{\hat{\sigma}})$-optimal approximations. Let

$$\{u_f\}_C = \{u_f \in H: \hat{u}_i^2 \le C_i^2, \quad \sum_{i=1}^{\infty} C_i^2 < +\infty\}.$$

Then $(\{T\},\{u_f\}_C,\{\xi\}_{\hat{\sigma}})$-optimal approximations are defined by

$$\hat{u}_{opt} = \sum_{i=1}^{\infty} \frac{\lambda_i C_i^2}{\hat{\sigma}^2+\lambda_i C_i^2}\frac{\tilde{f}_i}{\sqrt{\lambda_i}}\, e_i$$

and

$$\varepsilon^2_{opt}(\{u_f\}_C,\{\xi\}_{\hat{\sigma}}) = \sum_{i=1}^{\infty} \frac{\hat{\sigma}^2 C_i^2}{\hat{\sigma}^2+\lambda_i C_i^2} \to 0, \quad \hat{\sigma} \to 0.$$

$(\{T\},\{u_f\}_N,\{\xi\}_{\hat{\sigma}})$-optimal approximations. Let the system $\{e_i\}$ be finite-dimensional (of dimension N), and let $\{u_f\}_N = \{u_f \in H: \hat{u}_i^2 \le C^2, u = 1,\ldots,N\}$.

The following assertion holds.

The $(\{T\},\{u_f\}_N,\{\xi\}_{\hat{\sigma}})$-optimal solution of the problem (1) is a solution of the equation

$$(\alpha E + A^*A)u = A^*\tilde{f},$$

for $\alpha = \alpha_{opt} = \hat{\sigma}^2/C^2$. In this case

$$\varepsilon^2_{opt}(\{u_f\}_N,\{\xi\}_{\hat{\sigma}}) = \sum_{i=1}^{N} \frac{\hat{\sigma}^2 C^2}{\hat{\sigma}^2+\lambda_i C^2}$$

Proofs of these assertions are immediate and therefore we omit them here.

4. Estimates of Accuracy of Optimal Approximations. Let

$$\Delta^2(\{u_f\},\{\xi\}) = \sup_{u_f\in\{u_f\},\xi\in\{\xi\}} \varepsilon^2_{opt}(u_f,\xi)$$

where $\{\xi\} \subseteq \{\xi\}_{\hat{\sigma}}$. In a similar way we define the quantity

$$\Delta^2(\{u_f\},\{\xi\}_{\hat{\sigma}}) = \sup_{u_f\in\{u_f\}} \varepsilon^2_{opt}(u_f,\{\xi\}_{\hat{\sigma}}).$$

Theorem 96. Let

$$\{u_f\} = \{u_f\}^p_\zeta = \{u_f: \sum_{i=1}^{\infty} (\frac{\hat{u}_i}{\lambda_i^\zeta})^{2p/(2\zeta+1)} \le R^{2p/(2\zeta+1)}$$

$$\{\xi\} = \{\xi\}^q_\sigma = \{\xi \in \{\xi\}_{\hat{\sigma}}: \sum_{i=1}^{\infty} \sigma_i^{4\zeta q/(2\zeta+1)} \le \sigma^{4\zeta q/(2\zeta+1)}\}$$

where $\alpha > 0$, $p \ge 1$, $q \ge 1$, are some constants, with $1/p+1/q = 1$.

Then

$$\Delta(\{u_f\}^p_\zeta, \{\xi\}^q_\sigma) \le C_\zeta \sigma^{2\zeta/(2\zeta+1)} R^{1/(2\zeta+1)} \tag{13}$$

where $C^2_\zeta = (2\zeta)^{2\zeta/(2\zeta+1)}/1+2\zeta)$. The estimate (13) is unimprovable.

Proof: In conjunction with (9),

$$\varepsilon^2_{opt}(u_f,\xi) = \sum_{i=1}^{\infty} \frac{\sigma_i^2 g_i^2 \lambda_i^{2\zeta}}{\sigma_i^2 + g_i^2 \lambda_i^{2\zeta+1}}$$

where $g_i^2 = \hat{u}_i^2/\lambda_i^{2\zeta}$. Let

$$\phi(\lambda) = \frac{\bar{\sigma}^2 g^2 \lambda^{2\zeta}}{\bar{\sigma}^2 + g^2\lambda^{2\zeta+1}}, \quad \lambda \geq 0, \quad \bar{\sigma}^2, \quad g^2 \neq 0.$$

Then it is easy to see that

$$\max_{\lambda>0} \phi(\lambda) = \phi(\lambda_{max}) = C_\zeta^2 (\bar{\sigma}^2)^{2\zeta/(2\zeta+1)} (g^2)^{1/(2\zeta+1)}$$

where $\lambda_{max} = (2\zeta\bar{\sigma}^2/g^2)^{1/(2\zeta+1)}$. Therefore,

$$\varepsilon^2_{opt}(u_f,\xi) \leq C_\zeta^2 \sum_{i=1}^{\infty} \sigma_i^{4\zeta/(2\zeta+1)} g_i^{2/(2\zeta+1)}.$$

Applying the Hölder inequality, we obtain

$$\varepsilon^2_{opt}(u_f,\xi) \leq C_\zeta^2 \left[\sum_{i=1}^{\infty} \sigma_i^{4\zeta q/(2\zeta+1)}\right]^{1/q} \left[\sum_{i=1}^{\infty} g_i^{2p/(2\zeta+1)}\right]^{1/p}$$

yielding (13).

Let $\bar{u}_f = \hat{u}_i e_i$, where $\hat{u}_i = \lambda_i^\zeta R^2$, $\bar{\xi} = \xi_i \omega_i$, where ξ_i is such that $M\xi_i^2 = \sigma^2 = (1/2\zeta)\lambda_i^{2\zeta+1} R^{2(2\zeta+1)}$. It is easy to verify that $\bar{u}_f \in \{u_f\}_\zeta^p$, $\bar{\xi} \in \{\xi\}_\sigma^q$, with

$$\varepsilon_{opt}(\bar{u}_f,\bar{\xi}) = C_\zeta \sigma^{2\zeta/(2\zeta+1)} R^{1/(2\zeta+1)}.$$

that is, the estimate (13) is unimprovable. □

Remark. 1. Let $p = 1$, $q = \infty$. Then $\{\xi\}_\sigma^\omega = \{\xi\}_{\hat{\sigma}}$ and therefore

$$\hat{\Delta}(\{u_f\}_\zeta^1,\{\xi\}_{\hat{\sigma}}) \leq C_\zeta \hat{\sigma}^{2\zeta/(2\zeta+1)} R^{1/(2\zeta+1)}. \tag{14}$$

2. Let $p = 2\zeta+1$, $q = (2\zeta+1)/2\zeta$. Also, let

$$\{u_f\}_\zeta = \{u_f\}_\zeta^p = \left\{u_f\colon \sum_{i=1}^{\infty} \left(\frac{\hat{u}_i}{\lambda_i^\zeta}\right)^2 \leq R^2\right\}$$

$$\{\xi\}_\sigma = \{\xi\}_\sigma^q = \{\xi\colon \sum_{i=1}^{\infty} \sigma_i^2 \leq \sigma^2\}.$$

Then

$$\Delta(\{u_f\}_\zeta,\{\xi\}_\zeta) \leq C_\zeta \sigma^{2\zeta/(2\zeta+1)} R^{1/(2\zeta+1)}. \tag{15}$$

We list here sufficient conditions for pseudosolutions of (1) to belong to the class $\{u_f\}_\zeta$.

Lemma 51. If $u_f = A^*h$, $h \in F$, $||h||_F \le R$, then $u_f \in \{u_f\}_\zeta$ for $\zeta = 1/2$. If $u_f = (A^*A)^\zeta w$, $w \in H$, then $u_f \in \{u_f\}_\zeta$ for any $\zeta > 0$.

The conditions of the lemma are satisfied, for example, if the operator A is normally solvable (Section 24). To prove this lemma one should use the decompositions (5) and (6) for the elements w and h, respectively.

Remark 3. Let $\{u_f\}^*_\zeta = \{u_f\colon u_f = (A^*A)^\zeta_w, w \in H, ||w||_H \le R\}$. It is easy to see that the class $\{u_f\}^*_\zeta$ is compact in H for any $\zeta > 0$. We define the quasi-solution $\tilde{u}_\zeta$ of the problem (1) to be a solution of the problem of minimizing

$$||A\tilde{u}_\zeta - \tilde{f}||^2_F, \quad u \in \{u_f\}^*_\zeta.$$

(Here we assume that ξ is not random.) Assuming

$$\tilde{\Delta}(\{u_f\}^*_\zeta, \{\xi\}^*_\sigma) = \sup_{u_f \in \{u_f\}^*_\zeta, \xi \in \{\xi\}^*} ||\tilde{u}_\zeta - u_f||_H$$

where $\{\xi\}^*_\sigma = \{\xi \in F\colon ||\xi||_F \le \sigma$, it is possible to show that

$$\tilde{\Delta}(\{u_f\}^*_\zeta, \{\xi\}^*_\sigma) \le 2^{1/(2\zeta+1)}\sigma^{2\zeta/(2\zeta+1)}R^{1/(2\zeta+1)} \tag{16}$$

where $\tilde{C}_\zeta = 2^{1/(2\zeta+1)}$.

Comparing (15) with (16), we note that the orders of the estimates coincide. However, the constant in (15) is smaller than the one in (16), at least for $\zeta = 1/2, 1, 2$.

5. On Realization of Optimal Approximations. Realization of the $(\{T\}, u_f, \{\xi\}_{\hat{\sigma}})$-optimal approximations (12). We define now the element

$$\tilde{u}_{kopt} = \sum_{c=1}^{\infty} \left(\frac{\tilde{f}_i^2}{\hat{\sigma}^2 + \tilde{f}_i^2}\right) \frac{\tilde{f}_i}{\sqrt{\lambda_i}} e_i \tag{17}$$

which we will call a $(\{T\}, u_f, \{\xi\}_{\hat{\sigma}})$-*quasi-optimal approximation*. We note that in order to realize (17), we need to know only $\hat{\sigma}$; that is, the element f must be given exactly.

Further, we make an assumption which we will call the *linearization principle*. Let

$$F(\xi_i) = \frac{\tilde{f}_i^3}{\sqrt{\lambda_i}(\hat{\sigma}^2 + \tilde{f}_i^2)} - \frac{f_i}{\sqrt{\lambda_i}}$$

Then we assume that in the decomposition

$$F(\xi_i) = F(0) + F'(0)\xi_i + \cdots$$

we have a "sufficient" degree of accuracy if we restrict ourselves to linear terms.

Theorem 97. Let the linearization principle be satisfied. Then

$$M||\tilde{u}_{k\ opt} - u_f||^2 \leq g\varepsilon^2_{opt}(u_f,\{\xi\}_{\hat{\sigma}}). \tag{18}$$

Proof: We have

$$F(\xi_i) \approx \frac{1}{\lambda_i}\left[-\frac{\hat{\sigma}^2 f_i^2}{\hat{\sigma}^2+f_i^2} + \frac{3f_i^2\hat{\sigma}^2+f_i^4}{(\hat{\sigma}^2+f_i^2)^2}\xi_i\right], \qquad i = 1,2,\ldots$$

Therefore,

$$M||\tilde{u}_{k\ opt} - u_f||^2 \approx M\sum_{i=1}^{\infty} F^2(\xi_i) = \sum_{i=1}^{\infty}\left[\frac{\hat{\sigma}^2 f_i^2}{\lambda_i(\hat{\sigma}^2+f_i^2)} + \frac{4f_i^4\hat{\sigma}^4}{\lambda_i(\hat{\sigma}^2+f_i^2)^3}\right.$$

$$\left. + \frac{4f_i^4\hat{\sigma}^6}{\lambda_i(\hat{\sigma}^2+f_i^2)^4}\right] \leq \sum_{i=1}^{\infty}\frac{\hat{\sigma}^2 f_i^2}{\lambda_i(\hat{\sigma}^2+f_i^2)}.$$

The theorem is proved.

Corollary. If u_f $\{u_f\}^*$, taking into account (14), we can obtain the estimate of accuracy of quasi-optimal approximations:

$$M||\tilde{u}_{k\ opt} - u_f||^2 \leq gC_\zeta^2\hat{\sigma}^{4\zeta/(2\zeta+1)}R^{2/(2\zeta+1)}.$$

The regularization method. Let $u_f \in \{u_f\}_\zeta$, $\xi \in \{\xi\}_\sigma$, $\{T\} = \{T_\alpha^\zeta\} = \{T\colon Te_i = \alpha\lambda^{-2\zeta}e_i\}$, $\alpha > 0$ a parameter. Then

$$\varepsilon^2_{T_\alpha^\zeta}(u_f,\xi) = \sum_{i=1}^{\infty}\frac{\lambda_i^{4\zeta+1}\sigma_i^2 + \alpha^2\lambda_i^{2\zeta}g_i^2}{(\alpha+\lambda_i^{2\zeta+1})^2}$$

where,,as earlier, $g_i^2 = \hat{u}_i^2\lambda_i^{2\zeta}$. Let

$$\phi(\lambda) = \frac{\lambda^{4\zeta+1}}{(\alpha+\lambda^{2\zeta+1})^2}, \qquad \psi(\lambda) = \frac{\lambda^{2\zeta}}{(\alpha+\lambda^{2\zeta+1})^2}, \qquad \lambda > 0.$$

We have

$$\max_\lambda \phi(\lambda) = \phi(\lambda') = C'_\zeta\alpha^{-1/(2\zeta+1)},$$

$$\max_\lambda \psi(\lambda) = \psi(\lambda'') = C''_\zeta\alpha^{2\ /(2\zeta+1)},$$

$$C'_\zeta = \frac{(4\zeta+1)^{(4\zeta+1)/(2\zeta+1)}}{(4\zeta+2)^2}, \qquad C''_\zeta = \left(\frac{\zeta+1}{2\zeta+1}\right)^2\left(\frac{\zeta}{\zeta+1}\right)^{2\zeta/(2\zeta+1)}$$

and

$$\lambda' = \left(\frac{1}{4\zeta+1}\right)^{1/(2\zeta+1)}\alpha^{1/(2\zeta+1)}, \qquad \lambda'' = \left(\frac{\zeta}{\zeta+1}\right)^{1/(2\zeta+1)}\alpha^{1/(2\zeta+1)}.$$

Then

$$\varepsilon^2_{T^\zeta_\alpha}(u_f,\xi) \le C'_\zeta\sigma^2\alpha^{-1/(2\zeta+1)} + C''_\zeta R^2\alpha^{2\zeta/(2\zeta+1)} = \varepsilon^2_\zeta(\alpha) \tag{19}$$

for any $u_f \in \{u_f\}_\zeta$, $\xi \in \{\xi\}_\sigma$. Hence

$$\inf_{\alpha>0} \varepsilon^2_{T^\zeta_\alpha}(\{u_f\}_\zeta,\{\xi\}_\sigma) \le \inf_{\alpha>0} \varepsilon^2_\zeta(\alpha).$$

But

$$\inf_\alpha \varepsilon^2_\zeta(\alpha) = \varepsilon^2_\zeta(\alpha_\zeta) = C^{*2}_\zeta\sigma^{4\zeta/(2\zeta+1)}R^{2/(2\zeta+1)},$$

where

$$C^{*2}_\zeta = C'_\zeta\left(\frac{C'_\zeta}{2\zeta C''}\right)^{-1/(2\zeta+1)} + C''_\zeta\left(\frac{C'_\zeta}{2\zeta C''}\right)^{2\zeta/(2\zeta+1)}$$

and

$$\alpha_\zeta = \frac{C'_\zeta}{2_\zeta C''_\zeta}\sigma^2R^{-2} \approx \sigma^2R^{-2}. \tag{20}$$

Therefore

$$\inf_\alpha \varepsilon_{T^\zeta_\alpha}(\{u_f\}_\zeta,\{\xi\}_\sigma) \le C^*_\zeta\sigma^{2\zeta/(2\zeta+1)}R^{1/(2\zeta+1)}. \tag{21}$$

The estimate thus obtained is unimprovable in the order of the variables σ and R. In order to see that this is true, it suffices to assume that $\overline{u}_f = \hat{u}_i e_i$, $\hat{u}_i = \lambda_i^{2\zeta}R^2$, $\overline{\xi} = \xi_i\omega_i$, $M\xi_i^2 = \sigma^2$ and $\alpha = \sigma^2R^{-2}$ and, furthermore, to determine $\varepsilon_{T^\zeta_\alpha}(\overline{u}_f,\overline{\xi})$.

Therefore, the following theorem holds.

Theorem 98. The error of the $(\{T^\zeta_\alpha\},\{u_f\}_\zeta,\{\xi\}_\sigma)$-optimal approximation does not exceed the right side of the estimate (21) which is unimprovable in the order of the variables σ and R. To obtain quasi-optimal approximations in a $\{T^\zeta_\alpha\}$-regularizing algorithm it suffices to choose $\alpha = \alpha_\zeta$ using the formula (20). The accuracy of these approximations coincides in the order of the variables σ and R with the accuracy of optimal approximations.

Remark. The computation of $(\{T_\alpha^\zeta\},\{u_f\}_\zeta,\{\xi\}_\sigma)$-quasi-optimal approximations can be difficult because of the need to compute the spectral structure of the operator A^*A. Let the operator G be such that $Ge_i = \lambda_i^{-2\zeta}e_i$, u' is the null space of A^*A, $Gu' = 0$. Then the $\{T_\alpha^\zeta\}$-regularization is reduced to solving the equation

$$(\alpha G + A^*A)u = A^*f,$$

which can be satisfied sufficiently effectively.

Next we consider the case of a $\{T_\alpha\}$-regularization where $T_\alpha = \alpha E$. It is easy to see that

$$\varepsilon_{T_\alpha}^2(u_f,\xi) = \sum_{i=1}^{\infty} \frac{\lambda_i\sigma_i^2}{(\alpha+\lambda_i)^2} + \sum_{i=1}^{\infty} \frac{\alpha^2\hat{u}_i^2}{(\alpha+\lambda_i)^2} .$$

We assume that $u_f \in \{u_f\}_\zeta$, $\xi \in \{\xi\}_\sigma$. Then

$$\varepsilon_{T_\alpha}^2(u_f,\xi) = \sum_{i=1}^{\infty} \frac{\lambda_i\sigma_i^2}{(\alpha+\lambda_i)^2} + \alpha^2 \sum_{i=1}^{\infty} \frac{\lambda_i^{2\zeta}g_i^2}{(\alpha+\lambda_i)^2}$$

where, as before, $g_i^2 = \hat{u}_i^2/\lambda_i^{2\zeta}$.

Lemma 52. Let $\phi(\lambda) = \lambda/(\alpha+\lambda)^2$, $\psi(\lambda) = \lambda^{2\zeta}/(\alpha+\lambda)^2$, $\lambda > 0$. Then

$$\phi(\lambda) \le \phi(\lambda') = \frac{1}{4\alpha}, \quad \psi(\lambda) \le \psi(\lambda'') = \begin{cases} \overline{C}\,\alpha^{2\zeta-2}, & 0 < \zeta \le 1 \\ \lambda^{2(\zeta-1)}, & \zeta \ge 1 \end{cases}$$

$$\lambda' = \alpha, \quad \overline{C}_\zeta = \left(\frac{\zeta}{1-\zeta}\right)^{2\zeta}(1-\zeta)^2$$

$$\lambda'' = \begin{cases} \left(\frac{\zeta}{1-\zeta}\right), & 0 < \zeta \le 1 \\ +\infty, & \zeta \ge 1. \end{cases}$$

Using this lemma, we obtain

$$\sum_{i=1}^{\infty} \frac{\lambda_i\sigma_i^2}{(\alpha+\lambda_i)^2} \le \frac{\sigma^2}{4\alpha}, \quad \alpha^2 \sum_{i=1}^{\infty} \frac{\lambda_i^{2\zeta}g_i^2}{(1+\lambda_i)^2} \le \begin{cases} \alpha^{2\zeta}\overline{C}_\zeta R^2, & 0 < \zeta \le 1 \\ \alpha^2 R^2, & \zeta \ge 1. \end{cases}$$

The estimates thus obtained are unimprovable. In fact, assuming $\xi = \xi_i\omega_i$, $M\xi_i^2 = \sigma^2$, $\alpha = \lambda_i$, we obtain

$$\sum_{i=1}^{\infty} \frac{\lambda_i\sigma_i^2}{(\alpha+\lambda_i)^2} = \frac{\sigma^2}{4\alpha} .$$

Similarly, for $0 < \zeta < 1$, assuming that $u_f = u_i e_i$, $u_i^2 = \lambda_i^{2\zeta} R^2$, $\alpha = (1-\zeta)/\zeta)\lambda_i$, we have

$$\alpha^2 \sum_{i=1}^{\infty} \frac{\lambda_i^{2\zeta} g_i^2}{(\alpha+\lambda_i)^2} = \alpha^{2\zeta} \bar{C}_\zeta R^2.$$

It remains to note that for $\zeta \geq 1$

$$\lim_{\alpha \to 0} \sum_{i=1}^{\infty} \frac{\lambda_i^{2\zeta} g_i^2}{(\alpha+\lambda_i)^2} = R^2.$$

Using the estimates obtained above, we have

$$\varepsilon_{T_\alpha}^2(\{u_f\}_\zeta, \{\xi\}_\sigma) \leq \frac{\sigma^2}{4\alpha} + \begin{cases} \alpha^{2\zeta} \bar{C}_\zeta R^2, & 0 < \zeta \leq 1 \\ \alpha^2 R^2, & \zeta \geq 1. \end{cases}$$

The minimum of the right side of the foregoing estimate is attained for

$$\alpha = \alpha_\zeta = \begin{cases} \dfrac{\sigma^2}{2\zeta \bar{C}_\zeta} R^{-2/(2\zeta+1)}, & 0 < \zeta \leq 1 \\ \dfrac{\sigma^2}{8} R^{-2/3}, & \zeta \geq 1 \end{cases} \tag{22}$$

which yields

$$\varepsilon_{T_\alpha\,opt}^2(\{u_f\}, \{\xi\}_\sigma) \leq \begin{cases} \left[\dfrac{2\zeta+1}{8\zeta \frac{2\zeta+1}{2\zeta}}\right] \sigma^{4\zeta/(2\zeta+1)} (\bar{C}_\zeta R^2)^{1/(2\zeta+1)}, & 0<\zeta\leq 1 \\ \dfrac{3}{\sqrt[3]{64}} \sigma^{4/3} R^{2/3}, & \zeta \geq 1 \end{cases} \tag{23}$$

Theorem 98. Let $\tilde{u}_\alpha$ be a solution of the equation

$$(\alpha E + A^*A)u = A^*\tilde{f}$$

and let the parameter α be defined by (22). Then the accuracy of the approximation obtained is given by the right side of the estimate (23).

Remark. It is not hard to see that the orders of the estimates (21) and (23) for $0 < \zeta \leq 1$ coincide. Although the estimate (23) is of lower order with respect to σ, for $\zeta \geq 1$, the constant involved does not depend on ζ, which makes it easier to apply. Furthermore, the $\{T_\alpha\}$-algorithm is easier to realize than the $\{T_\alpha^\zeta\}$-algorithm.

6. Example. The results given above are applicable to many problems. As an example, we conisder the problem of solving the evolution equatoon

$$\frac{du}{dt} = Lu, \quad u(0) = f, \quad 0 < t \leq T \tag{24}$$

where L is a positive definite self-adjoint linear operator on H into H with discrete unbounded spectrum. We assume that the problem (24) has a solution.

Let $A_t = e^{-Lt}$. Then Eq. (24) is reduced to solving the parameter operator equation

$$A_t u = f. \tag{25}$$

Let $\{e_i\}$ be a complete system of eigenfunctions of the operator L:

$$Le_i = \Lambda_i e_i, \quad i = 1,2,\ldots, \quad \lim_{i\to\infty} \Lambda_i = +\infty.$$

Then

$$u_f(t) = \sum_{i=1}^{\infty} \exp(\Lambda_i t) f_i e_i,$$

where $f_i = (f,e_i)_H$.

The $(\{T\},u_f,\xi)$-optimal solution of the problem (25) is seen to be

$$\tilde{u}_{opt}(t) = \sum_{i=1}^{\infty} \frac{f_i^2}{\sigma_i^2 + f_i^2} \exp(\Lambda_i t)\tilde{f}_i e_i .$$

which easily yields a $(\{T\},u_f,\{\xi\}_{\hat{\sigma}})$-optimal and, respectively, quasi-optimal solutions.

Let $\zeta = (T-t)/2t$. Then, obviously,

$$u_f(t) = \sum_{i=1}^{\infty} \exp[-2\Lambda_i \zeta t + \Lambda_i T] f_i e_i = (A_t^* A_t)^{\zeta} u_f(T) = A_t^{2\zeta} u_f(T);$$

that is, the conditions of Lemma 51 are satisfied.

The $\{T_\alpha^\zeta\}$-regularized approximations are defined by the formula

$$\tilde{u}_\alpha(t) = \sum_{i=1}^{\infty} \exp(A_i t)(1 + \alpha \exp(2\Lambda_i T))^{-1}\tilde{f}_i e_i .$$

Noting that $\{u_f\}_\zeta = \{u_f\colon ||u_f(T)||_H^2 \leq R^2\}$ and, in addition, using (21), we obtain

$$M||\tilde{u}_\zeta^\alpha(t) - u_f(t)||_H^2 \leq C_{\zeta(t)}^{*2} \sigma^{2(1-(t/T))} ||u_f(T)||_H^{2t/T}, \quad 0 \leq t \leq T.$$

As we proved earlier, the order of this estimate is unimprovable.

To conclude, we note that our analysis is valid also in the case where the element f is assumed to be random; we need only to require that $M\xi_i f_i = 0$ and replace everywhere f_i^2 by Mf_i^2. In the case where $A = E$, Problem 1 coincides with the problem of optimal filtering in the random process f.

Section 26. Numerical Algorithms for Regularization Parameters

1. Numerical solution of ill-posed problems by Tikhonov's regularization method is usually reduced to the problem of minimizing the quadratic form

$$\phi_\alpha[u] = ||Au-f||_m^2 + \alpha||u||_C^2, \quad \alpha > 0 \tag{1}$$

on the space of vectors $u = (u_1,u_2,\dots,u_n) \in R_n$ as well as to the problem of determining the regularization parameter α from an additional condition of the type

$$\psi(u_\alpha) = \ell_\psi,$$

where ψ is a nonlinear functional defined on solutions of the problem (1), and ℓ_ψ is the a priori given level of admissible values of the functional ψ.

In (1) A denotes a rectangular $m \times n$ matrix. We denote the Euclidean norms in R_n and R_m by $||\cdot||_n$ and $||\cdot||_m$, respectively; C is the a priori given positive definite $n \times n$ matrix defining the degree of "consistency" of components of the vector u, and $||u||_C = (Cu,u)_n^{1/2}$.

Next we consider the following three techniques for choosing the regularization parameter α:

$$\rho(\alpha) \equiv ||Au_\alpha - f||_m = e \tag{2}$$

$$\gamma(\alpha) \equiv ||u_\alpha||_C = e_\gamma \tag{3}$$

$$\phi(\alpha) \equiv \rho^2(\alpha) + \alpha\gamma^2(\alpha) = e_\phi. \tag{4}$$

The quantity e_ρ characterizes the given level of residual and it can be found by the residual method. The quantity e_γ is determined by the method of quasi-solutions and it depends on the a priori information on the dimension of a C-sphere which contains the required solution. The quantity e_ϕ is chosen on the basis of the same considerations as the quantity e_ρ.

We proved before that the functions ρ, γ, and ϕ are continuous on $(0,\infty)$, ρ and ϕ being strictly monotone increasing and γ being strictly monotone decreasing. At the same time we determined the limit values for $\rho(\alpha)$ and $\phi(\alpha)$ as $\alpha \to 0, +\infty$. Therefore, if e_ρ, e_γ, e_ϕ belong to sets of values of the functions ρ, γ and ϕ, (2)-(4) have a unique solution. We assumed this condition to be satisfied. We denote the roots of (2)-(4) by α_ρ, α_γ and α_ϕ, respectively; we also assume that $0 < \alpha_\psi < +\infty \ \forall\psi$.

We now note that the properties of the functions ρ, γ and ϕ mentioned above do not ensure unconditional applicability of rapidly convergent algorithms for finding roots of (2)-(4) of the Newton type. We shall indicate below appropriate modifications of these equations, to which Newton-type methods can be applied.

2. Let

$$R(\alpha) = \rho(1/\alpha), \quad F(\alpha) = \phi(1/\alpha), \quad 0 < \alpha < \infty.$$

We have the following lemma.

Lemma 53. The functions R^s, γ^s, and F^s are decreasing convex downward functions for any $s > 0$ and, furthermore, they are increasing convex upward functions for $-1 \leq s < 0$. If $s < -1$, these functions change the direction of convexity.

Proof: Decomposing C as K^TK, where K is upper triangular, and also assuming that $x_\alpha = Ku_\alpha$, $N = AK^{-1}$, we obtain

$$\rho(\alpha) = ||Nx_\alpha - f||_m, \quad \gamma(\alpha) = ||x_\alpha||_n, \quad \phi(\alpha) = ||Nx_\alpha - f||_m^2 + \alpha||x_\alpha||_n^2 \tag{5}$$

Now let $N = P_1\Lambda P_2$ be the *singular-value decomposition* of the matrix N, that is, P_1 and P_2 are orthogonal and Λ is "diagonal". Using the fact that x_α satisfies the equation

$$(N^TN + \alpha E)x_\alpha = N^Tf$$

and, furthermore, making simple transformations in Eq. (5), we arrive at the final formulas

$$\begin{gathered} R(\alpha) = \left(\sum_{i=1}^m \left(\frac{g_i}{\alpha\lambda_i^2+1}\right)^2\right)^{1/2}, \quad \gamma(\alpha) = \left(\sum_{i=1}^m \left(\frac{\lambda_i g_i}{\lambda_i^2+\alpha}\right)^2\right)^{1/2} \\ F(\alpha) = \sum_{i=1}^m \frac{g_i^2}{\alpha\lambda_i^2+1} \end{gathered} \tag{6}$$

where λ_i is the ith eigenvalue of the matrix N for $i \leq n$ and $\lambda_i = 0$ for $i > n$, g_i is the ith component of the vector P_1^Tf.

From the representation (6) the monotonicity of the functions R^s,

γ^s and F^s is obvious. The convexity properties formulated in the lemma follow from the fact that the sign of the second derivative of

$$y(\alpha) = \left(\sum_{i=1}^{m} \frac{a_i}{(b_i\alpha+c_i)^k}\right)^{\ell}$$

(where a_i, b_i, c_i are nonnegative, $k > 0$ for $\alpha > 0$ and $\ell \geq -(1/k)$) coincides with the sign of ℓ.

In the sequel it will be convenient to call the values $s \geq -1$, $s \neq 0$ *admissible*. We also note that the functions R^s, γ^s, and F^s can easily be defined more precisely using right continuity at $\alpha = 0$ to preserve smoothness.

Next, we consider the equations

$$R^s(\alpha) = e_\rho^3, \quad \gamma^s(\alpha) = e_\gamma^s, \quad F^s(\alpha) = e_\phi^s. \tag{7}$$

The following theorem is an immediate corollary of the lemma as well as of the solvability condition of (2)-(4).

Theorem 99. Newton's method of tangents for finding roots of Eq. (7) is convergent for admissible values of s for any initial approximation $\alpha \geq 0$.

3. We choose now values of degrees s yielding the maximal increment of the parameter α at each step (starting from the second step) while solving Eq. (7) by Newton's method of tangents. For the increment $\Delta\alpha$ we obtain

$$\Delta\alpha = \frac{e_y^s - y^s}{sy^{s-1}y'} = -\frac{y}{y'}\left[\frac{1-(e_y/y)^s}{s}\right] \tag{8}$$

where y is one of the functions R, γ or F; $e_y = e_\rho$, e_γ or e_ϕ, respectively.

We shall investigate $\Delta\alpha$ as the function s. Using the lemma for $s = 1$, we obtain that the multiplier $-y/y'$ in (8) is positive and, in addition, the positive quantity $q = e_y/y$ is less than unity, at least starting from the second step. This implies that for any $s \neq 0$ the expression in the square brackets is positive. Since the function $(1-q^s)/s$ is the decreasing function of s, the maximum of the former for $s \geq -1$ is attained for $s = -1$. We note that for $s < -1$ the unconditional convergence of Newton's method for Eqs. (7) is not guaranteed. Therefore, the value $s = -1$ ensures the maximal increment $\Delta\alpha$ for every function at each step in Newton's method; that is, the maximal rate of convergence of the method.

Next we estimate the rate of convergence of Newton's method while solving (7). In this case we confine ourselves only to the case $s = -1$, which is the best in the sense indicated.

We shall write (7) as

$$Y(\alpha) = y^{-1}(\alpha) - e_y^{-1} = 0. \tag{9}$$

Here $y(\alpha)$ has the same meaning as in (8). As is well known, the rate of convergence of Newton's method of tangents is given by the formula

$$\alpha_y - \alpha_{n+1} = M_y(\alpha_y - \alpha_n)^e,$$

where α_n and α_{n+1} are respectively, the nth and (n+1)th approximations to the root α_y of Eq. (9), and

$$M_y = -\frac{1}{2}(Y''(\xi_n)/Y'(\alpha_n)),$$

where for the signs Y' and Y'' found above we have:

$$\alpha_n \le \xi_n \le \alpha_y, \quad \alpha_n \le \alpha_{n+1} \le \alpha_y, \qquad n \ge 2.$$

For α_n sufficiently close to the value of the root of the corresponding equation we have the following estimates for M_y for $y = R, \gamma$, and F:

$$M_R \le \frac{3}{2}\left[\frac{R(\alpha_n)}{R(\xi_n)}\right]^3 \lambda_{max}^2 \le \frac{3}{2}\,\frac{\lambda_{max}^2}{[1-(\alpha_\rho-\alpha_n)\lambda_{max}^2]^3}$$

$$M_\gamma \le \frac{3}{2}\left[\frac{\gamma(\alpha_n)}{\gamma(\xi_n)}\right]^3 \lambda_{min}^{-2} \le \frac{3}{2}\,\frac{1//_{min}}{[1-(\alpha_\gamma-\alpha_n)/\lambda_{min}^2]^3}$$

$$M_F \le \left[\frac{F(\alpha_n)}{F(\xi_n)}\right]^2 \lambda_{max}^2 \le \frac{\lambda_{max}^2}{[1-(\alpha_\phi-\alpha_n)\lambda_{max}^2]^2}$$

where $\lambda_{max} = \max_i |\lambda_i|$, $\lambda_{min} = \min_{i,\lambda_i \ne 0} |\lambda_i|$, and the λ_i, as before, are the eigenvalues of the matrix N. Therefore, the quantities M_R, M_γ, and M_F can, asymptotically, be approximately bounded by

$$\frac{3}{2}\lambda_{max}^2, \qquad \lesssim \frac{3}{2}\lambda_{min}^{-2}, \qquad \le \lambda_{max}^2 \tag{10}$$

respectively.

4. We proceed to describing the computational scheme for determining roots of (7). As is seen from (7), for each Newton step, we need the values $y(\alpha_n)$ and $y'(\alpha_n)$. Since solving the minimization problem (1) is equivalent to solving the system of linear equations

$$(A^T A + \alpha C)u_\alpha = A^T f, \tag{11}$$

it is possible to compute the value of the function $y(\alpha)$ for $\alpha = \alpha_n$ after we have determined the solution u_α of the system (11) using one of the formulas defining the function $y(\alpha)$.

To compute the values of $y'(\alpha)$ we shall need the values of derivatives $\rho'(\alpha)$, $\gamma'(\alpha)$ or $\phi'(\alpha)$. Using the definitions of these functions, we obtain

$$\rho'(\alpha) = -\alpha \frac{(u'_\alpha, Cu_\alpha)}{\rho(\alpha)}, \quad \gamma'(\alpha) = \frac{(u'_\alpha, Cu_\alpha)}{\gamma(\alpha)}, \quad \phi'(\alpha) = (u_\alpha, Cu_\alpha), \tag{12}$$

where the vector $u'_\alpha = du_\alpha/d\alpha$ can be found by solving the system

$$(A^T A + \alpha C)u'_\alpha = -Cu_\alpha \tag{13}$$

with the same matrix as that for the computation of u_α but with a different right-hand side. We obtain (13) by direct differentiation of (11). It is easy to compute derivatives for the functions R and F using the relation (12); that is,

$$R'(\alpha) = p^3 \frac{(u'_p, Cu_p)}{\rho(p)}, \quad \gamma'(\alpha) = \frac{(u'_\alpha, Cu_\alpha)}{\gamma(\alpha)}, \quad F'(\alpha) = -p^2(u_p, Cu_p), \tag{14}$$

where $p = 1/\alpha$. We find the vectors u_p and u'_p from the equations

$$(A^T A + pC)u_p = A^T f, \quad (A^T A + pC)u'_p = -Cu_p. \tag{15}$$

To reduce the amount of computations, one can use further transformation. For example, the matrix C can be decomposed into the product $C = K^T K$, where K is upper triangular, by the square root method. Next, one reduces the matrix $N = AK^{-1}$ to a "bidiagonal" form: $N = QDR$, where Q and R are orthogonal matrices and D is a "bidiagonal" matrix. Letting $z_\alpha = RKu_\alpha$, we reduce (11) to a system with a tridiagonal matrix $D^T D$:

$$(D^T D + \alpha E)z_\alpha = D^T \phi, \quad \phi = Q^T f. \tag{16}$$

The solution u_α of the initial system (11) is found as a solution z_α of the transformed system (16) from the relation

$$Ku_\alpha = R^T z_\alpha.$$

It is easy to obtain the following expressions for the functions ρ, γ, and ϕ:

$$\rho(x) = ||Dz_\alpha - \phi||_m, \quad \gamma(\alpha) = ||z_\alpha||_n, \quad \phi(\alpha) = \rho^2(\alpha) + \alpha\gamma^2(\alpha), \tag{17}$$

For their derivatives,

$$\rho'(\alpha) = \alpha \frac{(z'_\alpha, z_\alpha)}{\rho(\alpha)}, \quad \gamma'(\alpha) = \frac{(z'_\alpha, z_\alpha)}{\gamma(\alpha)}, \quad \phi'(\alpha) = (z_\alpha, z_\alpha),$$

where z'_α is the solution of the system

$$(\alpha E + D^T D) z'_\alpha = -z_\alpha. \tag{18}$$

Then

$$R'(\alpha) = p^3 \frac{(z'_p, z_p)}{\rho(p)}, \quad \gamma'(\alpha) = \frac{(z'_\alpha, z_\alpha)}{\gamma(\alpha)}, \quad F'(\alpha) = -p^2 (z_p, z_p)$$

where $p = 1/\alpha$ and z_p and z'_p are the solutions of the tridiagonal systems

$$(pE + D^T D) = D^T \phi, \quad (D^T D + pE) z'_p = -z_p \tag{19}$$

with the same matrix of coefficients.

The formulas (17), (18) and (19) involve considerably less computations than the formulas (2)-(4), (14) and (15).

5. We have considered earlier three techniques for determining a regularization parameter, which we have called for convenience the ρ, ϕ and γ-methods. However, from the computational point of view, these methods are not equally effective.

For the ϕ-method the system of equations has to be solved at each step of Newton's method; for the other two methods, it has to be solved twice at each step.

Next, the equations approximate the respective roots on the left, at least starting from the second step. For the methods ρ and ϕ this is a favorable factor because the conditioning of the systems of (15) or (19) is best for $\alpha = 0$ and deteriorates as α increases. For the method γ this factor is unfavorable since for $\alpha = 0$ the systems (11) and (13) or (16) and (18) are conditioned in the worst way (it is also assumed that these systems are degenerate). This can lead to a large loss of accuracy or even to overflow of the discharing mesh when u_α (z_α)

is being determined if α turns out to be sufficiently close to zero at the start or as a result of the first step of Newton's method.

The inequalities (10) show that with respect to the rate of convergence the γ-method is better than the other methods. This happens because the quantity λ_{min}^{-2} depends on the arrangement of eigenvalues of the matrix N, while the quantity λ_{max}^{2} is always bounded by the number $||N^TN||$.

We note that in the methods ρ and ϕ, one can always take $\alpha = 0$ as the initial approximation to the root. As can easily be calculated,

$$R(0+) = ||f||_m = ||\phi||_m, \quad R'(0_+) = -\frac{(C^{-1}A^Tf, A^Tf)_n}{||f||_m}$$

$$= -\frac{||N^Tf||_m^2}{||f||_m} = -\frac{||D^T||_m^2}{||\phi||_m}$$

$$F(0+) = ||f||_m^2 = ||\phi||_m^2, \quad F'(0+) = -(C^{-1}A^Tf, A^Tf)_n = -||N^Tf||_m^2$$

$$= ||D^T\phi||_m^2.$$

The method ρ for choosing the regularization parameter as described above is given in FORTRAN (Numerical Analysis on the FORTRAN, 6, 7, ed. by V. V. Vojevodin, Moscow, Izd. MGU, 1974), where matrix decompositions were used, reducing the initial regularized system of equations to the system (16) with a tridiagonal matrix. The regularized solution u is restored only after the root $\alpha = \alpha_\rho$ of Eq. (2) has been determined.

One takes always $\alpha = 0$ as the initial approximation to the root. The computations performed showed unexpectedly rapid convergence of the ρ-method. For example, the normal solution of a system of equations of the order eight with a singular matrix was determined in two iterations with relative accuracy of computation of the root of (2) equal to 10^{-3}. This fact suggests that if high accuracy of the root of (2) is not needed, then in solving the system (15) with normal accuracy there is no need to make matrix decompositions since they require considerable computer time which is justified only for repeated solutions of the regularized system of equations. In the present case a regularized solution satisfying (2) can be obtained in quite reasonable time immediately on the basis only of the formulas (15).

6. Let L be an $s \times n$ matrix such that the quadratic form $||Au||_m^2 + ||Lu||_s^2$ is positive definite. This assumption ensures, obviously, that the completeness condition is satisfied. Let u_α denote

the minimum of the functional

$$\Phi_\alpha[u] = ||Au-f||_m^2 + \alpha||Lu-g||_s^2, \quad u \in R_n,$$

where $g \in R_s$ is a given vector.

For $\alpha = 1/\lambda$, we denote by u^λ the vector minimizing the quadratic form

$$\lambda||Au-f||_m^2 + ||Lu-g||_s^2 \quad \text{for} \quad u \in R_n. \tag{20}$$

Obviously, $u^\lambda = u_{1/\lambda}$. Let $\rho_i(\lambda) = ||Au^\lambda-f||$. Then $\rho_0(\lambda) = \rho(1/\lambda) = \rho(\alpha)$.

Theorem 100. If the quadratic form $||Au||_m^2 + ||Lu||_s^2$, $u \in R_n$, is positive definite, for all $\lambda > 0$ the function $\rho_0(\lambda)$ is twice continuously differentiable and $\rho_0'(\lambda) \le 0$, $\rho_0''(\lambda) \ge 0$; that is, the function $\rho_0(\lambda)$ is convex downward for $\lambda > 0$.

Proof: The solutions u^λ of the problem (20) are found from the Euler equation

$$L^T(Lu^\lambda-g) + \lambda A^T(Au^\lambda-f) = 0,$$

the vector-derivative $du^\lambda/d\lambda$ is found from the equation

$$(L^TL + A^TA)\frac{du^\lambda}{d\lambda} = -A^T(Au^\lambda-f),$$

and, finally, the vector of second derivatives $d^2u^\lambda/d\lambda^2$ is found from the equation

$$(L^TL + A^TA)\frac{d^2u^\lambda}{d\lambda^2} = -2A^TA\frac{du^\lambda}{d\lambda}. \qquad \square$$

Using the definition of $\rho_0(\lambda)$ and the foregoing equations, we obtain

$$\begin{aligned}\rho_0(\lambda),\rho_0'(\lambda) &= (Au^{\lambda'},Au^\lambda-g)_m = (u^{\lambda'},A^T(Au^\lambda-f))_n \\ &- -((L^TL + \lambda A^TA)^{-1}A^T(Au^\lambda-f),A^T(Au^\lambda-f))_n \le 0\end{aligned} \tag{21}$$

because the matrix $(L^TL + \lambda A^TA)^{-1}$ is positive definite; that is, $\rho_0'(\lambda) \le 0$ (here $u^{\lambda'} = du^\lambda/d\lambda$). Next, differentiating (21) again, we obtain:

$$\rho_0(\lambda)\rho_0''(\lambda) + \rho_0'^2(\lambda) = (u^{\lambda''},A^T(Au^\lambda-f))_m + (Au^{\lambda'},Au^{\lambda'})_m$$

where $u^{\lambda''} = d^2u^\lambda/d\lambda^2$.

Using the Cauchy-Schwarz inequality, we have:

$$\rho_0'^2(\lambda) = \frac{1}{\rho_0^2(\lambda)} |(Au^{\lambda'}, Au^{\lambda}-f)_m|^2 \leq ||Au^{\lambda'}||_m^2.$$

Hence

$$\rho_0''(\lambda)\rho_0(\lambda) \geq (u^{\lambda''}, A^T(Au^{\lambda}-f))_n$$
$$= 2((L^TL + A^TA)^{-1}A^TA(L^TL + \lambda A^TA)^{-1}A^T(Au^{\lambda}-f)),$$
$$A^T(Au^{\lambda}-f))_n \geq 0, \quad \lambda > 0.$$

Remark. The matrix $C = L^TL$ need not be positive definite; that is, in the general case it is only nonnegative definite. If $C > 0$, we may assume in the theorem that $\lambda \geq 0$.

7. Let $C > 0$ and let for definiteness the vector $g = 0$. Making decompositions, as in Subsection 4, we can write

$$\rho_0(\lambda) = ||Dz^{\lambda}-\phi||_m,$$

where z^{λ} is found from the system of equations analogous to (16):

$$(E + \lambda D^TD)z^{\lambda} = \lambda D^T\phi \tag{22}$$

whose matrix is tridiagonal. It is easily seen that $z^{\lambda} = z_{1/\lambda} = z_{\alpha}$, $\alpha = 1/\lambda$.

The employment of (22) for computation of $\rho_0(\lambda)$ is connected with the explicit definition of the solution z^{λ} of this equation. In fact, it suffices to know only the vector $\varepsilon^{\lambda} = Dz^{\lambda} - \phi$. We shall give below a simple method to compute this vector. Also, it ensures the computation with higher accuracy of the values of the function $\rho_0(\lambda)$ as well as its derivatives, necessary for numerical solution of the equation

$$\rho_0(\lambda) = e_{\rho} = \Delta. \tag{23}$$

In fact, multiplying (22) on the left by the matrix D, we obtain that ε^{λ} satisfies the equation

$$(E + \lambda DD^T)\varepsilon^{\lambda} = -\phi.$$

It is easy to see that the solution z^{λ} of (22) is related to ε^{λ} as follows:

$$z^{\lambda} = -\lambda D^T\varepsilon_{\lambda}.$$

We also note that $||(E + \lambda DD^T)^{-1}|| \le 1$ uniformly in $\lambda \ge 0$.

Sequential approximations to the root λ_Δ of the equation

$$1/\rho_0(\lambda) = 1/\Delta,$$

equivalent to (23), are defined according to the formula

$$\lambda_{n+1} = \lambda_n + \frac{\rho_0(\lambda_n)}{\Delta} \cdot \frac{\Delta - \rho_0(\lambda_n)}{\rho'(\lambda_n)}, \qquad n = 0,1,\ldots,$$

where the initial value λ_0 is chosen so that $0 \le \lambda_0 \le \lambda_\Delta$.

Let us examine the method for choosing an initial approximation. Obviously, in the given case one can assume that $\lambda_0 = 0$. However, with some insignificant additional computational work one can indicate a more precise approximation to the root of (23).

The idea is based on the construction of a simple function r, which is a minorant for the function ρ_0. It is then natural to take λ_0 as the solution of the equation $r(\lambda) = \Delta$. We have

$$\rho_0'(\lambda)\rho_0(\lambda) = -((E + \lambda DD^T)^{-1}DD^T\varepsilon_\lambda, \varepsilon_\lambda)_m.$$

But

$$((E + \lambda DD^T)^{-1}DD^T\varepsilon_\lambda, \varepsilon_\lambda)_m \le ||(E + \lambda DD^T)^{-1}DD^T||\ ||\varepsilon_\lambda||_m^2$$

$$\le \frac{||D||^2}{1 + \lambda||D||^2}\rho_0^2(\lambda).$$

Therefore,

$$\rho_0'(\lambda) \ge -\frac{||D||^2}{1 + \lambda||D||^2}\rho_0(\lambda), \qquad \rho_0(0) = ||\phi||_m$$

Solving the inequality obtained for $\lambda \ge 0$, we have $\rho_0(\lambda) \ge r(\lambda) \equiv \rho_0(0)(1 + \lambda||D||^2)^{-1}$, and λ_0 is defined by

$$0 \le \lambda_0 = \frac{\rho_0(0) - \Delta}{||D||^2} = \frac{||\phi_m|| - \Delta}{\Delta||D||^2} \le \lambda_\Delta$$

We note that all the quantities in this expression are computable.

8. We now discuss some particular computational schemes. Noting that the matrix D is seen to be

$$D = \begin{bmatrix} \hat{D} \\ \cdots \\ 0 \end{bmatrix},$$

where $\hat{D}$ is the upper bidiagonal matrix of the order n n, we obtain

$$DD^T = \left[\begin{array}{c|c} \hat{D}\hat{D}^T & 0 \\ \hline 0 & 0 \end{array}\right].$$

Let

$$\varepsilon^\lambda = (\hat{\varepsilon}, \varepsilon_{n+1}, \ldots, \varepsilon_m), \quad \phi = (\hat{\phi}, \phi_{n+1}, \ldots, \phi_m), \quad \frac{d\varepsilon^\lambda}{d\lambda} = (\hat{\varepsilon}', \varepsilon'_{n+1}, \ldots, \varepsilon'_m)$$

where $\hat{\varepsilon}$, $\hat{\varepsilon}'$ and $\hat{\phi}$ are n-dimensional vectors. It is not hard to see that

$$\varepsilon_i = -\phi_i, \quad \varepsilon'_i = 0, \quad i \geq n+1,$$

and the vectors $\hat{\varepsilon}$ and $\hat{\varepsilon}'$ are defined by the "truncated" systems of equations:

$$(E + \lambda\hat{D}\hat{D}^T)\hat{\varepsilon} = -\hat{\phi}, \quad (E + \lambda\hat{D}\hat{D}^T)\hat{\varepsilon}' = -\hat{D}\hat{D}^T\hat{\varepsilon} = \frac{\hat{\varepsilon} + \hat{\phi}}{\lambda}. \tag{24}$$

The advantages obtained by going to (24) are obvious. Eq. (24) can be solved using the piping technique. We compute $\rho_0(\lambda)$ and $\rho'_0(\lambda)$ using the formulas

$$\rho_0^2(\lambda) = \sum_{i=n+1}^{m} \phi_0^2 + (\hat{\varepsilon}, \hat{\varepsilon})_n, \quad \rho'_0(\lambda)\rho_0(\lambda) = (\hat{\varepsilon}', \hat{\varepsilon})_n.$$

We find the solution z^λ of (22) using the formula

$$z^\lambda = -\lambda\hat{D}^T\hat{\varepsilon}$$

or directly from the system of equations

$$(E + \lambda\hat{D}^T\hat{D})z^\lambda = \lambda\hat{D}^T\hat{\phi}$$

for the value of the parameter $\lambda = \lambda_\Delta$ defined by (23).

9. The employment of generalized criterion $\tilde{\rho}$ for choosing the regularization parameter is due to the need to solve the following equation

$$\rho(\alpha) = h\gamma(\alpha) + m, \tag{25}$$

where $h > 0$, $m > 0$ are fixed values.

We now describe the procedure of finding the root of (25) if this root exists.

We take an $\alpha_0 > 0$ and, in addition, we replace $\gamma(\alpha)$ by the straight line $\gamma(\alpha_0) + \gamma'(\alpha_0)(\alpha - \alpha_0)$. Next we find α_1 from the condition of intersection of the graph of the function $\rho(\alpha)$ with the line $h[\gamma_0 + \gamma_0'(\alpha-\alpha_0)] + m$, where $\gamma_0 = \gamma(\alpha_0)$, $\gamma_0' = \gamma'(\alpha_0)$; that is, as a solution of the equation

$$\rho(\alpha) = h\gamma_0'(\alpha-\alpha_0) + m + h\gamma_0. \tag{26}$$

It is seen from Figure 5 that this point exists and is unique. In order to solve (26) it is natural to use the property of downward convexity of the function ρ_0.

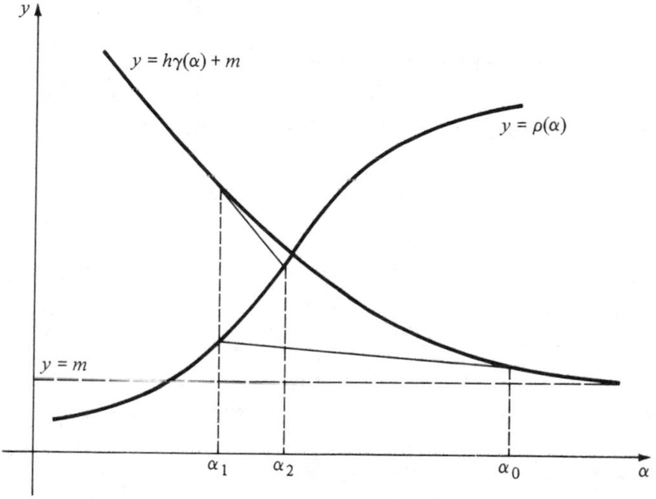

Figure 5

For $\alpha = 1/\lambda$ in (26), we have the equation

$$\rho_0(\lambda) = h\gamma_0' 1/\lambda + m + h\gamma_0 - h\alpha_0\gamma_0',$$

or

$$\rho_0(\lambda) - (h\gamma_0'/\lambda) = m + h\gamma_0 - h\alpha_0\gamma_0'.$$

The function $\rho_0(\lambda) - (h\gamma_0'/\lambda)$ is obviously convex downward since $\rho_0(\lambda)$ is convex downward and furthermore $\gamma_0' < 0$. Hence the root λ_1 of (27) and therefore the root $\alpha_1 = 1/\lambda_1$ of (26) can be found using Newton's method. As the zero approximation while we are finding λ_1, it is natural to take the root of the equation $\rho_0(\lambda) = m$.

After α_1 is defined, we can repeat the procedure described. It is easy to see that α_1 convertes to the root of (25).

10. Although we have dealt only with finite-dimensional operators A and L in this subsection, the results are valid in the general case as well. We omit them here because they are cumbersome.

Section 27. Heuristic Methods for Choosing a Parameter

Under the conditions of the basic problem let $G = H$ and $L = E$, and let $\overset{1}{u}_\alpha$ be the solution of the problem of minimizing

$$\phi_\alpha[u;A,f;u^*] \equiv ||Au-f||_F^2 + \alpha||u-u^*||_H^2, \quad \alpha > 0 \quad \text{for} \quad u \in H. \tag{1}$$

We assume that the operator A is bounded and defined on all of H.

Along with the problem (1) defining the regularized solutions $\overset{1}{u}_\alpha$, we consider the problem of minimizing the same functional for $u^* = \overset{1}{u}_\alpha$; that is, the minimizing problem of

$$\Phi_\alpha[u;A,f;\hat{u}_\alpha], \quad u \in H. \tag{2}$$

We denote by $\overset{2}{u}_\alpha$ solutions of the problem (2). We call them *two-fold regularized families*. We define a p-fold regularized family in a similar way.

We now express the p-fold regularized family in terms of the primary family. We have obviously, by the Euler equation,

$$\hat{u}_\alpha = (\alpha E + A^*A)^{-1}(A^Tf + \alpha u^*),$$

$$\overset{2}{u}_\alpha = (\alpha E + A^*A)^{-1}(A^Tf + \alpha\hat{u}_\alpha) = (\alpha E + A^*A)^{-1}(A^Tf + \alpha u^* + \alpha(\overset{1}{u}_\alpha - u^*))$$

$$= \hat{u}_\alpha - \alpha(E + A^*A)^{-1}(u^* - \overset{1}{u}_\alpha) = \overset{1}{u}_\alpha - \alpha\frac{du}{d\alpha}$$

that is,

$$\overset{2}{u}_\alpha = \hat{u}_\alpha - \alpha\frac{d\hat{u}_\alpha}{d\alpha}. \tag{3}$$

Similarly,

$$\overset{3}{u}_\alpha = (\alpha E + A^*A)^{-1}(A^Tf + \alpha\overset{2}{u}_\alpha) = \hat{u}_\alpha - \alpha(\alpha E + A^*A)^{-1}(u^* - \overset{2}{u}_\alpha)$$

$$= \hat{u}_\alpha - \alpha(\alpha E + A^*A)^{-1}(u^* - \hat{u}_\alpha + \frac{d\hat{u}_\alpha}{d\alpha})$$

$$= \hat{u}_\alpha - \alpha\frac{d\hat{u}_\alpha}{d\alpha} - \alpha^2(\alpha E + A^*A)^{-1}\frac{d\hat{u}_\alpha}{d\alpha}$$

$$= \hat{u}_\alpha - \alpha\frac{d\hat{u}_\alpha}{d\alpha} + \frac{\alpha^2}{2}\frac{d^2\hat{u}_\alpha}{d\alpha^2}.$$

By induction we prove that

$$\overset{p}{u}_\alpha = \sum_{i=0}^{p-1} \frac{(-1)^s \alpha^s}{s!} \frac{d^3 \hat{u}_\alpha}{d\alpha^3}$$

that is, the p-fold family is closely connected with the primary regularized family.

It is not hard to see that for $\hat{u} = u^*$, the solution of the basic problem is given by

$$\overset{p}{u}_\alpha = \hat{u} \quad \forall p \geq 1.$$

Next we consider only the case $p = 2$. We choose a *geometric* mesh in the parameter α, that is, $\alpha_{j+1} = \tau\alpha_j$, $j = 0,1,\ldots,N$, $\tau \simeq 1$ in a neighborhood of 0; then, since for $\tau \simeq 1$

$$\alpha \frac{d\hat{u}_\alpha}{d\alpha} = \frac{\hat{u}_\alpha - \hat{u}_{\tau\alpha}}{\alpha - \tau\alpha} = \frac{\overset{1}{u}_\alpha - \overset{1}{u}_{\tau\alpha}}{1 - \tau},$$

by virtue of (3) we have

$$\overset{2}{u}_j = \overset{1}{u}_j - \frac{\hat{u}_j - \hat{u}_{j+1}}{1 - \tau} = \frac{\overset{1}{u}_{j+1} - \tau\overset{1}{u}_j}{1 - \tau} \tag{4}$$

where $\hat{u}_j = \hat{u}_{\alpha_j}$, $\overset{2}{u}_j = \overset{2}{u}_{\alpha_j}$, $j = 0,1,2,\ldots,N$.

The formula (4) shows that the elements of the two-fold regularized family on the geometric mesh can be computed approximately on the basis of elements of the initial family.

2. In (1) let approximate $\{A_h, f_\delta\}$ be given instead of the exact $\{A,f\}$. As was shown, there exists $\alpha = \alpha(\delta,h)$ such that $\hat{u}_{\delta,h} = \hat{u}_{\alpha(\delta,h)} \to \hat{u}$ in H as $\delta,h \to 0$.

Since in practice δ and h are finite, we can, given some "appropriate" value of the parameter $\alpha = \overline{\alpha}$, construct in its neighborhood a geometric mesh with respect to α and compute the elements $\hat{u}_j$, $j = 0,1,\ldots,N$. Intuitively, it seems reasonable to choose a value $j = j_0$ for which

$$\|\hat{u}_{j+1} - \hat{u}_j\|_H \text{ is minimized,} \tag{5}$$

which corresponds to the hypothesis of existence of the "plait" of regularized solutions concentrating near the required solution $\hat{u}$.

It is common to call the value α_{j_0} a *quasi-optimal value* of the regularization parameter [91]. Unfortunately, it has not been possible to justify this technique for choosing the parameter although it is widely used for solving unstable problems. This technique is close in principle to the Runge-Kutta technique in numerical integration.

We can say for sure only that this principle of choosing the parameter corresponds to approximate minimization of the second term of the decomposition (3). This remark allows us to make further generalizations of this technique using the representation of the p-fold regularized family of solutions.

Now we note that under certain conditions it is possible to prove the convergence of the two-fold regularized family to the solution of the basic problem. For more detail, see [53] and [55].

At the present time the method for choosing quasi-optimal values of the regularization parameter is used for making values of the parameter more precise, defined by justified criteria such as ρ, ϕ, γ, and others.

3. In (1) let $H = R_n$, $F = R_m$ and let $||u||_H^2 = (Cu,u)_n$, where C is a positive definite symmetric matrix. Then the regularized solutions

$$\tilde{u}_\alpha = (\alpha C + A^T A)^{-1}(A^T\tilde{f} + \alpha C u^*),$$

where $\tilde{f}$ is the approximation to f.

Let

$$R_\alpha = (\alpha C + A^T A)^{-1} A^T \tilde{f}.$$

We call the operator R_α a *processing operator* of $\tilde{f}$. If $\tilde{f}$ is a random Gaussian vector, $M\tilde{f} = f$, $M(\tilde{f}-f)^T(\tilde{f}-f) = \sigma^2 E_m$, where E_m is a unit matrix of the order m, it is advisable to satisfy the following principles for choosing the parameter:

1. $||A\tilde{u}_\alpha - \tilde{f}||_m$ is minimized in α;
2. $M||R_\alpha \xi||_C^2$ is minimized in α, where $\xi = \tilde{f}-f$.

The advantage of having condition 1 satisfied is obvious. Condition 2 implies that the processing operator must suppress the "numerical" error optimally. It is clear that it is impossible to satisfy conditions 1 and 2 at the same time. Therefore we suggest a compromi e criterion for choosing the parameter by minimizing the following functional with respect to α:

$$||A\tilde{u}_\alpha - \tilde{f}||_m^2 + M||R_\alpha \xi||_C^2.$$

We can show that the second term is monotone decreasing. Since the first term is monotone increasing, it is not difficult to devise an algorithm for the parameter satisfying the criterion suggested.

Section 28. The Investigation of Adequacy of Mathematical Models

1. Suppose we consider the problem of solving the equation

$$Au = f, \tag{1}$$

where $A: H \to F$ is a linear operator acting on the Hilbert space H into a Hilbert space F. We introduced earlier the measure of incompatibility (1):

$$\mu_A = \inf_{u \in H} ||Au-f||_F = \mu_A(f) \tag{2}$$

at the point $f \in F$. If we adopt the hypothesis that the element f corresponds exactly to measurable characteristics of a certain physical phenomenon, the incompatibility measure can be regarded as the measure of adequacy of the model (1) to the physical phenomenon under investigation (or a phenomenon of any other nature). Thus we say that the model (1) is *f-adequate* if

$$\mu_A(f) << ||f||_F.$$

In some cases it is worthwhile to investigate $\mu_A(f)$ for $f \in Q \subseteq F$. Then the quantity

$$\sup_{f \in Q} \mu_A(f) = \mu_A(Q)$$

characterizes the incompatibility measure of (1) on the class Q. If $\mu_A(f) = 0$, we say that the model (1) is *f-compatible*. If $\mu_A(Q) = 0$, we say that the model (1) is *Q-compatible*.

Obviously, the functional μ_A is defined on all of F (the existence of the element $\hat{u} \in H$ on which (2) is realized is not necessary). Let f be given approximately; that is, an element $\tilde{f} \in F$, $||\tilde{f}-f||_F < \delta$ is known. Let $u_\varepsilon \in H$ be such that

$$\mu_A(f) \leq ||Au_\varepsilon-f||_F \leq \mu_A(f) + \varepsilon \quad \text{for all} \quad \varepsilon > 0.$$

Then

$$\mu_A(\tilde{f}) \leq ||Au_\varepsilon-\tilde{f}||_F < ||Au_\varepsilon-f||_F + \delta \leq \mu_A(f) + \delta + \varepsilon.$$

Choosing $\tilde{u}_\varepsilon \in H$ such that

$$\mu_A(\tilde{f}) \le ||A\tilde{u}_\varepsilon - f|| \le \mu_A(\tilde{f}) + \varepsilon,$$

we obtain in a similar way that

$$\mu_A(f) \le \mu_A(\tilde{f}) + \delta + \varepsilon.$$

Therefore

$$||\mu_A(f) - \mu_A(\tilde{f})|| \le \delta + \varepsilon \quad \forall\varepsilon > 0;$$

that is,

$$|\mu_A(f) - \mu_A(\tilde{f})| \le \delta.$$

We have now the following theorem.

Theorem 101. The functional $\mu_A(f)$, $f \in F$, depends continuously on $f \in F$.

2. A. N. Tikhonov has introduced the following definition [92]:

Let $\tilde{f} \in F$, $||f-\tilde{f}||_F < \delta$ and let $U_\delta = \{u \in H: ||Au-\tilde{f}||_F \le \delta\}$. If $U_\delta \ne \emptyset$ for all sufficiently small $\delta > 0$, the model (1) is said to be *consistent* we shall speak about f-*consistency* meaning that f is fixed). We can speak about Q (or F)-consistency if the model (1) is consistent for any $f \in Q$ (or $f \in F$).

We examine the relationship between the concepts of f-compatibili y and f-consistency.

Theorem 102. For f-consistency of the model (1) it is necessary and sufficient that it is f-compatible.

Proof: Let the model (1) be f-consistent. Then there exist elements $u_\delta \in H$, $||Au_\delta-\tilde{f}||_F \le \delta$ $\forall\delta > 0$ sufficiently small. Then $||Au_\delta-f|| \le 2\delta$ and, therefore,

$$\mu(A) \le ||Au_\delta-f||_F \le 2\delta \quad \forall\delta > 0;$$

that is, $\mu_A(f) = 0$.

If $\mu_A(f) = 0$ (that is, the model (1) is f-compatible), for any $\varepsilon > 0$ there exists an element $u_\varepsilon \in H$ with $||Au_\varepsilon-f||_F$. Then

$$||Au_\varepsilon-f||_F \le ||Au_\varepsilon-f||_F + ||f-\tilde{f}||_F < \varepsilon + \delta \quad \text{for all} \quad \varepsilon > 0;$$

that is, for sufficiently small $\varepsilon > 0$, $u_\varepsilon \in U_\delta$ and therefore the set $U_\delta \ne \emptyset$ for all sufficiently small δ. Therefore, the model (1) is f-consistent. □

Theorem 102 shows that the f-compatibility and f-consistency are equivalent concepts.

3. We say that the model (1) is *f-solvable* if the set $U_A = \{u \in H: Au = f\} \neq \emptyset$. It is clear that the f-solvability implies f-compatibility (and f-consistency) of the model (1). The opposite is, however, not true, which can easily be shown. We shall speak also about the Q-solvability of the model (1), meaning by this the f-solvability for all $f \in Q \subseteq F$.

Theorem 103. Let Q be the set of solvability for the model (1). If the model (1) is F-compatible (or F-consistent) then the closure Q coincides with F; that is, $\overline{Q} = F$.

Proof: Suppose $\overline{Q} \neq F$. Then there exists an element $f \neq 0$, $f \perp \overline{Q}$, such that $||Au-f||_F \geq a > 0$ $\forall u \in H$. Then, however, $\mu_A(f) \geq a > 0$, which contradicts the F-compatibility of the model (1). □

4. Suppose we know that the model (1) is f-compatible. Is it possible to answer the question about the f-solvability of the model (1) within the frame work of an ideal experiment, that is, in the case where $\delta \to 0$?

We now consider the problem: find an element $u_\delta \in U_\delta$ such that

$$||u_\delta||_H = \inf_{u \in U_\delta} ||u||_H;$$

that is, the residual method is applicable.

The following theorem gives an answer to the question posed above.

Theorem 104. If the model (1) is f-compatible, in order for it to be f-solvable, it is necessary and sufficient that $||u_\delta||_H \leq C < +\infty$ for all sufficiently small δ.

Proof: If the model (1) is f-solvable, there exists a normal solution $\hat{u}$ of the problem (1). Then, obviously, $||u_\delta||_H \leq ||\hat{u}||_H = C$ for all δ.

Next we assume that $||u_\delta|| \leq C < +\infty$. Using this condition, we select from the family $\{u_\delta\}$ a weakly convergent sequence $\{u_n\}$; $u_n \xrightarrow{\text{weakly}} u_0 \in H$. Using the weak lower semi-continuity of the norm in a Hilbert space and the continuity of the operator A in the weak topology, we obtain:

$$||Au_0-f||_F \leq \underline{\lim}_{n\to\infty} ||Au_n-f||_F \leq \overline{\lim}_{n\to\infty} ||Au_n-\tilde{f}||_F + \delta \leq 2\delta, \quad \text{for all } \delta,$$

which implies that $Au_0 = f$; that is, the model (1) is f-solvable. □

5. The computation of the incompatibility measure $\mu_A(f)$ is necessary not only for estimating the adequacy of the mathematical model (1) to the physical phenomenon under investigation, but also in the cases where some regular methods of solving (1) are applied or, in the general case, solving the basic problem. Let $\{A_h, f_\delta\}$ be approximate quantities in the problem (1), where $A_h: H \to F$ is bounded and linear for any h, $0 < h \leq h_0$, and $||A-A_h|| \leq h$, also with $f_\delta \in F$, $||f-f_\delta||_F < \delta$.

Let

$$\mu_\sigma = \inf_{u \in H} ||A_h u - f_\delta||_F, \quad \sigma = (h,\delta).$$

It is easy to show that in the general case $(h \neq 0)$ the relation $\lim_{\sigma\to 0} \mu_\sigma = \mu_A$ does not hold. In fact, if $H = F = R_n$ is a Euclidean space of dimension n, $A = Q$ is the $n \times n$ zero matrix and $f \neq 0$, then $\mu_A = ||f||_m > 0$, whereas $\mu_\sigma \equiv 0$ for $A_h = hE$ and any $h > 0$.

As follows from Theorem 101, stability against perturbations of f is not maintained. For the perturbations of the operator A we can assert only that

$$\overline{\lim_{\sigma\to 0}}\, \mu_\sigma = \mu_A(f).$$

Indeed, choosing $u_\varepsilon \in H$ such that $||Au_\varepsilon - f||_F \leq \varepsilon$, $\varepsilon > 0$, we have:

$$\mu_\sigma \leq ||A_h u_\varepsilon - f_\delta||_F \leq \mu_A(f) + h||u_\varepsilon|| + \delta + \varepsilon;$$

that is,

$$\overline{\lim_{\sigma\to 0}}\, \mu_\sigma \leq \mu_A(f) + \varepsilon \quad \forall\varepsilon > 0,$$

which yields the required solution.

Next we consider the problem of constructing from any approximate $\{A_h, f_\delta\}$ values $\hat{\mu}_\sigma$ such that they satisfy the *convergence condition*

$$\lim_{\sigma\to 0} \hat{\mu}_\sigma = \mu_A(f)$$

and, therefore, can be regarded as approximations to the required incompatibility measure $\mu_A(f)$. We do not assume however that the problem (1) is f-solvable. This method can be recommended for preliminary estimation of an adequacy measure of the model (1), which is essential for many problems; in particular, for approximation problems.

6. We state now several auxiliary lemmas which are of interest per se. Assume, according to the regularization method, that $\Phi_\alpha[u] = ||Au-f||_F^2 + \alpha||u||_H^2$, $\alpha > 0$.

Then we have the following lemma.

Lemma 59. Let $u_\alpha \in H$ be the solution of the variational problem

$$\inf_{u \in H} \Phi_\alpha[u] = \Phi_\alpha[u].$$

Then

$$\mu_A = \lim_{\alpha \to 0} \mu_\alpha, \qquad \mu_\alpha = ||Au_\alpha - f||_F.$$

Proof: We have proved earlier that the variational problem has a unique solution. We choose $\varepsilon_n > 0$ such that $\lim_{n\to\infty} \varepsilon_n = 0$ and $u_n \in H$ satisfy the condition

$$\mu_A \le ||Au_n - f||_F \le \mu_A + \varepsilon_n.$$

Then, using the obvious properties of extremal problems, we obtain

$$\mu_A^2 \le \mu_\alpha^2 \le \Phi_\alpha[u_\alpha] \le \Phi_\alpha[u_n] \le (\mu_A + \varepsilon_n)^2 + \alpha||u_n||_H^2;$$

that is,

$$\overline{\lim_{\alpha\to 0}} |\mu_A - \mu_\alpha| \le \varepsilon_n, \qquad \lim_{n\to\infty} \varepsilon_n = 0,$$

proving thus the assertion. □

We note that the element u_α satisfies the equation

$$(\alpha E + A^*A)u_\alpha = A^*f,$$

where A^* is the adjoint of A. Assuming $Au_\alpha = f_\alpha$ and also applying the operator A to both sides of the foregoing equation, we obtain

$$(\alpha E + A^*A^*)f_\alpha = AA^*f$$

and therefore

$$\mu_A = \lim_{\alpha\to 0} \mu_\alpha, \qquad \mu_\alpha = ||f_\alpha - f||_m,$$

where f_α is defined uniquely by the above given equation. The element f_α can be edefined also by the following extremal problem:

$$\inf_{g \in F} \{||A^*g - A^*f||_H^2 + \alpha||g||_F^2\} = ||A^*f_\alpha - A^*f||_H^2 + \alpha||f_\alpha||_F^2.$$

Let $f' \in F$ be an element with minimal norm (a normal solution) among all solutions of the equation $A^*g = A^*f$. According to the remark made above,

$$||A^*f_\alpha - A^*f||_H^2 + \alpha||f_\alpha||_F^2 \le \alpha||f'||_F^2$$

and therefore

$$\lim_{\alpha\to 0} ||A^*f_\alpha - A^*f'||_H = 0, \qquad \overline{\lim_{\alpha\to 0}} ||f_\alpha||_F \le ||f'||_F,$$

which, in accord with the main results, yield that

$$\lim_{\alpha\to 0} f_\alpha = f'.$$

Thus the following lemma holds.

Lemma 55. We have the equality

$$\mu_A = \lim_{\alpha\to 0} ||f_\alpha - f||_F = ||f' - f||_F.$$

We now define the following functions of the parameter $\alpha > 0$:

$$\rho(\alpha) \equiv ||A^*f_\alpha - A^*f||_H, \quad \gamma(\alpha) = ||f_\alpha||_F.$$

The following lemma is a corollary of the basic results.

Lemma 56. The function ρ is a continuous, strictly increasing function for which

$$\lim_{\alpha\to 0} \rho(\alpha) = 0, \qquad \lim_{\alpha\to\infty} \rho(\alpha) = ||A^*f||_H.$$

The function γ is a continuous, strictly decreasing function for which

$$\lim_{\alpha\to 0} \gamma(\alpha) = ||f'||_F, \qquad \lim_{\alpha\to\infty} \gamma(\alpha) = 0.$$

Next we assume that $||A^*f||_H > 0$. Since

$$\lim_{\sigma\to 0} ||A_h^*f_\delta||_H = ||A^*f||_H,$$

for sufficiently small σ we have the inequality

$$||A_h^*f_\delta||_H > h||f_\delta||_F.$$

Let

$$\tilde{\mu}_\alpha = ||\tilde{f}_\alpha - f_\delta||_F,$$

where $\tilde{f}_\alpha$ satisfies the equation

$$(\alpha E + A_h A_h^*)\tilde{f}_\alpha = A_h A_h^* f_\delta$$

and, in addition,

$$\tilde{\rho}(\alpha) = ||A_h^*(\tilde{f}_\alpha - f_\delta)||_H, \qquad \tilde{\gamma}(\alpha) = ||\tilde{f}_\alpha||_F.$$

The foregoing assertions imply the following lemma.

Lemma 57. For sufficiently small σ the equation

$$\tilde{\rho}(\alpha) = 2h||f_\delta||_F$$

has a unique root $\alpha_\sigma > 0$.

7. We can solve the problem of computing approximations to the lower bound of the functional (2) using the following theorem.

Theorem 105. Let $f_\delta = f_{\alpha_\sigma}$, $\hat{\mu}_\sigma = ||f_\sigma - f_\delta||_F$, where α_σ was defined in Theorem 57. Then

$$\lim_{\sigma \to 0} \hat{\mu}_\sigma = \mu_A.$$

Proof: Let f_δ' be a solution of the equation $A^*g = A^*f_\delta$ with minimal norm. For any $\alpha > 0$,

$$||A_h^* f_\alpha - A_h^* f_\delta||_H^2 + \alpha||f_\alpha||_F^2 \leq h^2(||f_\delta'||_F + ||f_\delta||_F)^2 + \alpha||f_\delta'||_F^2.$$

Assuming $\alpha = \alpha_\sigma$ and taking into account (3), we obtain, after simple computations, that

$$||f_\sigma||_F \leq ||f_\delta'||_F, \qquad ||A^*f_\sigma - A^*f_\delta'||_H \leq 2h(||f_\delta'||_F + ||f_\delta||_F).$$

Since $||f_\delta' - f'||_F \leq \delta$, these estimates imply the relations

$$\lim_{\sigma \to 0} ||A^*f_\sigma - A^*f'||_H = 0, \qquad \overline{\lim_{\sigma \to 0}} ||f_\sigma||_F \leq ||f'||_F$$

and, in turn,

$$\lim_{\sigma \to 0} f_\sigma = f'.$$

It remains only to apply Lemma 55. □

8. We consider numerical aspects of computing lower bounds of the functional (1) on the basis of the approximate $\{A_h, f_\delta\}$.

We assume that A_h is the matrix operator acting on $R_n = H$ into $R_m = F$, where R_n and R_m are Euclidean spaces, $m \geq n$. Then u is a n-dimensional vector and f is a m-dimensional vector. For simplicity, we omit the subscripts h and δ in the matrix A_h and the vector f_δ.

As in Section 26, we expand A as

$$A = QDR,$$

where Q, R are orthogonal matrices, D is a "bidiagonal" matrix and the last $(m-n)$ rows are zeros.

Then the equation for determination of f_α becomes

$$\alpha(QQ^T + QDRR^TD^TQ^T)f_\alpha = QDRR^TD^TQ^Tf.$$

Assuming that

$$\phi^\alpha = Q^Tf_\alpha, \quad \phi = Q^Tf,$$

we obtain that ϕ^α satisfies the equation

$$(\alpha E + DD^T)\phi^\alpha = DD^T\phi. \tag{4}$$

Under these replacements, as can easily be verified,

$$\tilde{\rho}(\alpha) = ||D^T(\phi^\alpha - \phi)||_m.$$

According to the criterion (3) for choosing the parameter, we can find α_σ as the solution of the scalar equation

$$\tilde{\rho}(\alpha) = 2h||\phi||_m. \tag{5}$$

We define the required approximations $\hat{\mu}_\sigma$ using the formula

$$\mu_\sigma = ||\phi_{\alpha_\sigma} - \phi||_m.$$

We note that our approach ensures a considerable saving in time while solving the problem.

It is advisable to write Eq. (5) in equivalent form:

$$R^{-1}(\alpha) = (2h||\phi||_m)^{-1}, \qquad R(\alpha) \equiv \tilde{\rho}(1/\alpha). \tag{6}$$

The function R^{-1} is strictly convex upward and continuously differentiable on $[0,\infty)$ (see Section 26). In order to solve (6), we can apply Newton's method of tangents, which is convergent for any choice of the initial approximation $\alpha_0 \geq 0$.

Since $m \geq n$, it may seem that the vector ϕ^α from (4) will be defined from a system of sufficiently high order. Noting, however, that

$$D = \begin{bmatrix} \hat{D} \\ \cdots \\ 0 \end{bmatrix},$$

where $\hat{D}$ is an $m \times n$ matrix and 0 is the $n \times (m-n)$ zero matrix, we establish that the last $m-n$ components of the vector ϕ^α are equal to zero a priori. Furthermore, the first n components can be determined by solving the "truncated" system of equations $(\alpha E + \hat{D}\hat{D}^T)\hat{\phi}_\alpha = \psi$, where

$$\hat{\phi}_\alpha = (\phi_1^\alpha, \phi_2^\alpha, \ldots, \phi_r^\alpha), \qquad \psi = ((DD_\phi^T)_1), (DD_\phi^T)_2, \ldots, (DD_\phi^T)_r).$$

<u>Remark</u>. Instead of the criterion (3), it is possible to use the criterion

$$\tilde{\rho}(\alpha) = h\tilde{\gamma}(\alpha)$$

for choosing α_σ.

The proof of Theorem 105 will be somewhat modified in this case.

Bibliography

[1] V. Ya. Arsenin, Optimal summation of Fourier series with approximate coefficients, Soviet Math. Dokl., 9, 6 (1968), 1345-1348 (English translation).

[2] V. Ya. Arsenin, On methods of solving ill-posed problems (O metodakh reshenija nekorrectno postavlennykh zadach), Moscow, Print MIFI, 1973 (in Russian).

[3] A. B. Bakushinskii, Selected problems of approximate solution of ill-posed problems (Izbrannyje voprosy priblizhennogo reshenija nekorrektnykh zadach), Moscow, Computer Center, MGU, 1968 (in Russian).

[4] A. B. Bakushinskii, Some problems of the theory of regularization algorithms (Nekotoryje voprosy teorii reguljarizujushchikh algoritmov). In Sb.: Vychislitel'nyje metody i programmirovanije, Moscow, MGU, 12 (1969), 56-79 (in Russian).

[5] N. S. Bakhvalov, On the optimality of linear methods for operator approximation in convex classes of functions, USSR Comput. Math. and Math. Phys. 11, 4 (1971), 244-249 (English translation).

[6] V.V. Vasin, Relationship of several variational methods for the approximate solution of ill-posed problems, Math. Notes, 7, 3 (1970), 161-166 (English translation).

[7] V. V. Vasin, V. P. Tanana, Approximate solution of operator equations of the first kind (Priblizhennoje reshenije operatornykh uravnenii pervogo roda), Matematicheskii zametki, Ural'skii universitet, 6, 4 (1968), 27-37 (in Russian).

[8] V. V. Voevodin, The method of regularization, USSR Comput. Math. and Math. Phys., 9, 3 (1969), 228-229 (English translation).

[9] V. V. Voevodin, Approximation errors and stability in direct methods of linear algebra (Oshibki okruglenija i ustoichivost v prjamykh metodakh lineinoi algebry), Moscow, MGU, 1969 (in Russian).

[10] A. V. Goncharskii, A. S. Leonov, and A. G. Yagola, The generalized discrepancy principle, USSR Comput. Math. and Math. Phys., 13, 2 (1973), 25-37 (English translation).

[11] R. Denchev, The stability of linear equations on a compact set, USSR Comput. Math. and Math. Phys., 7, 6 (1967), 201-205 (English translation).

[12] V. I. Dmitriev, E. V. Zakharov, On numerical solution of Fredholm's integral equations of the first kind (O chislennom resheniji integral'nykh uravnenii Fredgol'ma pervogo roda). In Sb.: Vychislitel'nyje metody i programmirovanije, Moscow, MGU, 10 (1968), 49-54 (in Russian).

[13] G. F. Dokgopolova and V. R. Ivanov, On numerical differentiation, USSR Comput. Math. and Math. Phys., 6, 3 (1966), 223-232 (English translation).

[14] I. N. Dombrovskaya, On the solution of ill-posed linear equations in Hilbert spaces (O resheniji nekorrektnykh lineinykh uravnenii v gil'bertovykh prostranstvakh), Matematicheskije zapiski, Ural'skii universitet, 4, 4 (1964), 36-40 (in Russian).

[15] I. N. Dombrovskaya, On equations of the first kind with closed operator (Ob uravnenijakh pervogo roda s zamknutym operatorom), Izvestija Vysshykh uchebnykh zavedenii, Matematika 6, 1967, 39-72 (in Russian).

[16] I. N. Dombrovskaya, V. K. Ivanov, On the theory of linear equations in abstract spaces (K teoriji nekotorykh lineinykh uravnenii v abstraktnykh prostranstvakh), Sibirskii Matematicheskii zhurnal, 6, 3 (1960), 499-508 (in Russian).

[17] E. L. Zhukovskii and V. A. Morozov, On successive Bayes regularization of systems of algebraic equations, USSR Comput. Math. and Math. Phys., 12, 2 (1972), 222-223 (English translation).

[18] V. V. Ivanov, The theory of Approximate Methods and Their Application to the Numerical Solution of Singular Integral Equations, Leyden: Noordhoff International Publishing, 1976 (English translation).

[19] V. V. Ivanov and V. Yu. Kudrinskii, Approximate solution of linear operator equations in Hilbert space by the method of least squares, USSR Comput. Math. and Math. Phys., 6, 5 (1966), 60-75 (English translation).

[20] V. K. Ivanov, On linear problems which are not well-posed, Soviet Math. Dokl., 3, 4 (1962), 981-983 (English translation).

[21] V. K. Ivanov, Integral equations of the first kind and an approximate solution for the inverse problem of potential, Soviet Math. Dokl., 3, 1 (1962), 210-212 (English translation).

[22] V. K. Ivanov, A type of ill-posed linear equations in vector topological spaces (Ob odnom tipe nekorrektnykh lineinykh uravnenii v vektornykh topologicheskikh prostranstvakh), Sibirskii matematicheskii zhurnal, 6, 4 (1965), 832-839 (in Russian).

[23] V. K. Ivanov, The approximate solution of operator equations of the first kind, USSR Comput. Math. and Math. Phys., 6, 6 (1966), 197-205 (English translation).

[24] V. K. Ivanov, On ill-posed problems (O nekorrektno postavlennykh zadachakh), Matematicheskii sbornik, 61, 2 (1963), 211-223 (in Russian).

[25] V. K. Ivanov, On uniform regularization of unstable problems (O ravnomernoi reguljarizatsiji neustoichivykh zadach), Sibirskii matematisheskii zhurnal, 7, 3 (1966), 546-558 (in Russian).

[26] V. K. Ivanov, On integral equations of the first kind (Ob integral'nykh uravnenijakh pervogo roda), Differentsial'nyje uravnenija, 3, 3 (1967), 410-421 (in Russian).

[27] V. K. Ivanov, Incorrect problems in topological spaces, Siberian Math. Jour., 10, 5 (1969), 785-791 (English translation).

[28] V. K. Ivanov, On estimation of errors in solving operator equations of the first kind (Ob otsenke pogreshnostei pri resheniji operatornykh uravnenii pervogo roda). In Voprosy tochnosti i effektivnosti vychislitel'nykh algoritmov, Trudy simpoziuma, Vol. 2, Kiev, 1969 (in Russian).

[29] V. K. Ivanov, On the solution of operator equations not satisfying the correctness conditions (O resheniji operatornykh uravnenii ne udovletvorjajushchikh uslovijam korrektnosti), Trudy Matematicheskogo Instituta Akademii Nauk SSSR, Vol. 112, 1971, 232-240 (in Russian).

[30] V. K. Ivanov, Unstable linear problems with many-valued operators, Siberinn Math. Jour., 11, 5 (1970), 751-756 (English translation).

[31] V. K. Ivanov, T. I. Korolyuk, Error estimates for solutions of incorrectly posed linear problems, USSR Comput. Math. and Math. Phys., 9, 3 (1969), 35-49 (English translation).

[32] L. V. Kantorovich, On new approaches to computational methods and data processing (O novykh podkhodakh k vychislitel'nym metodam i obrabotke nablyudenii), Sibirskii matematicheskii zhurnal, 5 (1962), 701-709 (in Russian).

[33] L. F. Korkina, On regularization of operator equations of the first kind (O reguljarizatsiji operatornykh uravnenii pervogo roda), Izvestija vysshikh uchebnykh zavedenii, Matematika, 8 (1969), 26-29, (in Russian).

[34] S. G. Krein, On classes of correctness of some boundary value problems (O klassakh korrektnosti dlja nekotorykh granichnykh zadach), Doklady Akademii Nauk SSSR, 114, 6 (1957), 1162-1165 (in Russian).

[35] M. M. Lavrentjev, On the Cauchy problem for the Laplace equation (O zadache Cauchy dlja uravnenija Laplasa), Doklady Akademii Nauk SSSR, 102, 2 (1955), 205-206 (in Russian).

[36] M. M. Lavrentjev, On integral equations of the first kind (Ob integral'nykh uravnenijakh pervogo roda), Doklady Akademii Nauk SSSR, 127, 1 (1959), 31-33 (in Russian).

[37] M. M. Lavrentjev, On some ill-posed problems of mathematical physics (O nekotorykh nekorrektnykh zadachakh matematicheskoi fiziki), Novosibirsk, SO AN SSSR, 1962 (in Russian).

[38] M. M. Lavrentjev, Conditionally well-posed problems of differential equations (Uslovno korrektnyje zadachi dlja differentsial'nykh uravnenii), Novosibirsk, 1973 (in Russian).

[39] M. M. Lavrentjev, On the formulation of some ill-posed problems of mathematical physics (O postanovke nekotorykh nekorrektnykh zadach matematicheskoi fiziki), Sibirskii matematicheskii zhurnal, 7, 3 (1966), 559-576 (in Russian).

[40] R. Lattes and J.-L. Lions, The Method of Quasi-Reversibility; Applications to Partial Differential Equations, New York, American Elsevier Pub. Co., 1969 (Russian translation).

[41] J.-L. Lions, Optimal Control of Systems Governed by Partial Differential Equations, Berlin, New York, Springer-Verlag, 1971.

[42] V. I. Lebedev, On the solution of compact sets of some restoration problems, USSR Comput. Math. and Math. Phys., 6, 6 (1966), 79-102 (English translation).

[43] O. A. Liskovets, Incorrect problems with a closed irreversible operator, Differential Equations, 3, 4 (1967), 324-329 (English Translation).

[44] O. A. Liskovets, Regularization of equations with a closed linear operator, Differential Equations, 6, 7 (1970), 972-976 (English translation).

[45] O. A. Liskovets, Stability of quasisolutions of equations with a closed operator, Differential Equations, 7, 9 (1971), 1300-1302 (English translation).

[46] G. I. Marchuk, Formulation of some converse problems, Soviet Math. Dokl., 5, 3 (1964), 675-678 (English translation).

[47] G. I. Marchuk and V. G. Vasil'ev, On an approximate solution for operator equations of the first kind, Soviet Math. Dokl., 11, 6 (1970), 1562-1566 (English translation).

[48] V. P. Maslov, Regularization of incorrect problems for singular integral equations, Soviet, Math. Dokl., 8, 5 (1967), 1251-1254 (English translation).

[49] V. P. Maslov, The existence of a solution of an ill-posed problem is equivalent to the convergence of a regularizable process (Sushchestvovanije reshenija nekorrektnoi zadachi ekvivalentno skhodimosti reguljarizatsionnogo protsessa), Uspekhi matemat. nauk, 23, 3 (1968), 183-184 (in Russian).

[50] V. A. Morozov, Regularization of incorrectly posed problems and the choice of regularization parameter, USSR Comput. Math. and Math. Phys., 6, 1 (1966), 242-251 (English translation).

[51] V. A. Morozov, On the solution of functional equations by the method of regularization, Soviet Math. Dokl., 7, 2 (1966), 414-417 (English translation).

[52] V. A. Morozov, Choice of parameter for the solution of functional equations by the regularization method, Soviet Math. Dokl., 8, 4 (1967), 1000-1003 (English translation).

[53] V. A. Morozov, Methods for solving unstable problems (Metody reshenija neustoichivykh zadach), Print, Computer Center, MGU, 1967 (in Russian).

[54] V. A. Morozov, On restoring functions by the regularization method, USSR Comput. Math. and Math. Phys., 7, 4 (1967), 208-219 (English translation).

[55] V. A. Morozov, The error principle in the solution of operational equations by the regularization method, USSR Comput. Math. and Math. Phys., 8, 2 (1968), 63-87 (English translation).

[56] V. A. Morozov, Pseudosolutions, USSR Comput. Math. and Math. Phys., 9, 6 (1969), 196-203 (English translation).

[57] V. A. Morozov, On an effective numerical algorithm for constructing pseudosolutions, USSR Comput. Math. and Math. Phys., 11, 1 (1971), 339-343 (English translation).

[58] V. A. Morozov, The solution by regularization of incorrectly posed problems with unbounded nonlinear operators, Soviet Math. Dokl., 6, 8 (1970), 1107-1111 (English translation).

[59] V. A. Morozov, A stable method for computation of the values of unbounded operators, Soviet. Math. Dokl. 10, 2 (1969), 339-342 (English translation).

[60] V. A. Morozov, Optimal regularization of operator equations, USSR Comput. Math. and Math. Phys., 10, 4 (1970), 10-25 (English translation).

[61] V. A. Morozov, Error estimates for the solution of an incorrectly posed problem involving unbounded linear operators, USSR Comput. Math. and Math. Phys., 10, 5 (1970), 19-33 (English translation).

[62] V. A. Morozov, On stability of the problem of defining the parameters (Ob ustoichivosti zadach opredelenija parametrov). In Sb.: Vychislitel'nyje metody i programmirovannije, 1970, Vyp. XIV, Izd-vo MGU, pp. 63-67 (in Russian).

[63] V. A. Morozov, On regularizing families of operators (O reguljaizujustchikh semeistvakh operatorov). In Sb.: Vychislitel'nyje metody i programmirovanije, 19, Vyp. VIII, Izd-vo MGU, pp. 63-95 (in Russian).

[64] V. A. Morozov, Theory of splines and the problem of stable computation on the values of an unbounded operator, USSR Comput. Math. and Math. Phys., 11, 3 (1971), 1-19 (English translation).

[65] V. A. Morozov, On the approximate solution of operator equations by the method of splines, Soviet Math. Dokl., 12, 5 (1971), 1325-1329 (English translation).

[66] V. A. Morozov, Convergence of an approximate method of solving operator equations of the first kind, USSR Comput. Math. and Math. Phys., 13, 1 (1974), 1-20 (English translation).

[67] V. A. Morozov, On optimality of the discrepancy criterion in the problem of computing the values of unbounded operators, USSR Comput. Math. and Math. Phys., 11, 4 (1971), 250-257 (English translation).

[68] V. A. Morozov, On a new approach to solving linear equations of the first kind with an approximated operator (Ob odnom novom podkhode k resheniju lineinykh uravnenii pervogo roda s priblizhennym operatorom), Trudy pervoi konferentsii molodykh uchenykh f-ta VM i K, 1973, Izd-vo MGU (in Russian).

[69] V. A. Morozov, An optimality principle for the error when solving approximately equations with non-linear operators, USSR Comput. Math. and Math. Phys., 14, 4 (1974), 1-9 (English translation).

[70] V. A. Morozov, On a differentiation problem and some algorithms of approximation of experimental information. In Sb.: Vychislitel'nyje metody i programmirovanije, 1970, Vyp. XIV, Izd-vo MGU, pp. 46-62 (in Russian).

[71] V. A. Morozov, Calculation of the lower bounds of functionals from approximate information, USSR Comput. Math. and Math. Phys. 13, 4 (1973), 275-281 (English translation).

[72] V. A. Morozov, The error principle in the solution of incompatible equations by Tikhonov regularization, USSR Comput. Math. and Math. Phys., 13, 5 (1973), 1-16 (English translation).

[73] V. A. Morozov, Linear and nonlinear ill-posed problems (Lineinyje i nelineinyje nakorrektnyje zadachi), 1973, VINITI, Itogi nauki i tekniki, Ser. Matem. analiz, V. II, 129-178 (in Russian).

[74] V. A. Morozov, On regularization of some classes of problems (O reguljarizatsii nekotorykh klassov ekstremal'nykh zadach). In Sb.: Vychislitel'nyje metody i programmirovanije, 1969, Vyp. 12, MGU, 24-37 (in Russian).

[75] V. A. Morozov, On some general conditions for regularity of ill-posed variational problems (O nekotorykh obshchikh uslovijal reguljarizuemosti nekorrektnykh zadach), Trudy pervoi konferentsii molodykh uchenykh f-ta VM i K, Izd-vo MGU, 1973, 140-164 (in Russian).

[76] V. A. Morozov, On the determination of parameters of a linear model using the experimental data. In Trudy Vsesojuznogo simpoziuma: "Inverse problems for differential equations", Novosibirsk, 1972, Izd-vo SO AN SSSR (in Russian).

[77] V. A. Morozov, On stable methods for solving systems of linear algebraic equations (Ob ustoichivykh metodakh reshenija sistem lineinykh algebraicheskikh uravnenii), Trudy Vsesojuznogo soveshchanija "Vychislitel'nyje metody lineinoi algebry", Novosibirsk, 1974 (in Russian).

[78] V. A. Morozov, On particular characteristics of numerical realization of methods for solving unstable problems (Ob osobennostjakh chislennoi realizatsii metodov reshenija neustoichivykh zadach), Trudy Vsesojuznoi shkoly "Metody reshenija nekorrektnykh zadach i ikh primenenije", 1974, Izd-vo MGU, Moscow (in Russian).

[79] V. A. Morozov and V. I. Gordonova, Numerical parameter selection algorithms in the regularization method, USSR Comput. Math. and Math. Phys., 13, 3 (1973), 1-9 (English translation).

[80] V. A. Morozov and N. N. Kirsanova, On a generalization of the regularization method (Ob odnom obobshcheniji metoda reguljarizatsii). In Sb..: Vychislitel'nyje metody i programmirovanije, 1970, Vyp. XIV, Izd-vo MGU, 40-45 (in Russian).

[81] A. P. Petrov, Estimates of linear functionals for the solution of certain inverse problems, USSR Comput. Math. and Math. Phys. 7, 3 (1967), 241-249-c (English translation).

[82] A. A. Samarskii, Introduction to the theory of difference schemes (Vvedenije v teoriju raznostnykh skhem), Moscow, 1971 (in Russian).

[83] V. G. Romanov, An abstract inverse problem and questions of its uniqueness, Functional Analysis and Its Applications, 7, 3 (1973), 223-229 (English translation).

[84] S. B. Stechkin, Best Approximation of linear operators, Math. Notes 1, 1 (1967), 91-99 (English translation).

[85] V. N. Strakhov, Solution of incorrectly-posed linear problems in Hilbert space, Differential Equations, 6, 8 (1970), 1136-1140 (English translation).

[86] V. N. Strakhov, On methods of approximate solution of linear conditionally correct problems, Soviet Math. Dokl., 12, 1 (1971), 271-274, (English translation).

[87] V. P. Tanana, Incorrectly posed problems and the geometry of Banach spaces, Soviet Math. Dokl., 11, 4 (1970), 864-867 (English translation).

[88] S.L. Sobolev, Applications of functional analysis in mathematical physics, Amer. Math. Soc., Providence, R. I., 1963 (English translation).

[89] A.N. Tikhonov, On the stability of inverse problems, Doklady Akademii Nauk SSSR, 39, 5 (1943), 195-198 (in Russian).

[90] A.N. Tikhonov, On the solution of ill-posed problems and the method of regularization, Soviet Math. Dokl., 4 (1963), 1035-1038 (English translation).

[91] A.N. Tikhonov, On the regularization of ill-posed problems, Soviet Math. Dokl., 4 (1963), 1624-1627 (English translation).

[92] A. N. Tikhonov, On non-linear equations of the first kind, Soviet Math. Dokl., 6 (1965), 559-562 (English translation).

[93] A. N. Tikhonov, Incorrect problems of linear algebra and a stable method of their solution, Soviet Math. Dokl., 6, 4 (1965), 988-991 (English translation).

[94] A. N. Tikhonov, On ill-posed problems (O nekorrektno postavlennykh zadachakh). In Sb.: Vychislitel'nyje metody i programmirovanije, 1968, Vyp. VIII, Izd-vo MGU, 1967, 3-33 (in Russian).

[95] A. N. Tikhonov and V. B. Glasko, Use of the regularization method in nonlinear problems, USSR Comput. Math. and Math. Phys., 5, 3 (1965), 93-108 (English translation).

[96] A. N. Tikhonov, Methods for solving ill-posed problems (Metody reshenija nekorrektno postavlennykh zadach), Trudy Vsesojuznoi shkoly "Metody reshenija nekorrektnykh zadach i ikh primenenije", 1974, Izd-vo MGU, 6-11 (in Russian).

[97] V.N. Faddeeva, Shift for systems with badly posed matrices, USSR Comput. Math. and Math. Phys., 5, 5 (1965), 177-183 (English translation).

[98] Ahlberg, J. H., Wilson, E. W., Walsh, J. Z. The Theory of Splines and Their Applications, Academic Press, New York, 1967.

[99] Anselone, P. M., Laurent, P. J., A general method for the construction of interpolating or smoothing spline-functions, Num. Math., 1968, v. 12, 66-82.

[100] Arcangeli, R., Pseudosolutions de l'equation Ax = y. C. R. Acad. Sci., 1966, v. 263, A283-A285.

[101] Atteia, M., Fonction-spline généralisée, C. R. Acad. Sci., Paris, 1965, v. 261, 2149-2152.

[102] Bramble, J. H., Nitsche, J. A., A generalized Ritz-Least-Squares method for the Dirichlet problems, SIAM J. Num. Anal., 1973, v. 10, 81-93.

[103] Cooley, J. W., Tukey, J. W., An algorithm for the machine calculation of complex Fourier series, Math. of Comput., 1965, v. 19, 297-301.

[104] Douglas, J., Mathematical programming and integral equations, Symp. Numerical Treatment Ordinary Differential Equations, Integral and Integro-Differential Equations, Birkhäuser, 1960, 269-274.

[105] Franklin, J. N., Well-posed stochastic extensions of ill-posed linear problems, J. Math. Anal. and Appl., 1970, v. 33, 682-716.

[106] Hadamard, J., Sur les problèmes aux derivées partielies et leur signification physique, Bull. Univ. Princeton, 1902, v. 13, 49-52.

[107] Hadamard, J., Le probleme de Cauchy et les équations aux derivèes partielles linearies hyperboliques, Hermann, Paris, 1932.

[108] John, F., A note on "improper" problems in partial differential equations, Commun. Pure and Appl. Math., 1955, v. 8, 591-594.

[109] Munteanu, M.-J., Generalized smoothing spline functions for operators, SIAM J. Num. Anal., 1973, v. 10, 28-34.

[110] Nedelkov, I. P., Improper problems in computational physics, Computer Physics Commun., 1972, 157-164.

[111] Phillips, D. Z., A technique for the numerical solution of certain integral equations of the first kind, J. Assoc. Comput. Mach., 1962, v. 9, 84-97.

[112] Reinach, C. H., Smoothing by spline functions, Num. Math., 1967, v. 10, 177-187.